Illustrated Building Glossary

For my parents
with gratitude and love

Illustrated Building Glossary

Roxanna McDonald

OXFORD AUCKLAND BOSTON JOHANNESBURG MELBOURNE NEW DELHI

Butterworth-Heinemann
Linacre House, Jordan Hill, Oxford OX2 8DP
225 Wildwood Avenue, Woburn, MA 01801-2041
A division of Reed Educational and Professional Publishing Ltd

A member of the Reed Elsevier plc group

First published 1999

British Library Cataloguing in Publication Data
A catalogue record for this book is available from the British Library

Library of Congress Cataloguing in Publication Data
A catalogue record for this book is available from the Library of Congress

ISBN 0 7506 3643 2

Printed and bound in Great Britain by
Biddles Ltd, Guildford and King's Lynn

Contents

II. CONTROLS

Legal aspects

Administration

III. CONSTRUCTION PROCESS

Financial aspects

Project execution

IV THE BUILDING SITE

General

V. THE BUILDING FABRIC

Foundations

Superstructure – external walls

Superstructure – internal walls

Superstructure – roofs

Superstructure – stairs

Superstructure – chimneys

Superstructure – floors

Superstructure – openings in walls

Superstructure – fixings

Finishes

Glazing

Services – drainage and plumbing

Services – electrical

Services – heating

External works/landscaping

Foreword

Having been a practitioner, and now in my role as a teacher, I am acutely aware of the tendency to treat the young student and the practitioner as different beings. Of course the student's knowledge will be less rounded, but the act of learning is a continuing process, and to revisit words describing hard fact, presented in a new and illuminating way, is to be in a position not only to re-evaluate those facts but also to explore the ideas that stem from them.

The range of knowledge and skill required to operate maturely in architecture and construction is immense, and in studying one page of the *Illustrated Building Glossary*, this is all too obvious. Each *word* or *gloss* related to an item in an illustration, is simply a flag marking the tip of an iceberg. Each carefully chosen word is filled with potential; it introduces one piece, of one aspect, of one element of the process of building. In turn, the process of building is but the beginning of defining the place for a society to function; a place where the buildings gain a symbolic presence. So the volume you are holding is not only a glossary, it is a book with many resonances.

On being asked to prepare this forward, I took the precaution of searching for the root of the word *glossary.** *Gloss* (1548) from the Latin *glossa*, a word inserted between the lines, or in the margin, as an explanatory rendering of a word in the text. *Glossary* (1483) from the Latin *glossarium:* a list with explanations, of abstruse, antiquated, dialectal or technical terms; a collection of glosses.

Certainly this volume is a glossary, but it is a very unusual one. The historical strength of our society has been literary; text is used to describe the idea. Here, refreshingly, the explanation is visual, and through the clarity and completeness of the 'visual paragraph', meaning is given and a context described, in a form normally thought to require words. Indeed the only piece of written text by the author is her five hundred word preface: an admirable achievement.

Within the construction industry, there is still the sad tendency for us to focus strongly on our particular area of expertise at the expense of a broad understanding of the whole culture of design, construction and of our clients. Managers appear to reign and too few of them appear to have a deep knowledge of the culture they serve. Many interesting designers have become so esoteric in their approach, that they are out of touch with the construction industry they need to realize their ideas. Furthermore, the constructors and public alike find many designers difficult to understand. In consequence, the two sides are suspicious, one of the other. To Vitruvius and Alberti, to Lethaby and Loos, the idea and the making were of equal importance, one being seen as a corollary of the other. I would argue that the designers of importance have always understood this relationship and, from that understanding, their designs achieve their strength and authority. Client, architect, craftsman, engineer, fabricator and builder are each vital cogs in the process, interdependent and synergistic. Theory and practice, head and hand are still of equal value in making our surroundings. In Berthold Lubetkin's words 'Practice without theory is blind, theory without practice is dead'. It is from this strongly held belief that I welcome a book containing as much reference to the theoretical as to the every day. In our present situation, the division of the aesthetic from the pragmatic is destructive. The glossary of building opens up the full spectrum with commendable balance.

What of the presentation of this book? The structure and presentation themselves are worth study as a piece of design. A large quantity of material has been explored, digested and synthesized to present a core of information clearly without becoming simplistic. Each illustration in the glossary is from the same hand and must have taken hours to draw, never mind the weeks of research, assimilation and evaluation involved. Anyone who has tussled with a small design problem and who has attempted to present a solution in a simple line drawing is all too aware of the time taken in graphical study before undertaking the drawing itself.

So, as a general glossary, the *Illustrated Building Glossary* is useful for the public and the college student, but as many of us in architecture and construction learn as well or better through image than through the written word, it is of help to the advanced student and the hardened professional alike for its encyclopaedic quality. Here complex matters are explained with great simplicity. As an illustration of this, I refer the reader to the page summarizing 'planning approval' and 'construction control' (pp. 38 and 39). From these diagrams of the sequence involved, anyone with a background knowledge is able to picture the general shape of the legal structure and from there refer with confidence to detailed legislation. Needless to say, much of the diagrammatic representation is invaluable to any teacher of basic construction, construction management or professional and business practice.

A book of this kind is only useful if it presents the possibility of relating a term to a subject area and referring from there to an in depth bibliography. The author's bibliography is short but all the reference books are in common use, and from those, further channels can be explored.

In summary, the value of such a book for the experienced professional or crafts person is that it goes beyond its stated aim. It has much good straightforward information about the processes themselves, communicated in an attractive way. It contains much for the student to learn; and for the experienced, it contains much that we once knew and are ashamed to admit we have forgotten!

Richard Frewer,
Professor of Architecture, University of Bath
Director of Arup Associates (1977 – 91)

* Summarized from the *Shorter Oxford English Dictionary* .

Preface

It is not the intention of this book to provide an exhaustive list of building terms or to attempt a comprehensive teaching of building technology. There are many specialist encyclopedias, dictionaries and construction manuals which supply ample information in this respect. The book sets out to be primarily a *communication tool* using the *visual reference* as vocabulary.

The creation of a building is the result of a complex process of interaction between people of different professions, views, even nationalities, with varying technical knowledge and motivation. Architects, who at the centre of it all, often find themselves as 'interpreters' between the participants, use image as the safest interface.

The language we each use grows from our own personal experience and, sometimes, the same word can mean different things to different people depending on the circumstances in which they have learnt it. The same can apply to building terms.

Images on the other hand leave little room for ambiguity, and many a time a site query or dispute has been sorted out with the aid of a sketch scribbled on a wall! Words express ideas we have of tangible objects and can be classified into a system such as an alphabetical dictionary or be placed in a context as in a thesaurus. The same can apply to images – they can be attached to words arranged in alphabetical order or they can be placed in the context to which they are relevant.

It is the latter system this book has adopted, attempting to present the terms in the context in which they are likely to apply. The main building terms that form the language of construction are set out to follow the logical sequence of the building process. If one can't remember the right word or wants to know what a specific part is called, it should be simple enough to locate it on the sketch in the relevant section. Similarly, by placing something visually in context it should be much easier to learn terms rather than to memorize their abstract definition. At the same time, the index permits the reverse to take place making it possible to find the context of a given word.

The drawings are simple line sketches concerned mostly with descriptive clarity rather than comprehensive accuracy. The diagrams are intended to identify the sequence and relationships as well as particular terminology.

Compiled primarily as a visual checklist for students and early stages of practice building professionals, the book is also meant to help communication with the other participants to the building industry.

Its spirit, I hope, echoes the intentions of a much older introduction from which I quote below as it is as valid today as it was when it was first written.

IT is uſeful Knowledge only, that makes one Man more valuable than another, and eſpecially that part of Knowledge, which immediately concerns the Buſineſs he is to live by; and therefore, if this Work ſhould prove a Help to the Improvement of Knowledge in *Youth*, (for whoſe Sakes 'tis chiefly intended:) and be no Affront to the *ſage Workman*, by re-informing him of thoſe Rules which have ſlipt his Memory, and informing him of others which he never knew, it will anſwer the deſired End of their hearty Well-wiſher,

London, March 25th, 1741. THO. LANGLEY.

From the introduction to:

THE

BUILDER's JEWEL:

OR, THE

YOUTH's INSTRUCTOR,

AND

WORKMAN's REMEMBRANCER.

EXPLAINING

SHORT and EASY RULES,

Made familiar to the meaneſt Capacity,

For DRAWING and WORKING.

By B. and T. LANGLEY.

LONDON:

Printed for *R. Ware*, at the *Bible* and *Sun* in *Amen-Corner*, near *Pater-Noſter-Row*.
MDCCXLI. [Price 4*s*. 6*d*.]

Also by the same author; *The Fireplace Book*, Architectural Press (1984).

Acknowledgements

I am grateful to the following people for their supportive help during the preparation of this book:

Rob Dark, Architect, ARCHED, UK
B. Goilay, Engineer, Nord France Engineering
Dan S. Hanganu, Architect, Montreal, Canada
C. Lemaire and J. Capoulade, Architects, AREEL, France
A. Lemaire, Architecte, CNRS, France
Bibliotheque Centre Pompidou, Paris, France
The RIBA Library, London, UK
Veronique Thierry, Isabelle Mathieu, Monique Beranger,
Architects, Paris, France
Beatrice Jubien, France

My grateful thanks to my editors: Neil Warnock-Smith for his encouragement and advice, and Mike Cash for his understanding and enthusiasm.

My special thanks and gratitude to Jane Fawcett whose generous advice and personal example were an inspiration.

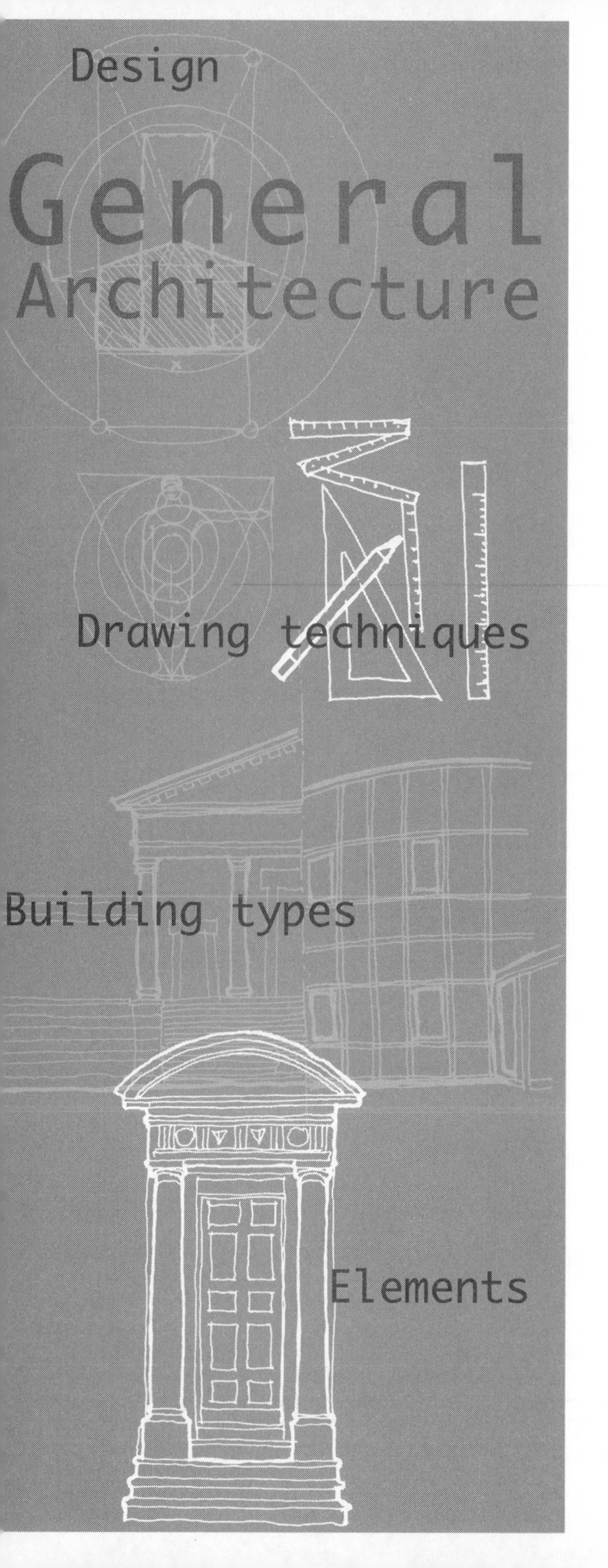

I. General Architecture

Golden number:
ratio of dimensions regarded as particularly harmonious from antiquity: 1.618

Golden section ⟶
application of golden number allows infinite sub-division to the same proportions

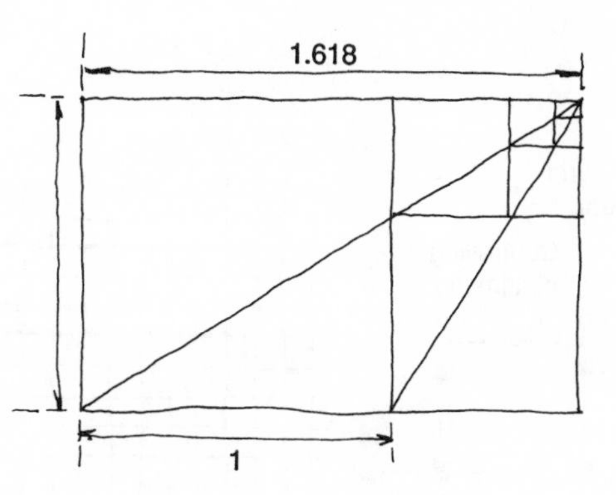

Modulor; Le Corbusier's application of the golden number proportions to human dimensions

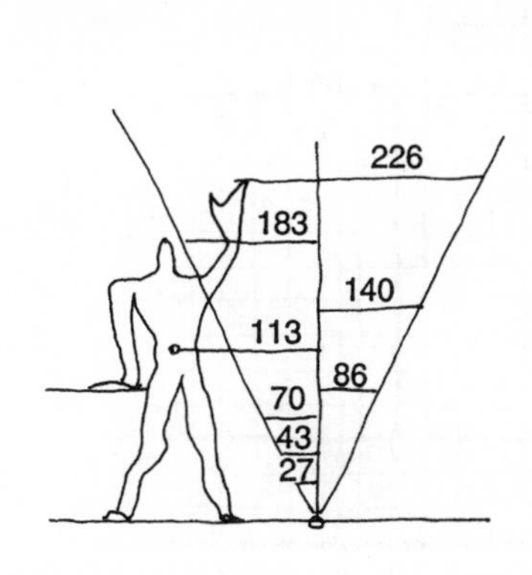

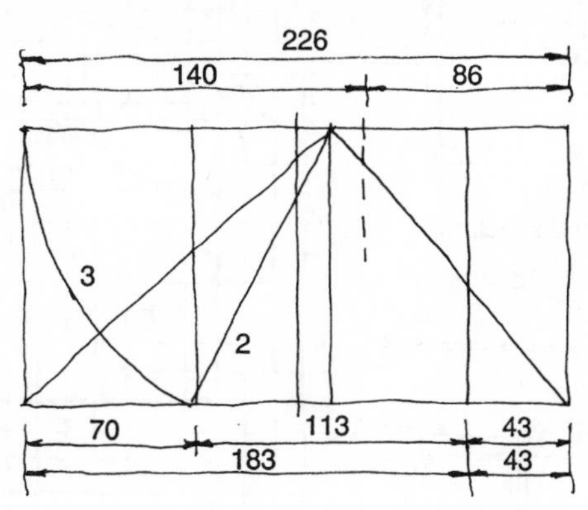

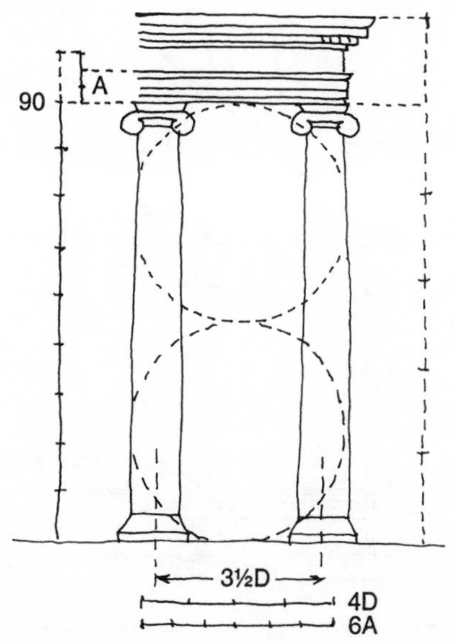

Classical orders (Vitruvius)

Module = half diameter of column base

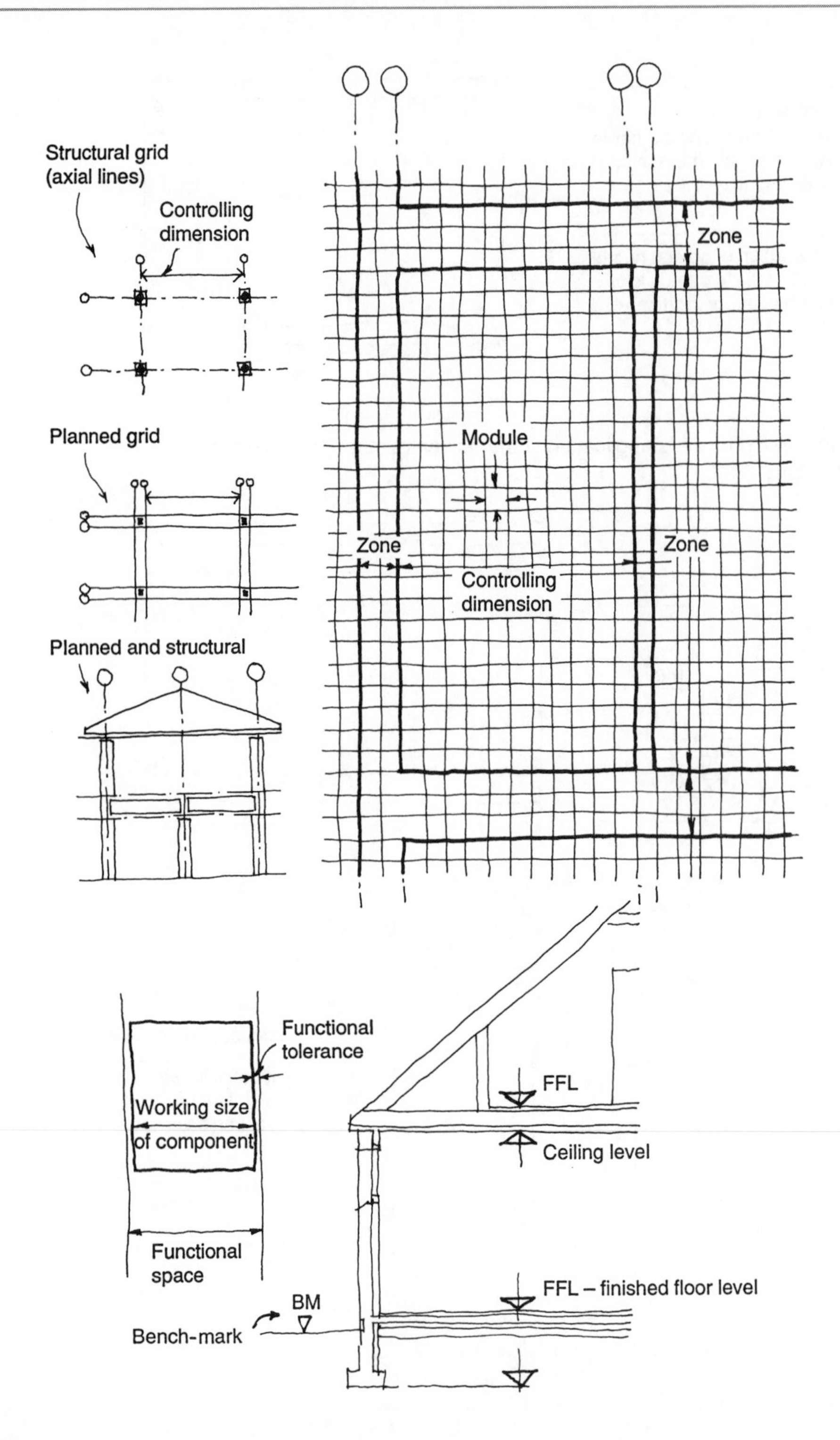
Structural grid
(axial lines)
Controlling
dimension
Planned grid
Planned and structural
Zone
Module
Zone
Controlling
dimension
Zone
Functional
tolerance
Working size
of component
Functional
space
FFL
Ceiling level
FFL – finished floor level
BM
Bench-mark

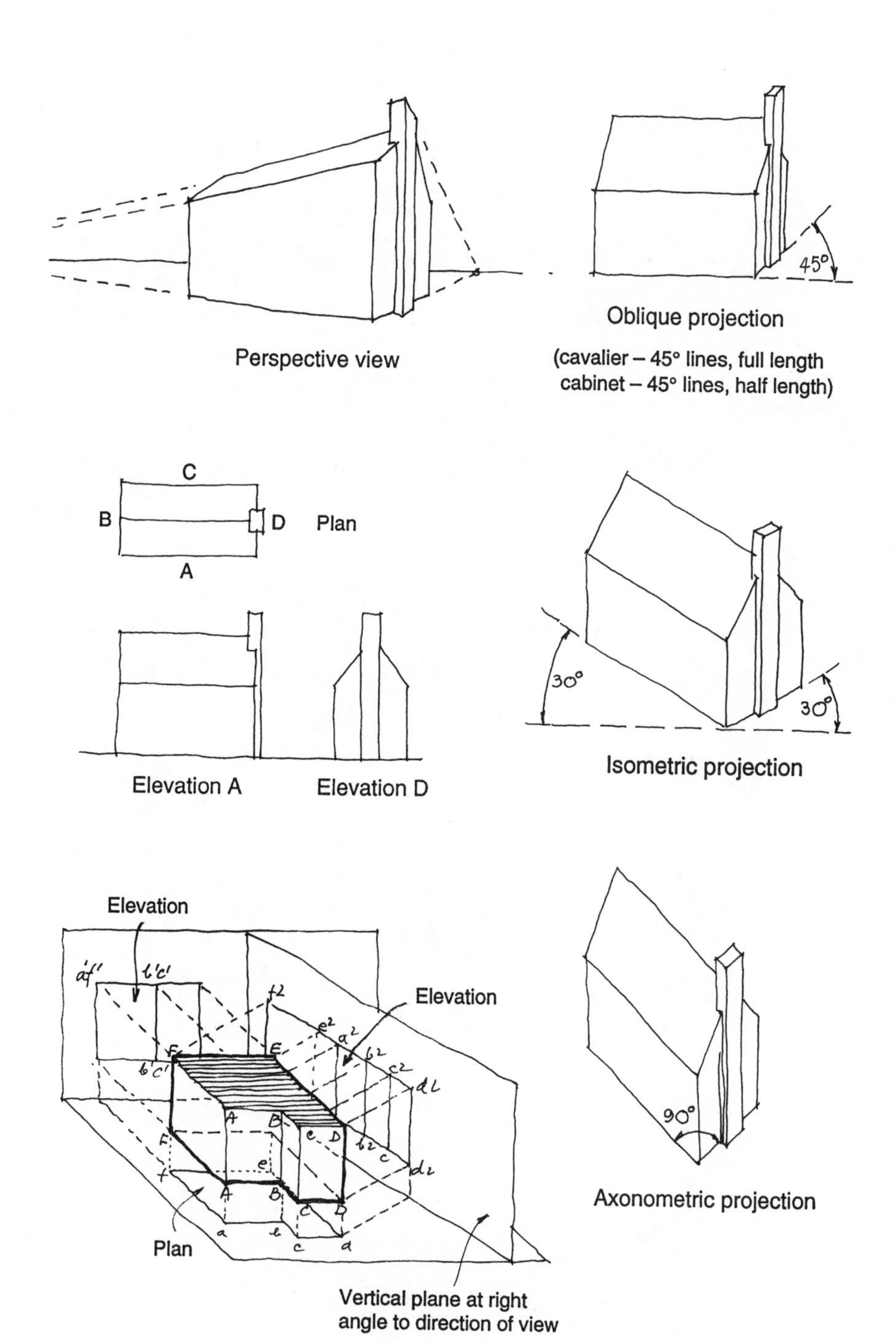

Orthographic projection

Volumes and shapes

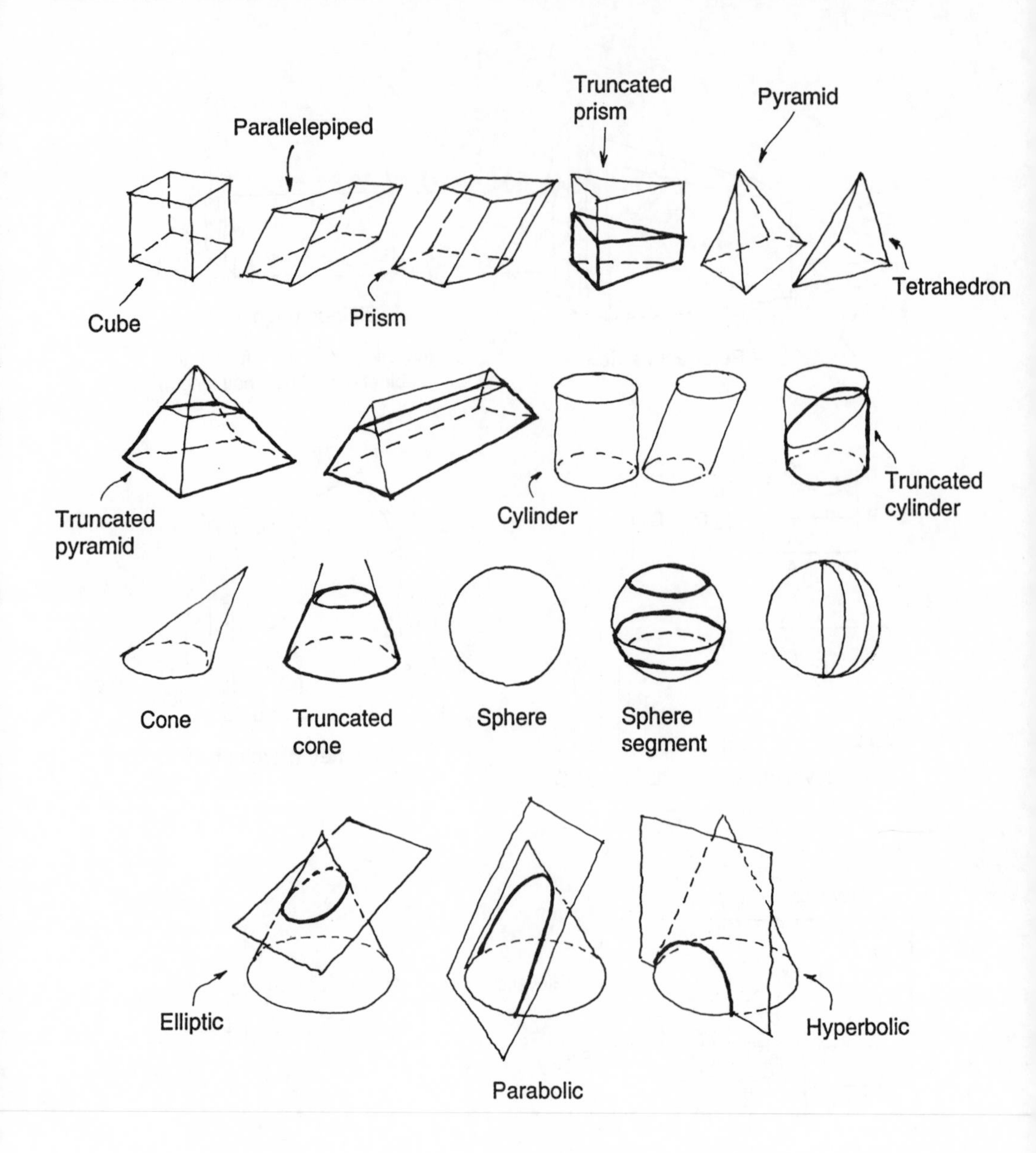

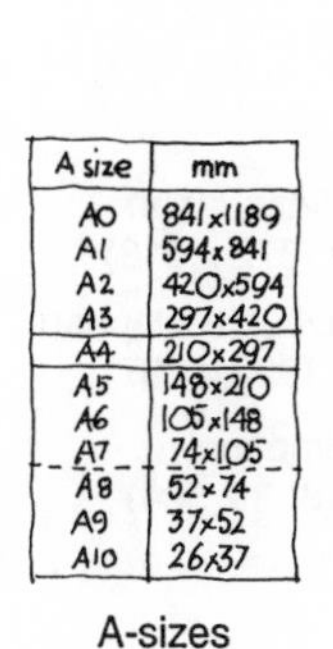

A size	mm
A0	841x1189
A1	594x841
A2	420x594
A3	297x420
A4	210x297
A5	148x210
A6	105x148
A7	74x105
A8	52x74
A9	37x52
A10	26x37

A-sizes

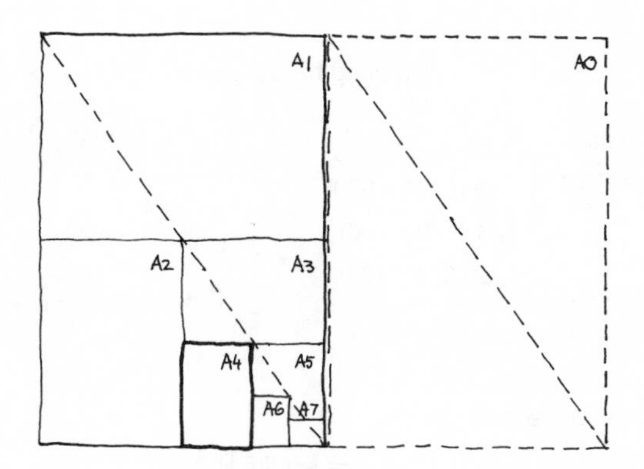

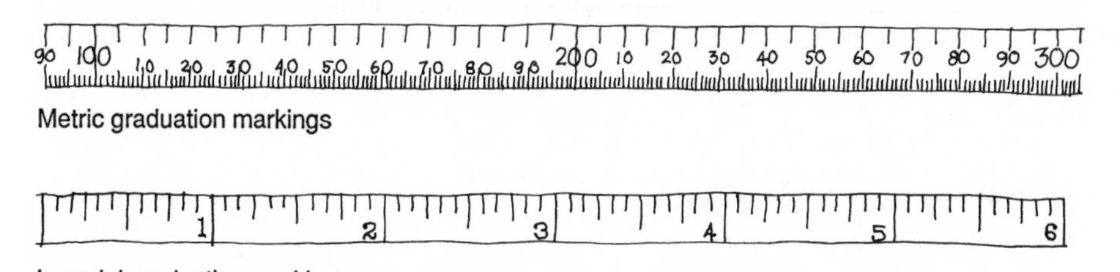

Metric graduation markings

Imperial graduation markings

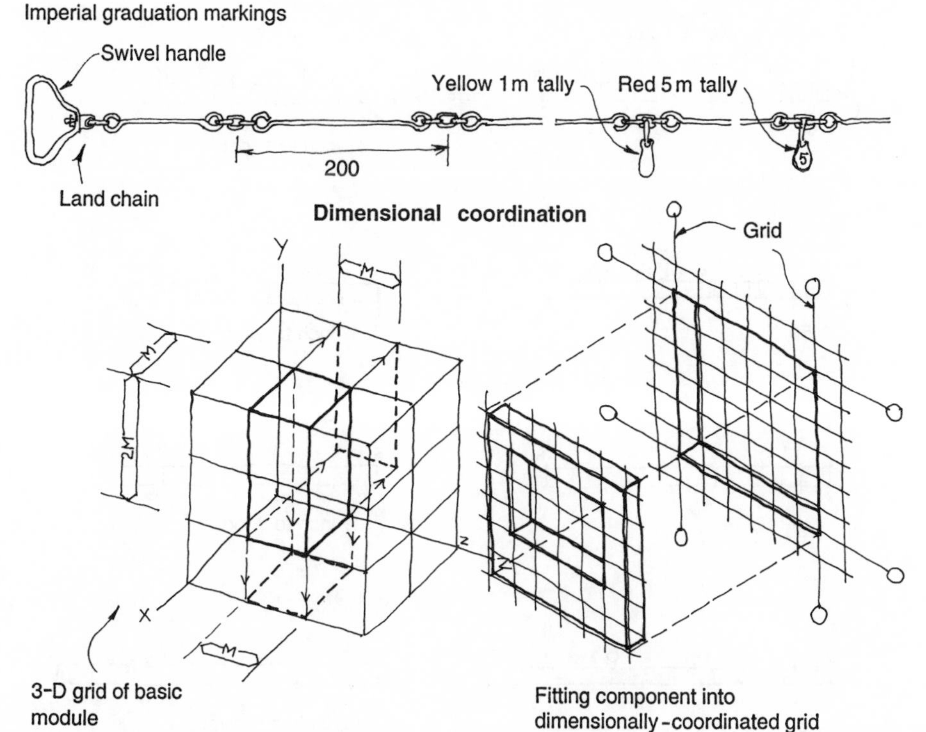

3-D grid of basic module

Fitting component into dimensionally-coordinated grid

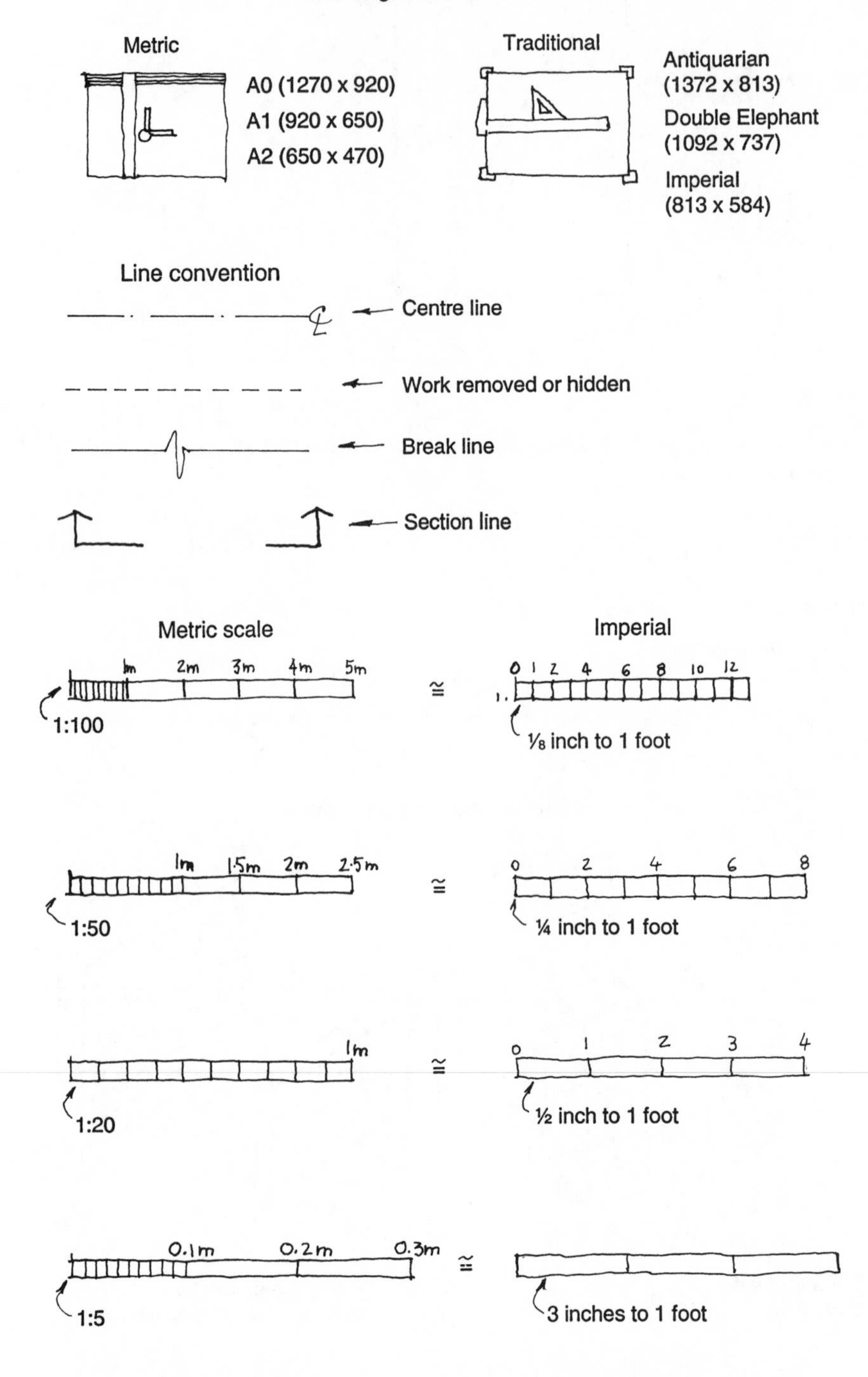
Drawing board sizes
Metric
A0 (1270 x 920)
A1 (920 x 650)
A2 (650 x 470)
Traditional
Antiquarian
(1372 x 813)
Double Elephant
(1092 x 737)
Imperial
(813 x 584)
Line convention
Centre line
Work removed or hidden
Break line
Section line
Metric scale
Imperial
1m 2m 3m 4m 5m
1:100
0 1 2 4 6 8 10 12
⅛ inch to 1 foot
1m 1.5m 2m 2.5m
1:50
0 2 4 6 8
¼ inch to 1 foot
1m
1:20
0 1 2 3 4
½ inch to 1 foot
0.1m 0.2m 0.3m
1:5
3 inches to 1 foot

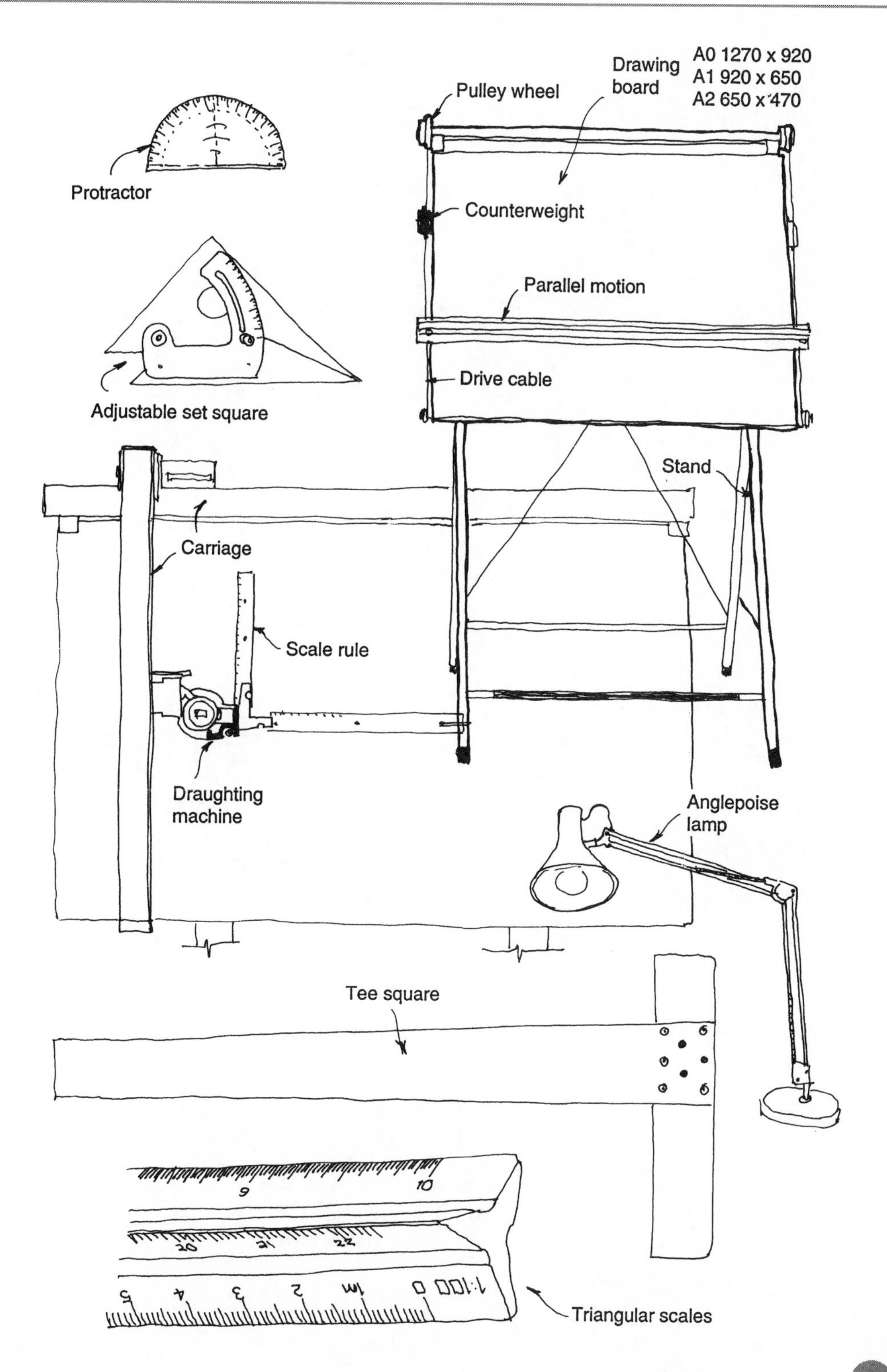
Drawing board
A0 1270 x 920
A1 920 x 650
A2 650 x 470
Pulley wheel
Protractor
Counterweight
Parallel motion
Adjustable set square
Drive cable
Stand
Carriage
Scale rule
Draughting machine
Anglepoise lamp
Tee square
9
10
20
21
22
1:100 0
1m
2
3
4
5
Triangular scales

Drawing instruments

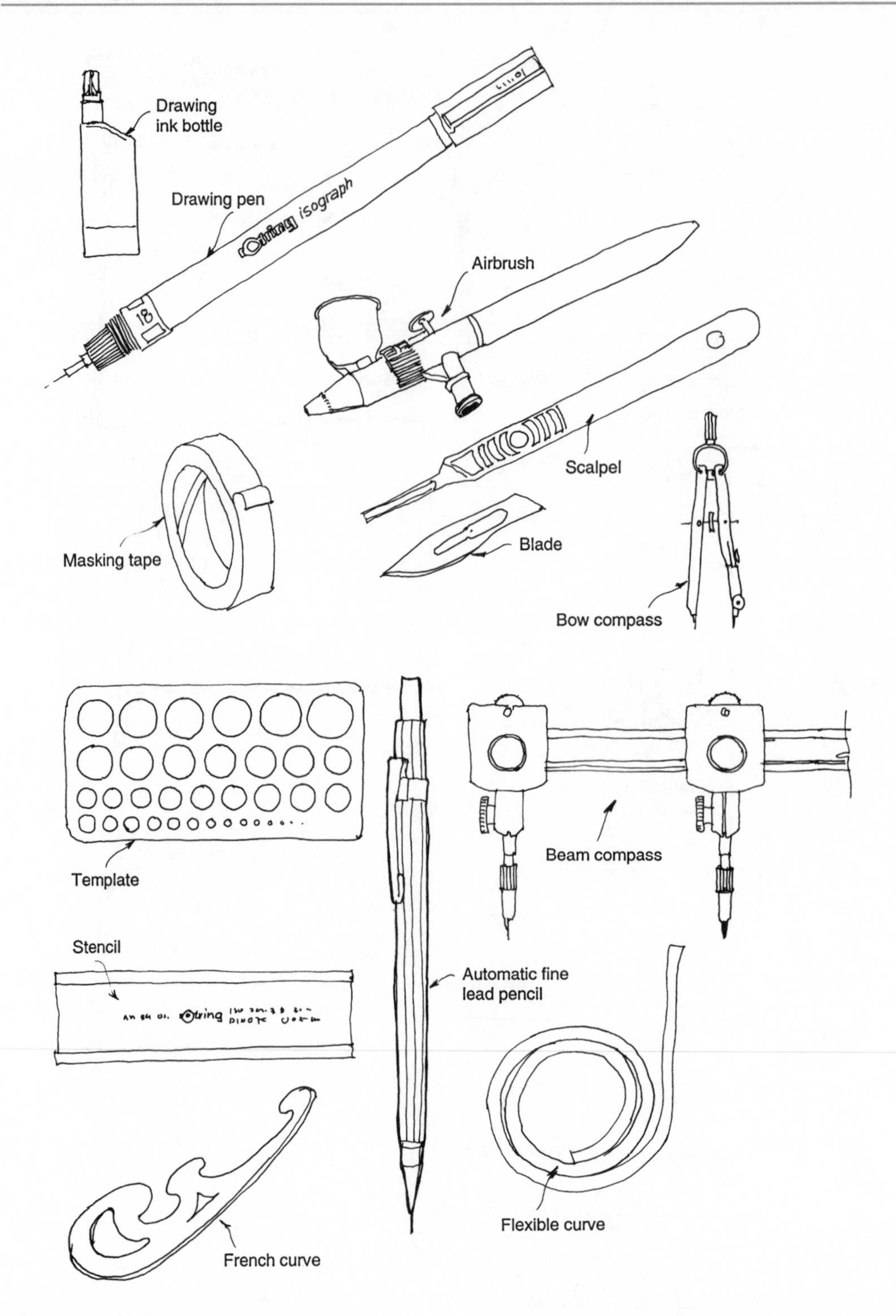

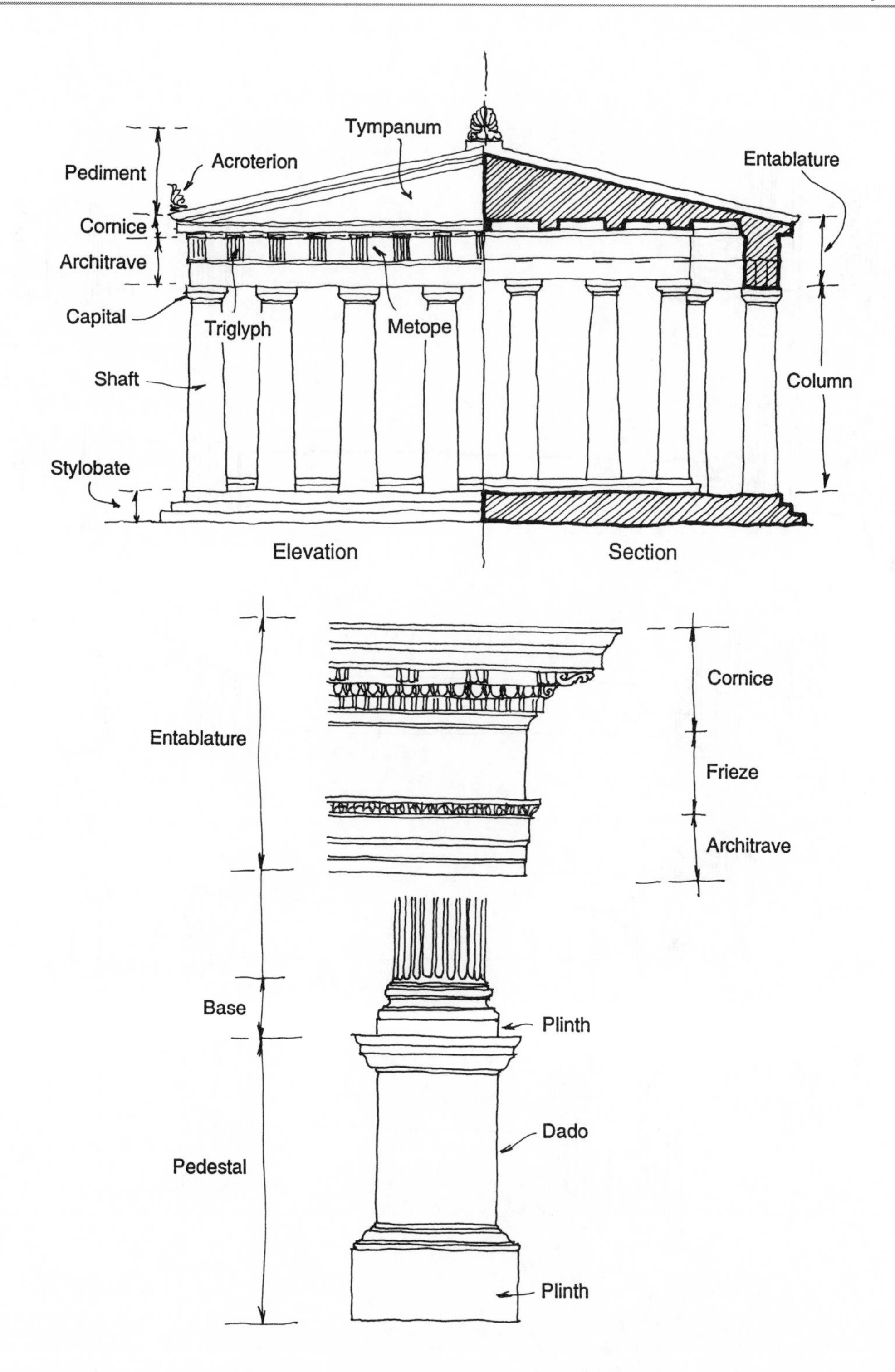
Tympanum
Pediment
Acroterion
Entablature
Cornice
Architrave
Capital
Triglyph
Metope
Shaft
Column
Stylobate
Elevation
Section
Entablature
Cornice
Frieze
Architrave
Base
Plinth
Dado
Pedestal
Plinth

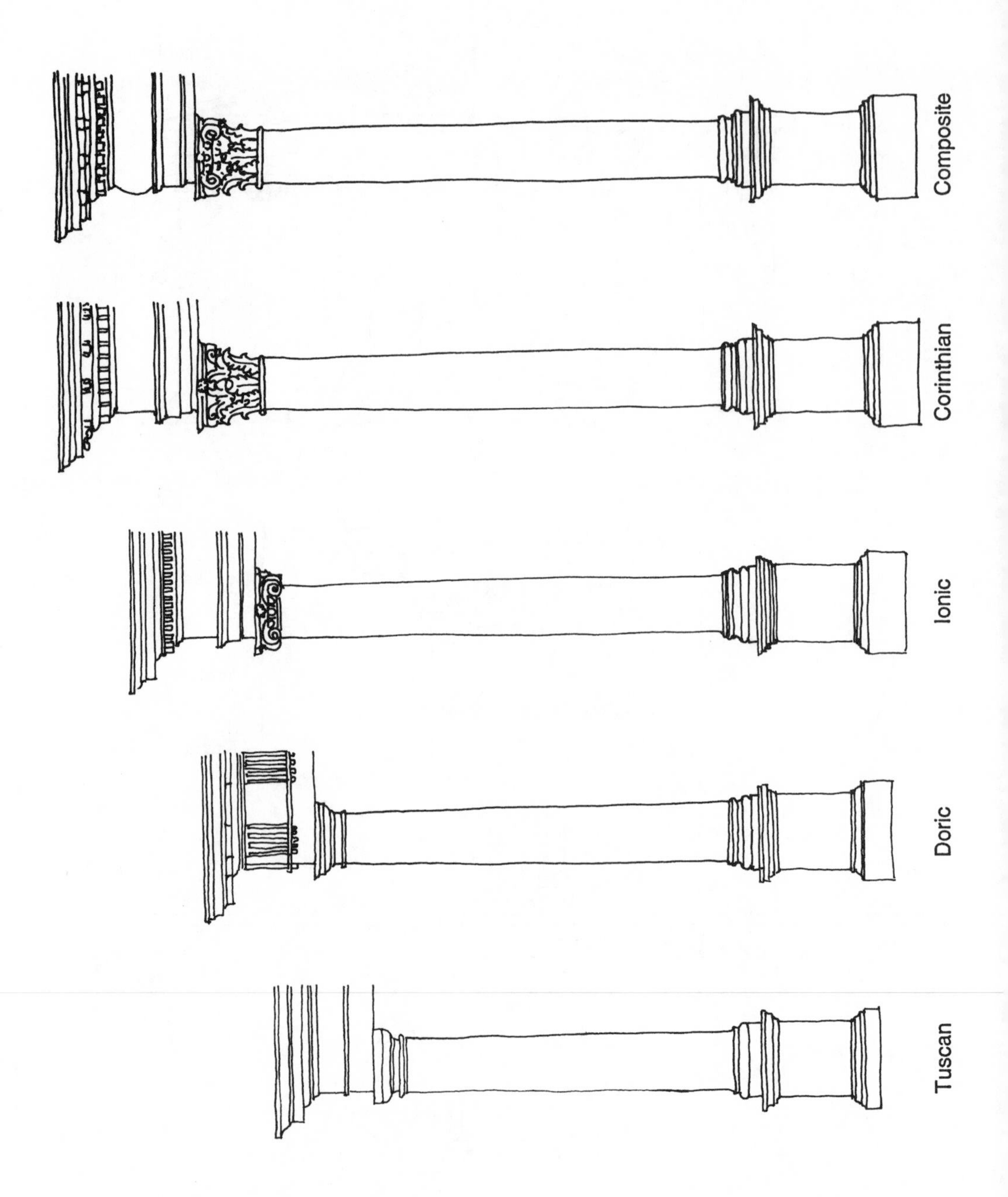
Composite
Corinthian
Ionic
Doric
Tuscan

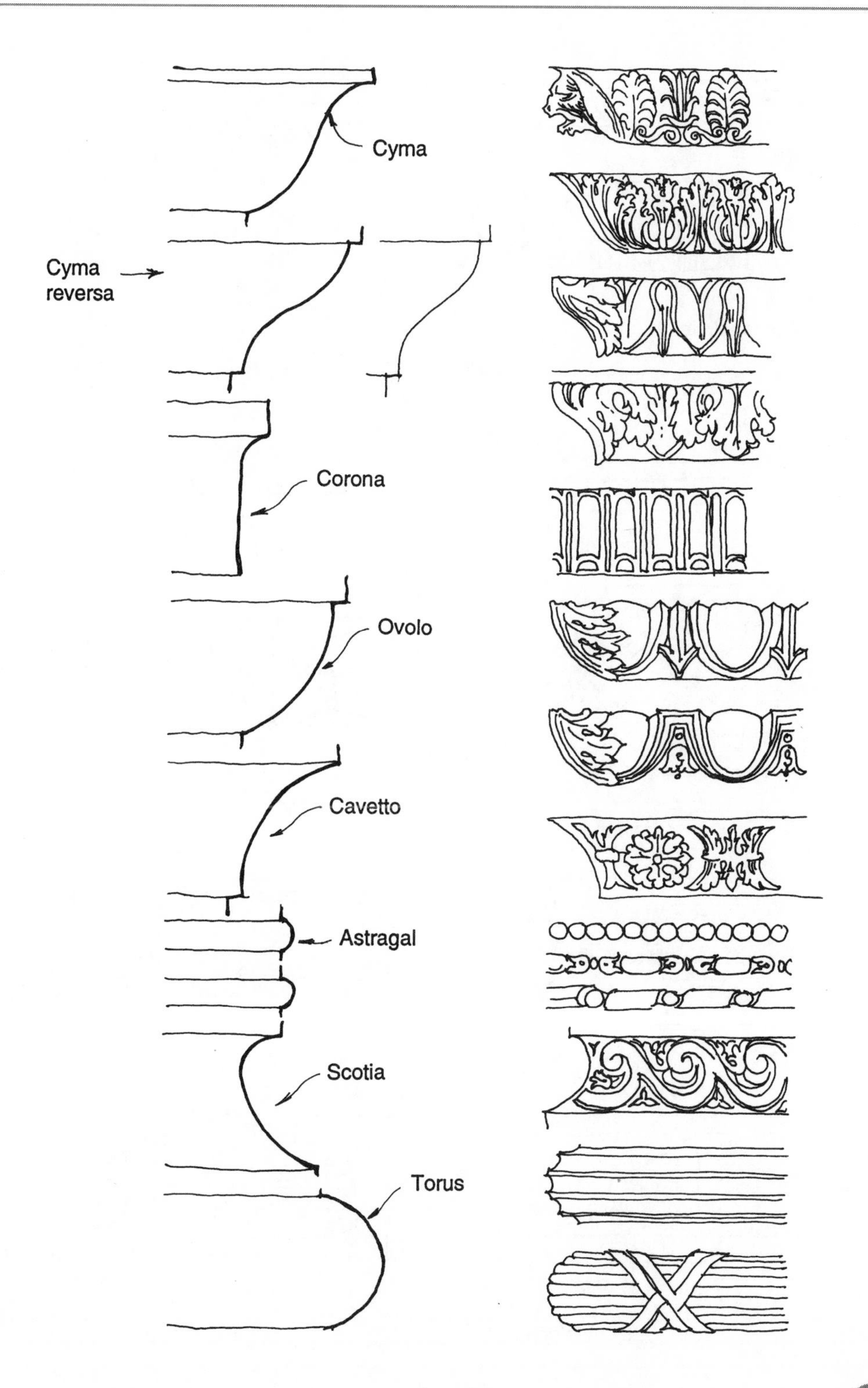
Cyma
Cyma reversa
Corona
Ovolo
Cavetto
Astragal
Scotia
Torus

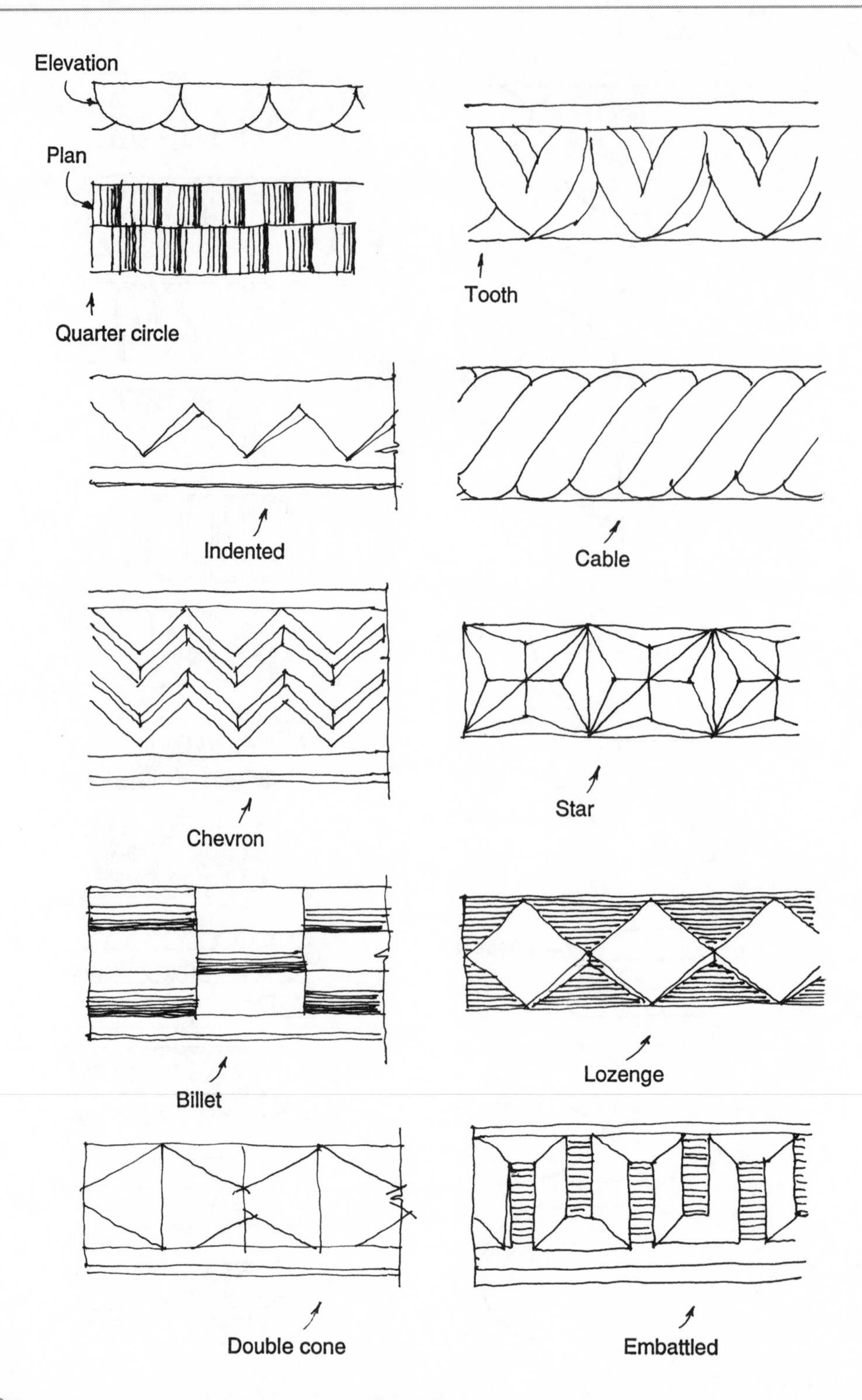
Elevation
Plan
Quarter circle
Tooth
Indented
Cable
Chevron
Star
Billet
Lozenge
Double cone
Embattled

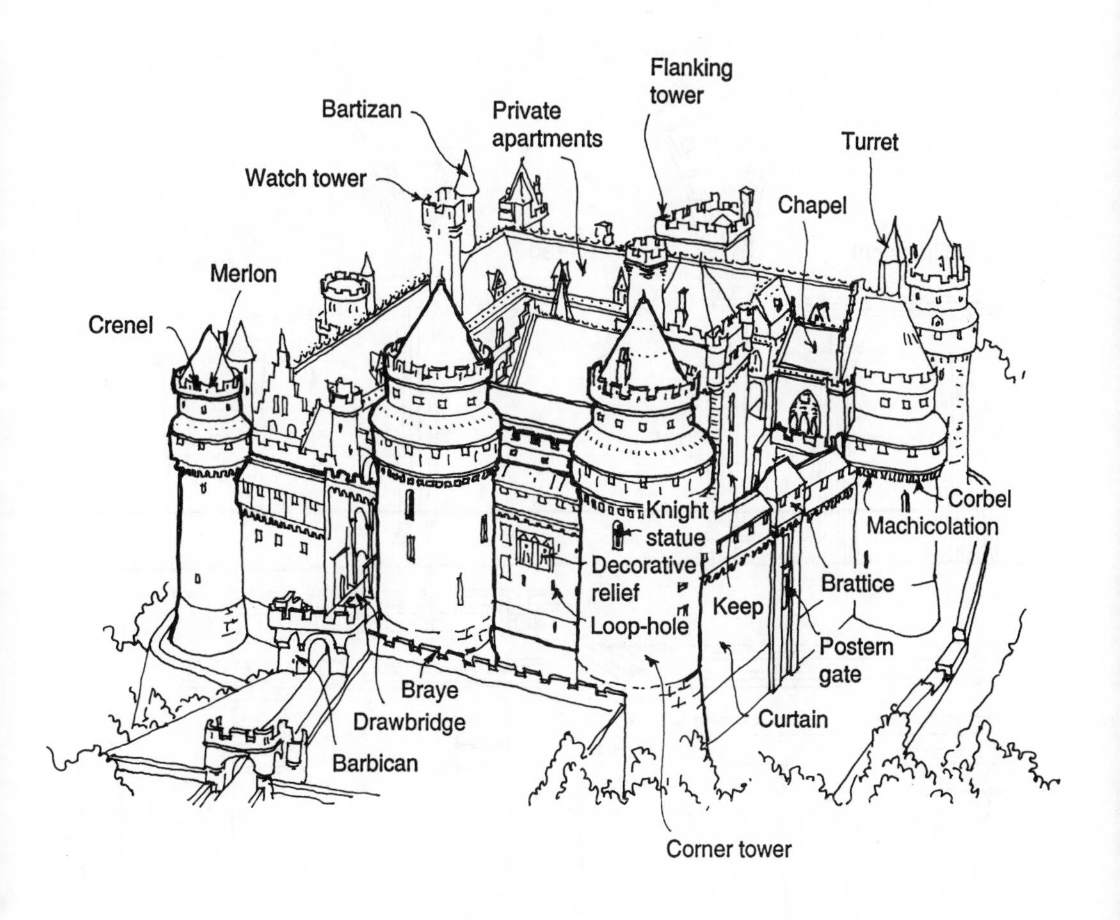
Flanking tower
Bartizan
Private apartments
Turret
Watch tower
Chapel
Merlon
Crenel
Knight statue
Corbel
Machicolation
Decorative relief
Brattice
Keep
Loop-hole
Postern gate
Braye
Curtain
Drawbridge
Barbican
Corner tower

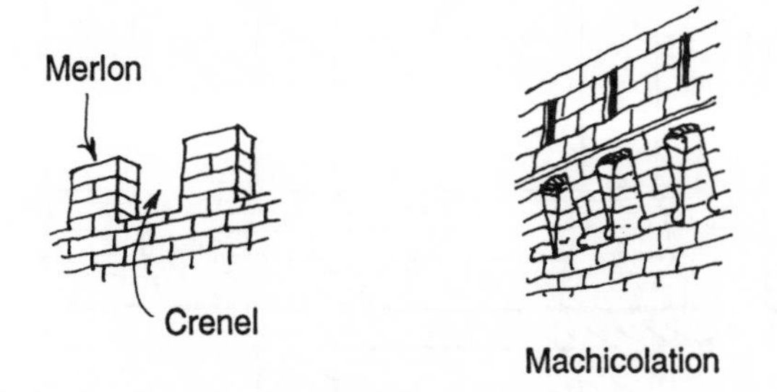
Merlon
Crenel
Machicolation

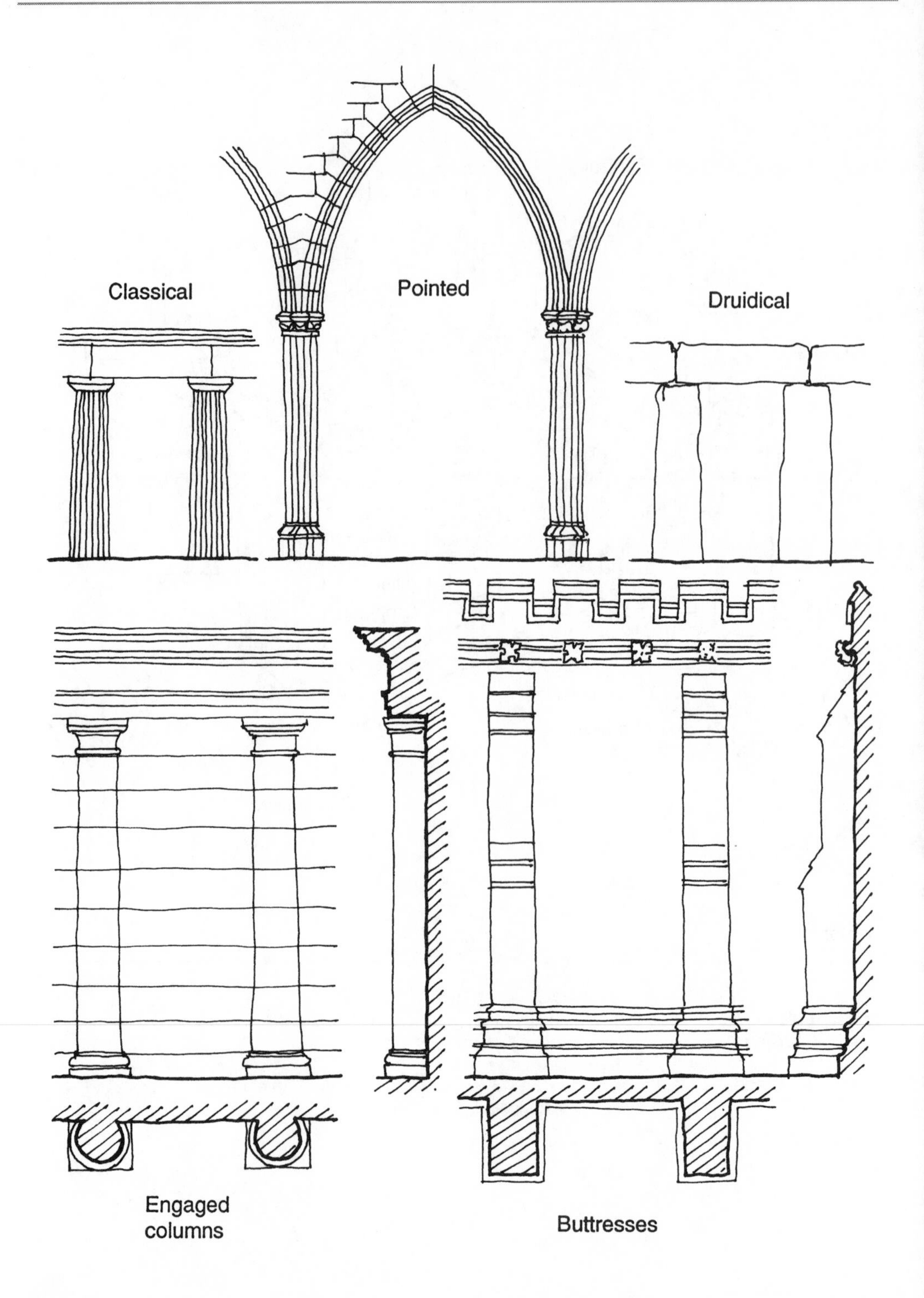
Classical
Pointed
Druidical
Engaged columns
Buttresses

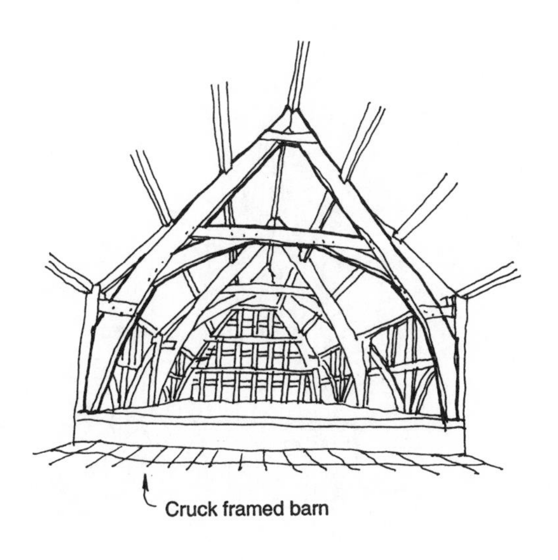
Cruck framed barn

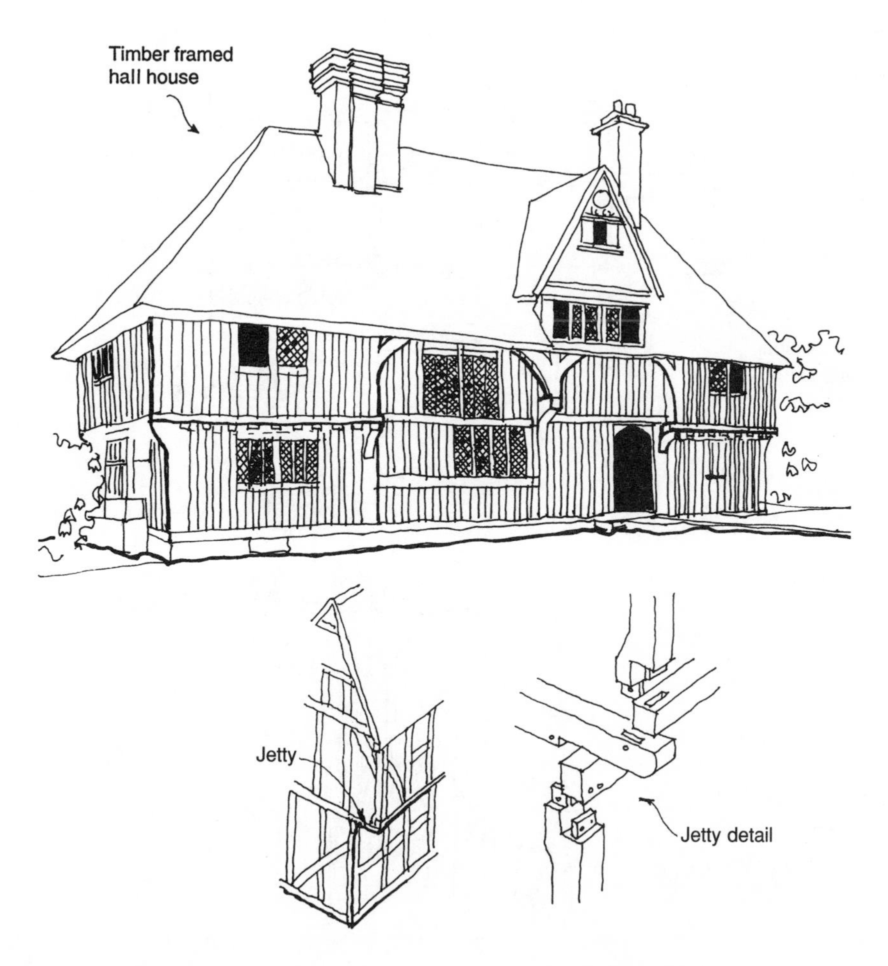
Timber framed
hall house
Jetty
Jetty detail

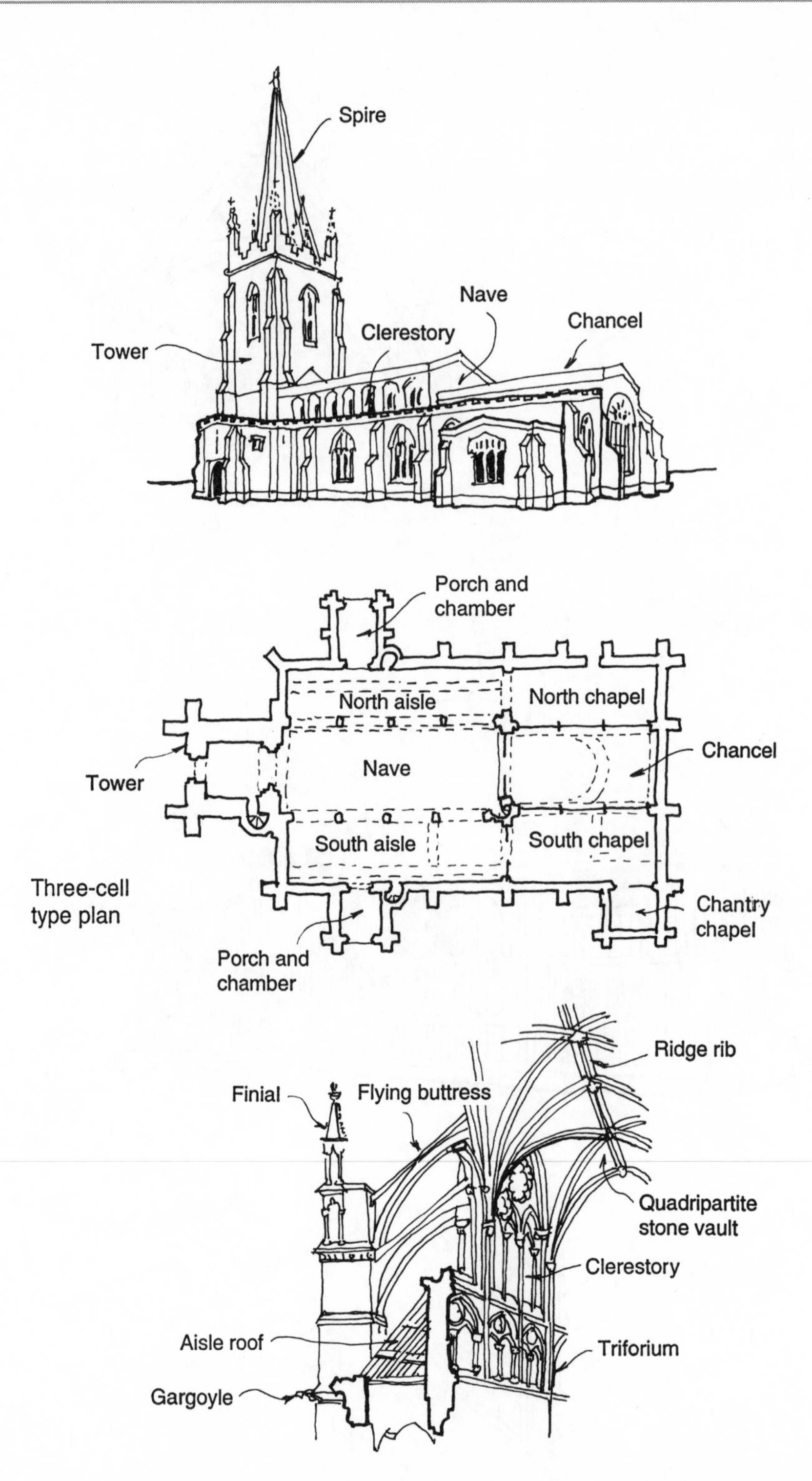
Spire
Nave
Chancel
Clerestory
Tower
Porch and chamber
North aisle
North chapel
Chancel
Tower
Nave
South aisle
South chapel
Three-cell type plan
Chantry chapel
Porch and chamber
Ridge rib
Finial
Flying buttress
Quadripartite stone vault
Clerestory
Aisle roof
Triforium
Gargoyle

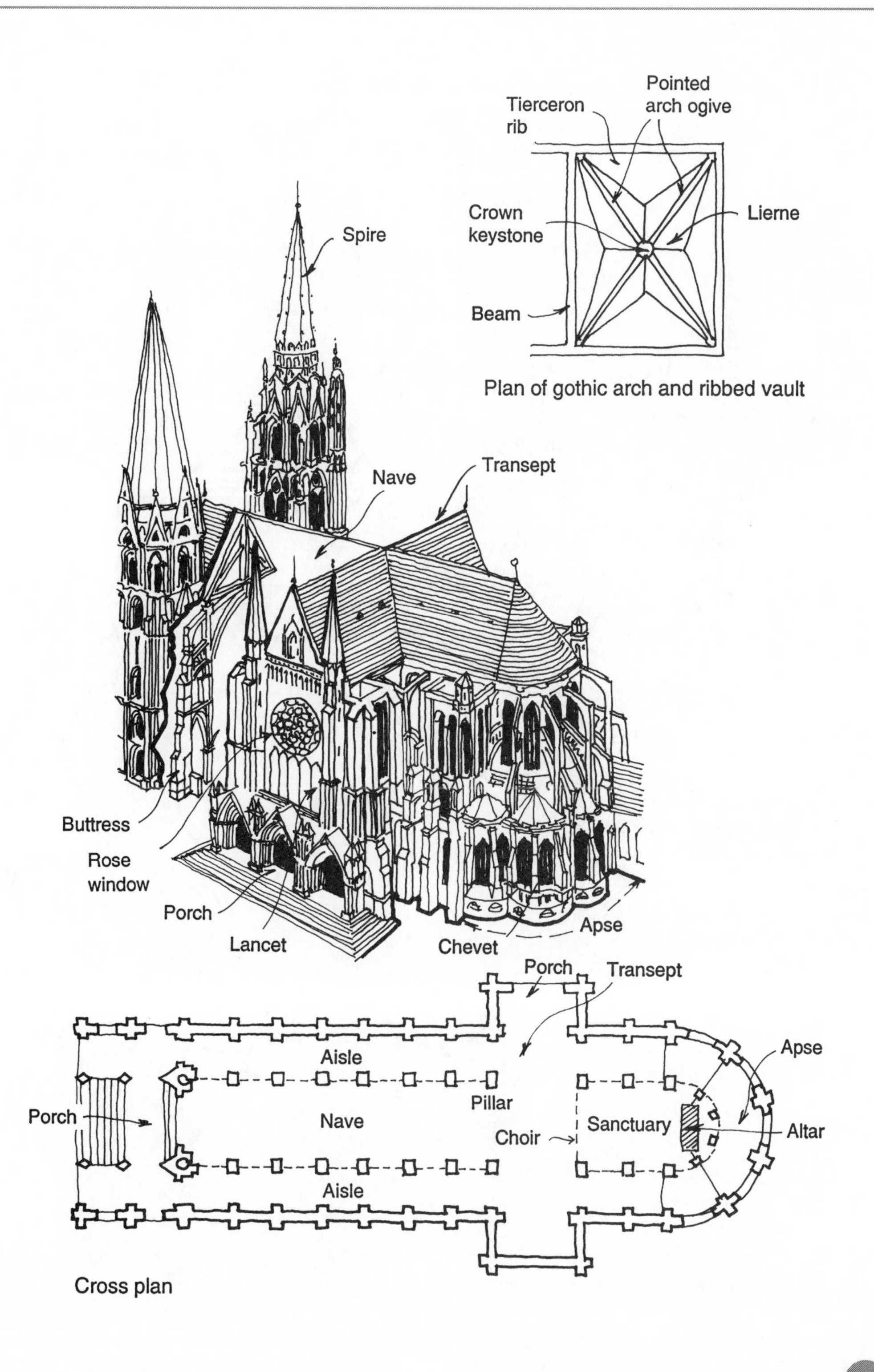

Plan of gothic arch and ribbed vault

Cross plan

Traditional house

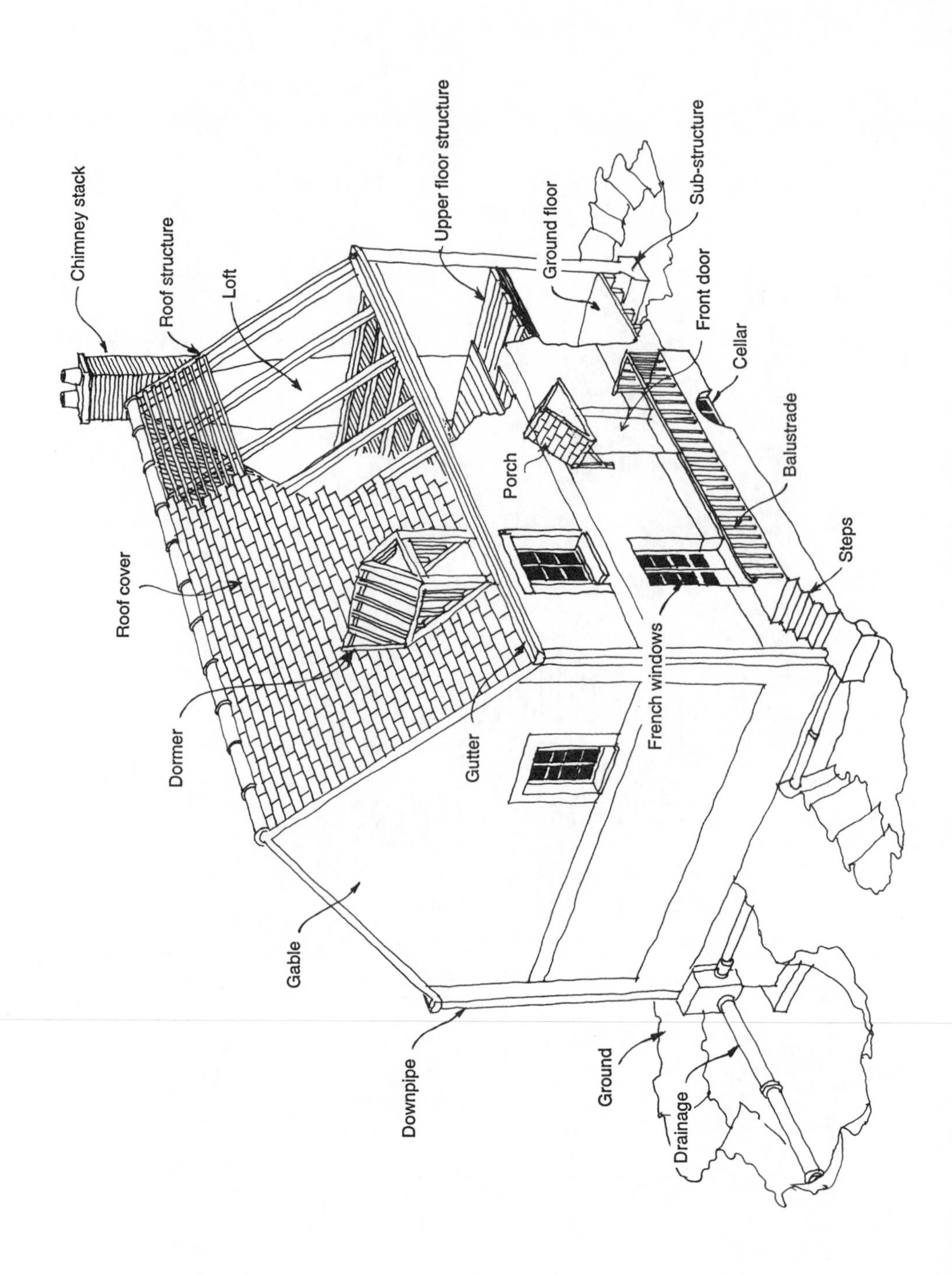

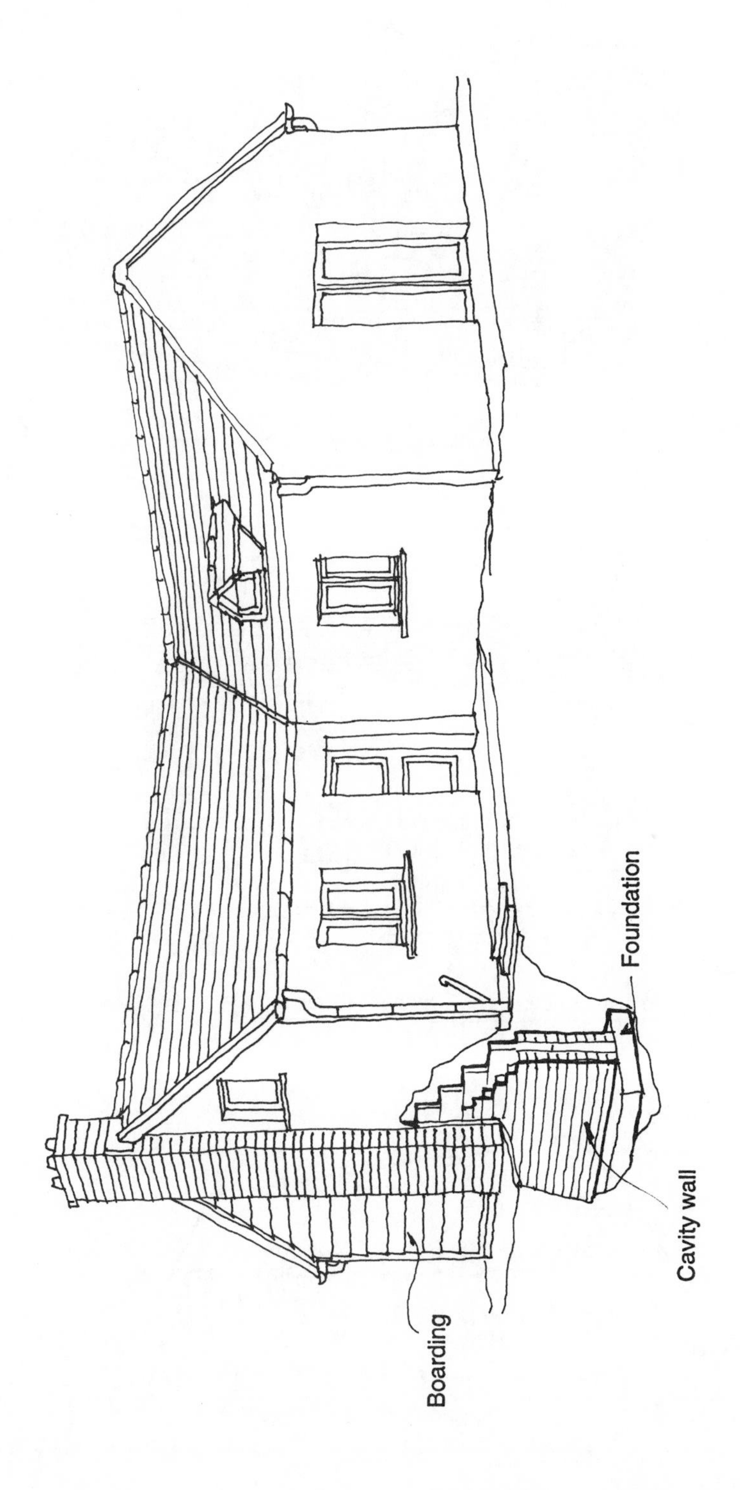
Foundation
Cavity wall
Boarding

Terraces

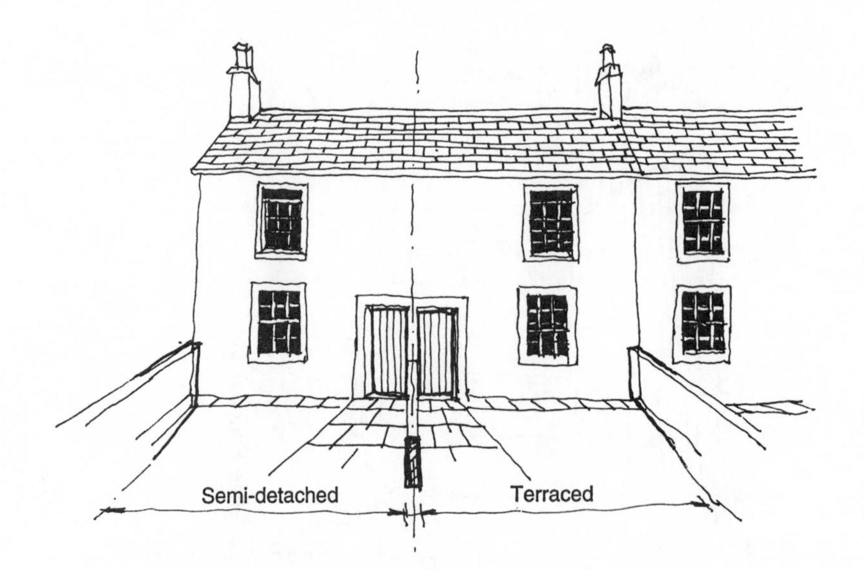

Semi-detached   Terraced

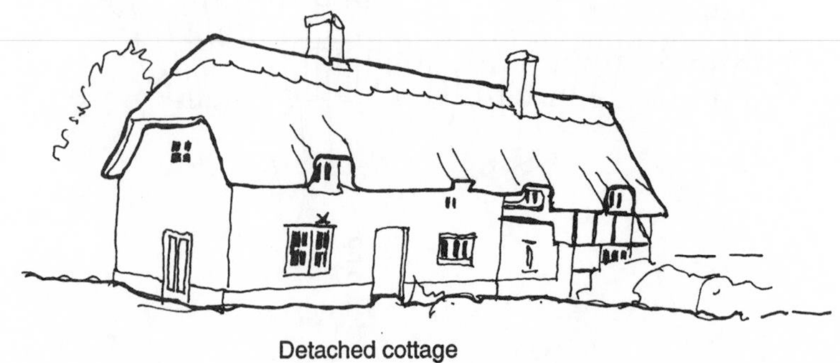

Detached cottage

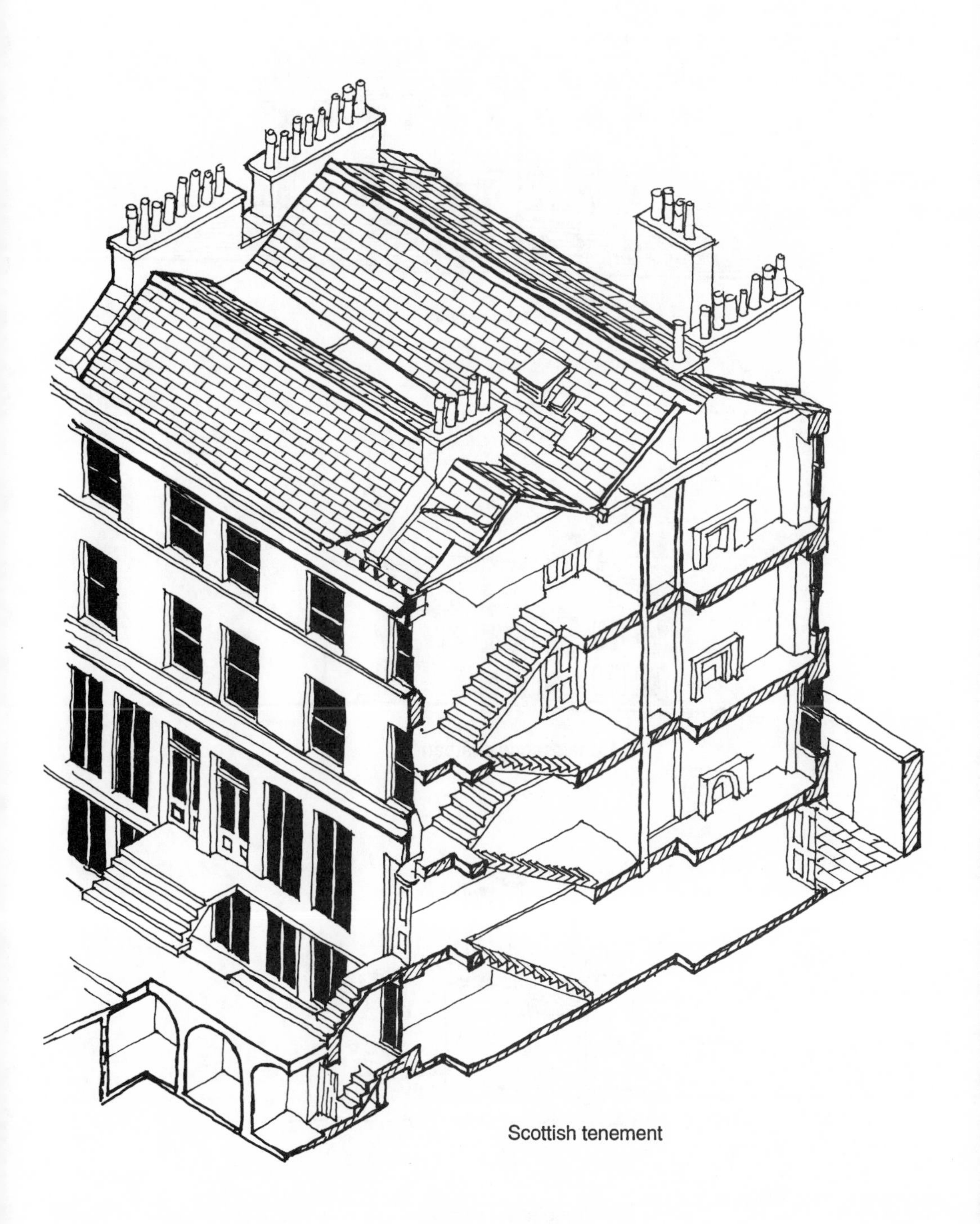

Scottish tenement

Stately home

Renaissance château

Mediaeval castle

Detached house

Bungalow

High-rise block of flats

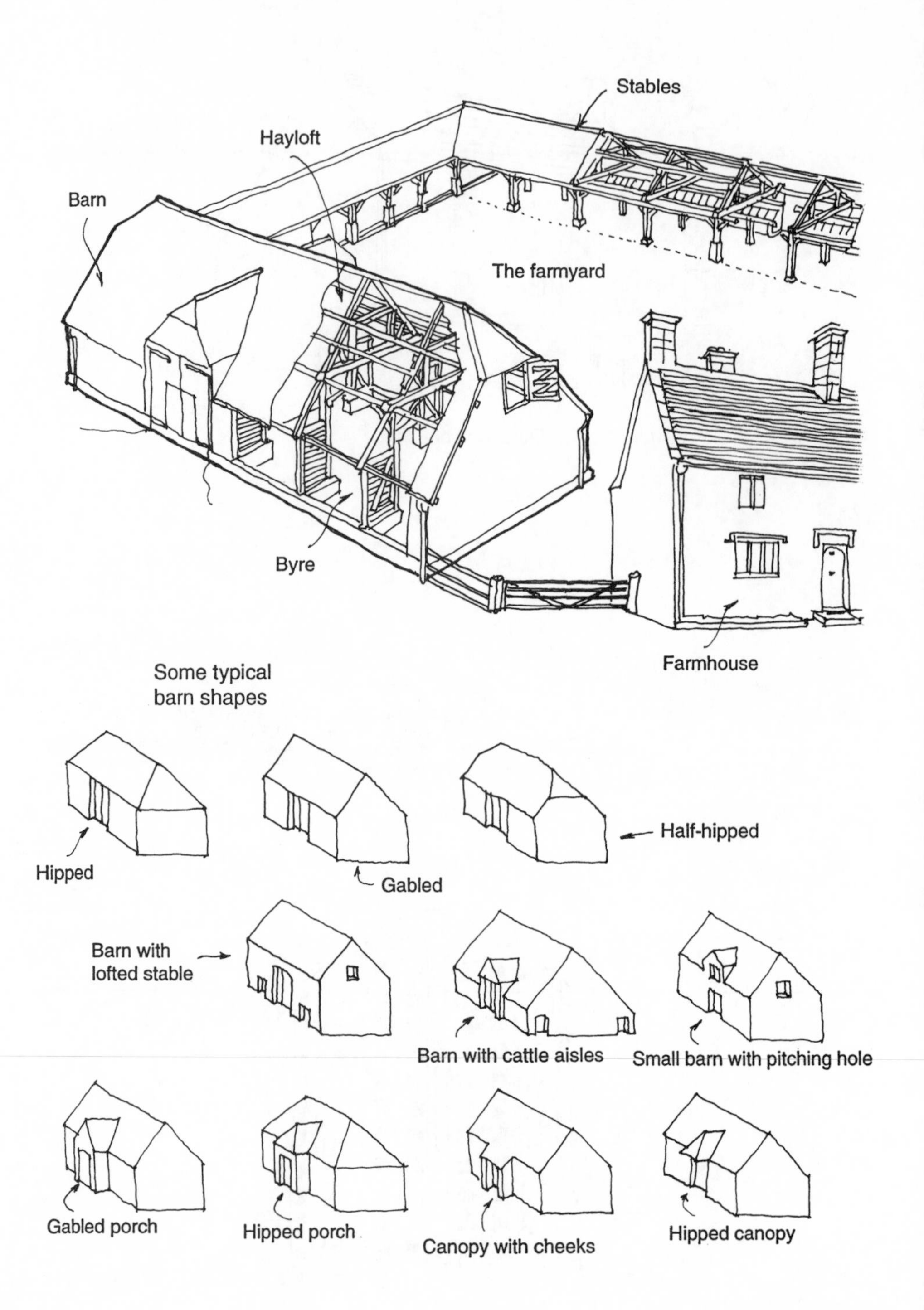
Stables
Hayloft
Barn
The farmyard
Byre
Farmhouse
Some typical barn shapes
Hipped
Gabled
Half-hipped
Barn with lofted stable
Barn with cattle aisles
Small barn with pitching hole
Gabled porch
Hipped porch
Canopy with cheeks
Hipped canopy

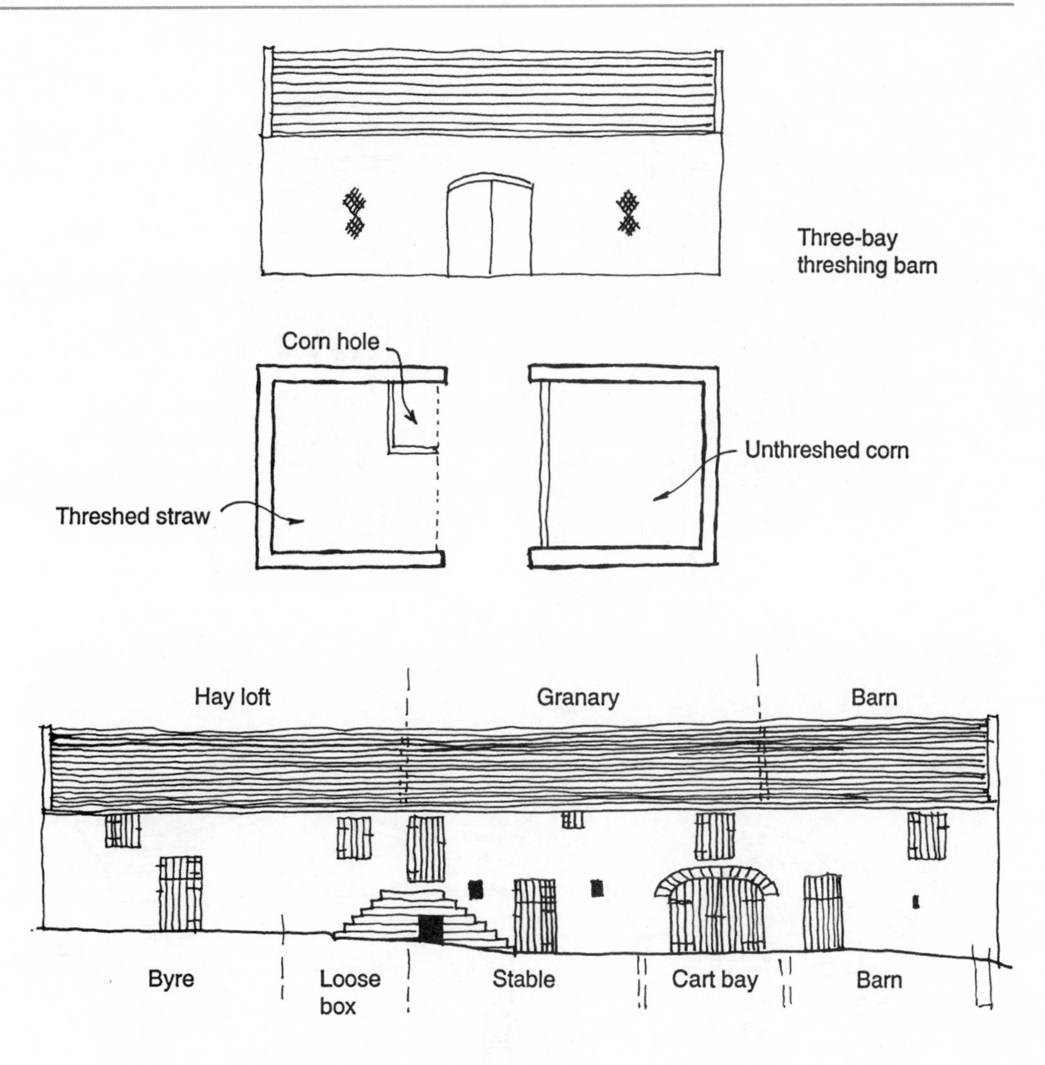

Field house

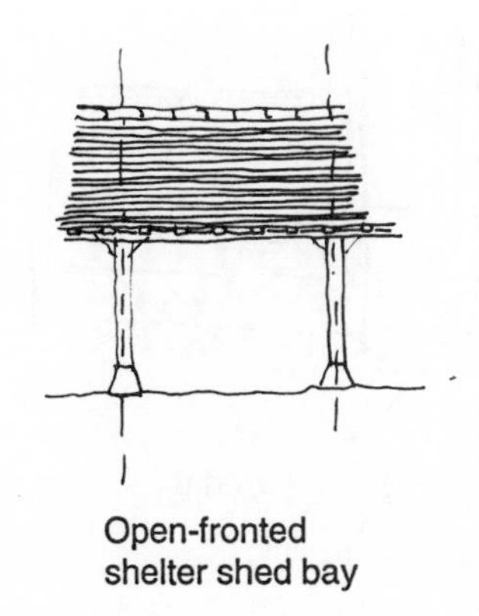

Open-fronted
shelter shed bay

Doorway
Doorcase
Door hood
Portal

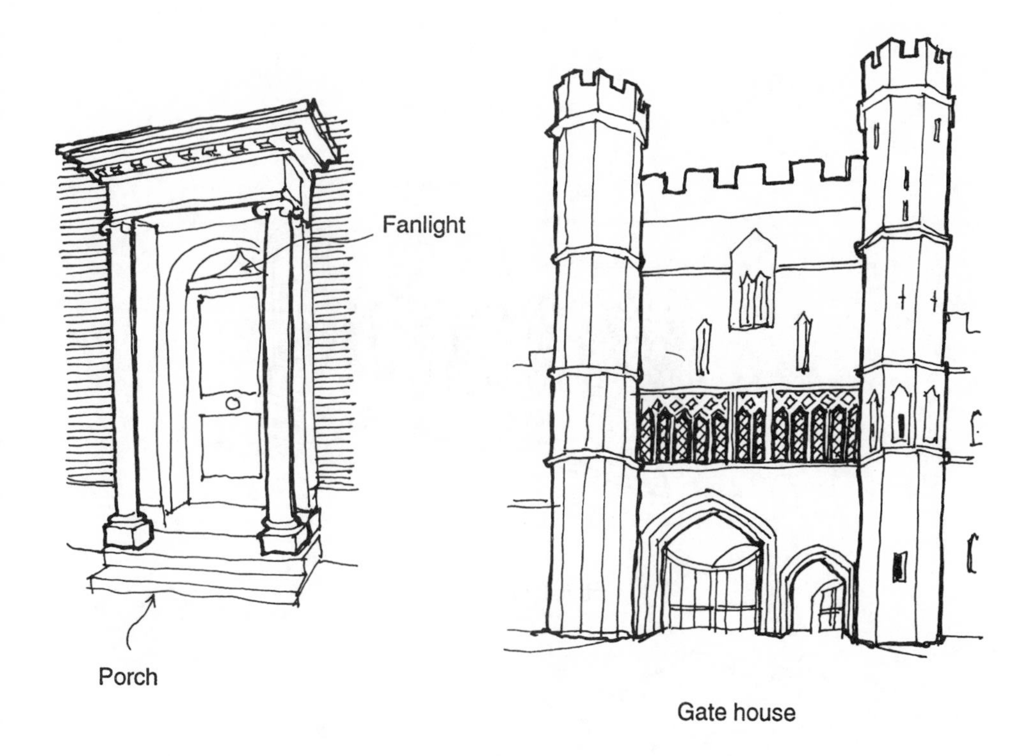

Gate house

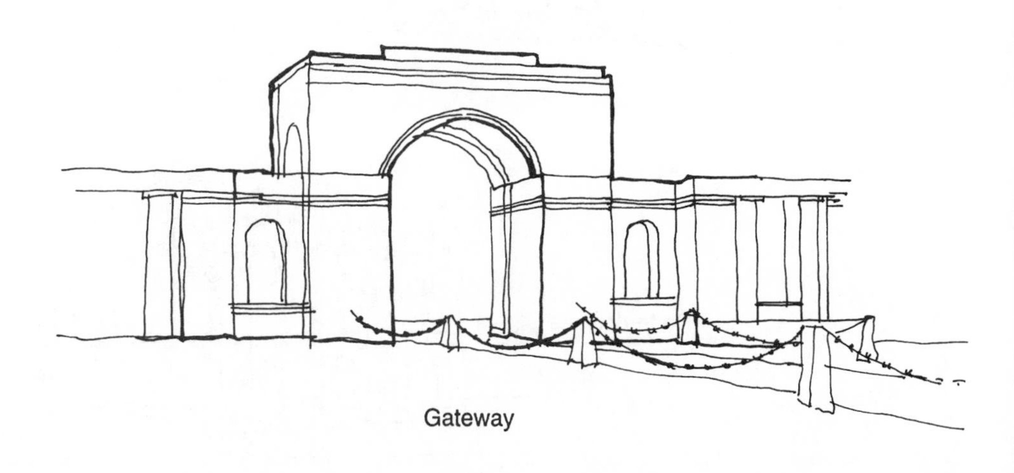

Gateway

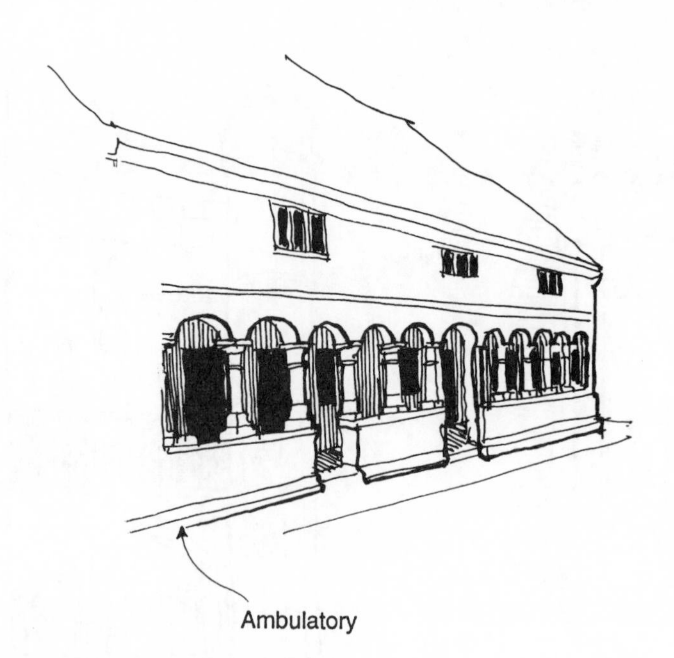
Ambulatory

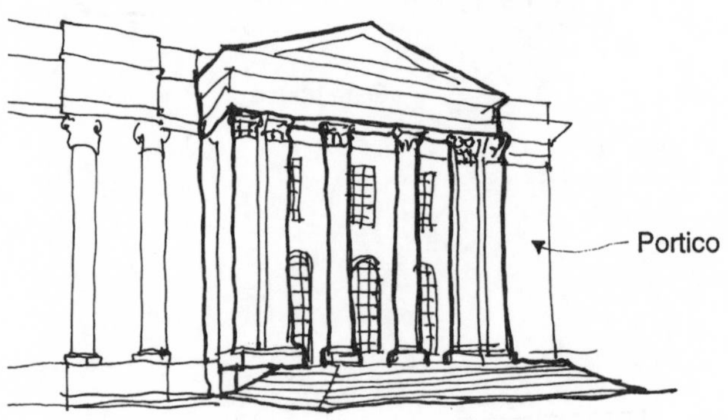
Portico

Porte-cochère

Lychgate

Loggia

Bay window
Bow window
Oriel window

II. Controls

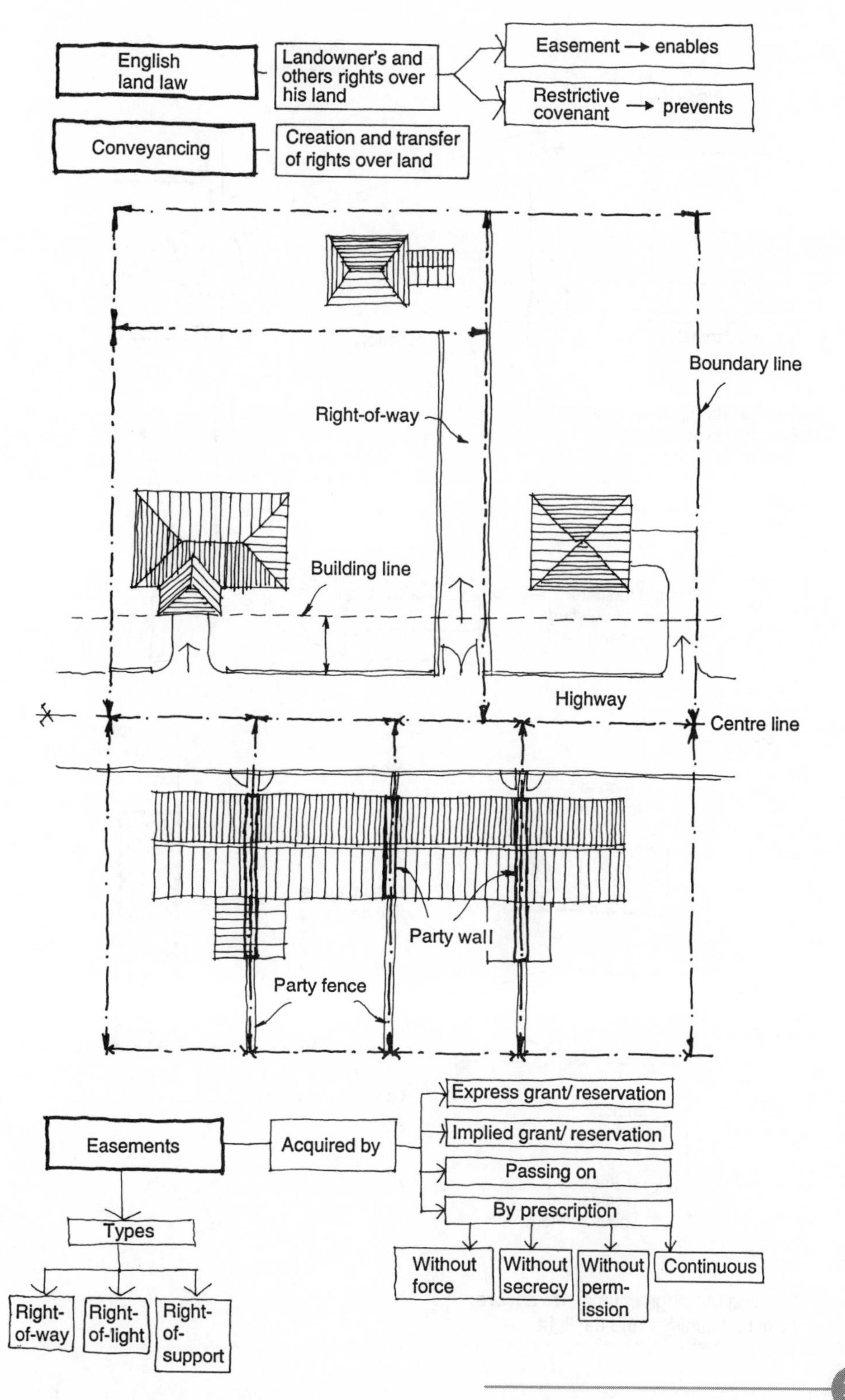
English land law
Landowner's and others rights over his land
Easement → enables
Restrictive covenant → prevents
Conveyancing
Creation and transfer of rights over land
Right-of-way
Boundary line
Building line
Highway
Centre line
Party wall
Party fence
Easements
Acquired by
Express grant/ reservation
Implied grant/ reservation
Passing on
By prescription
Without force
Without secrecy
Without perm-ission
Continuous
Types
Right-of-way
Right-of-light
Right-of-support

Sunlight and daylight

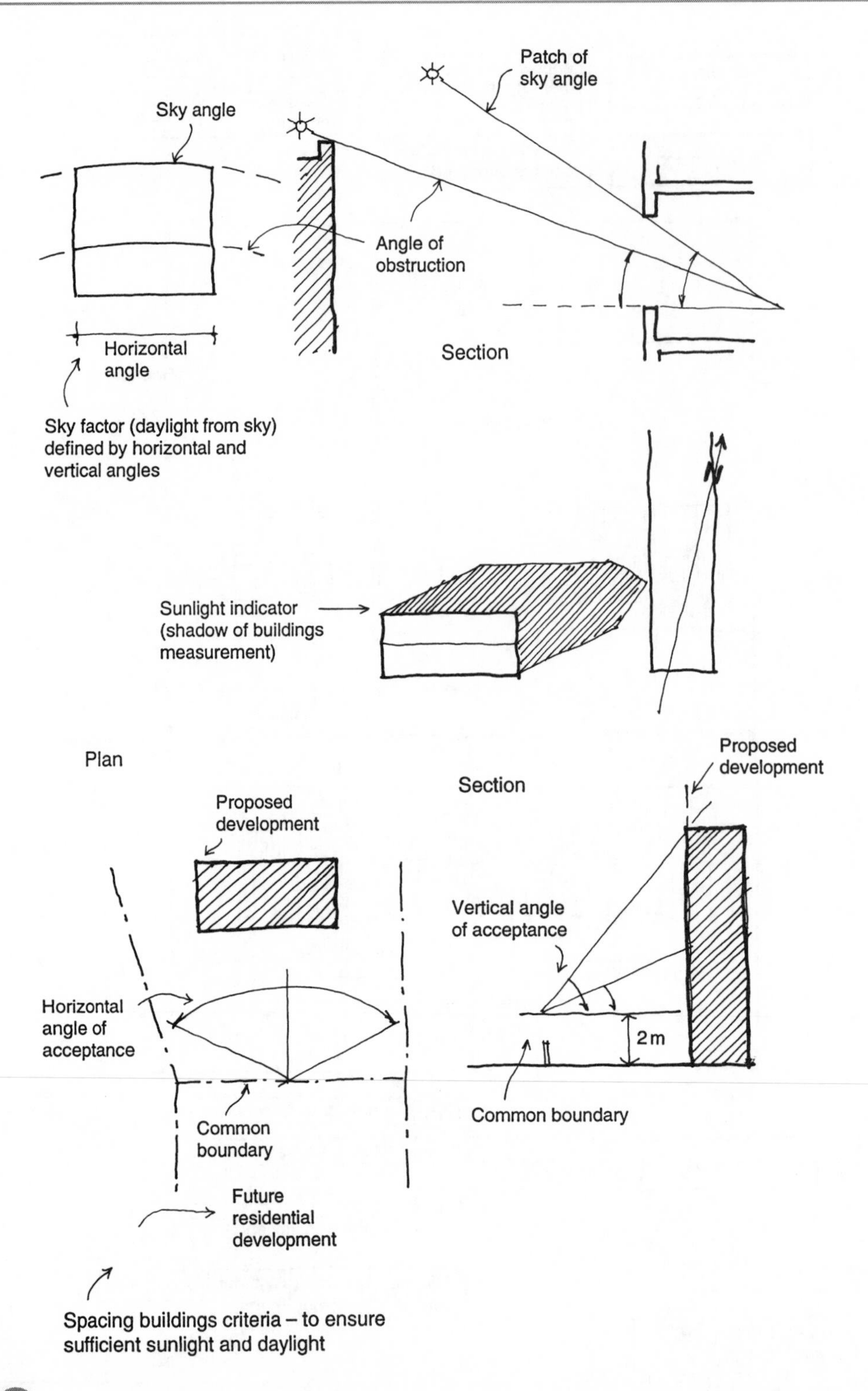

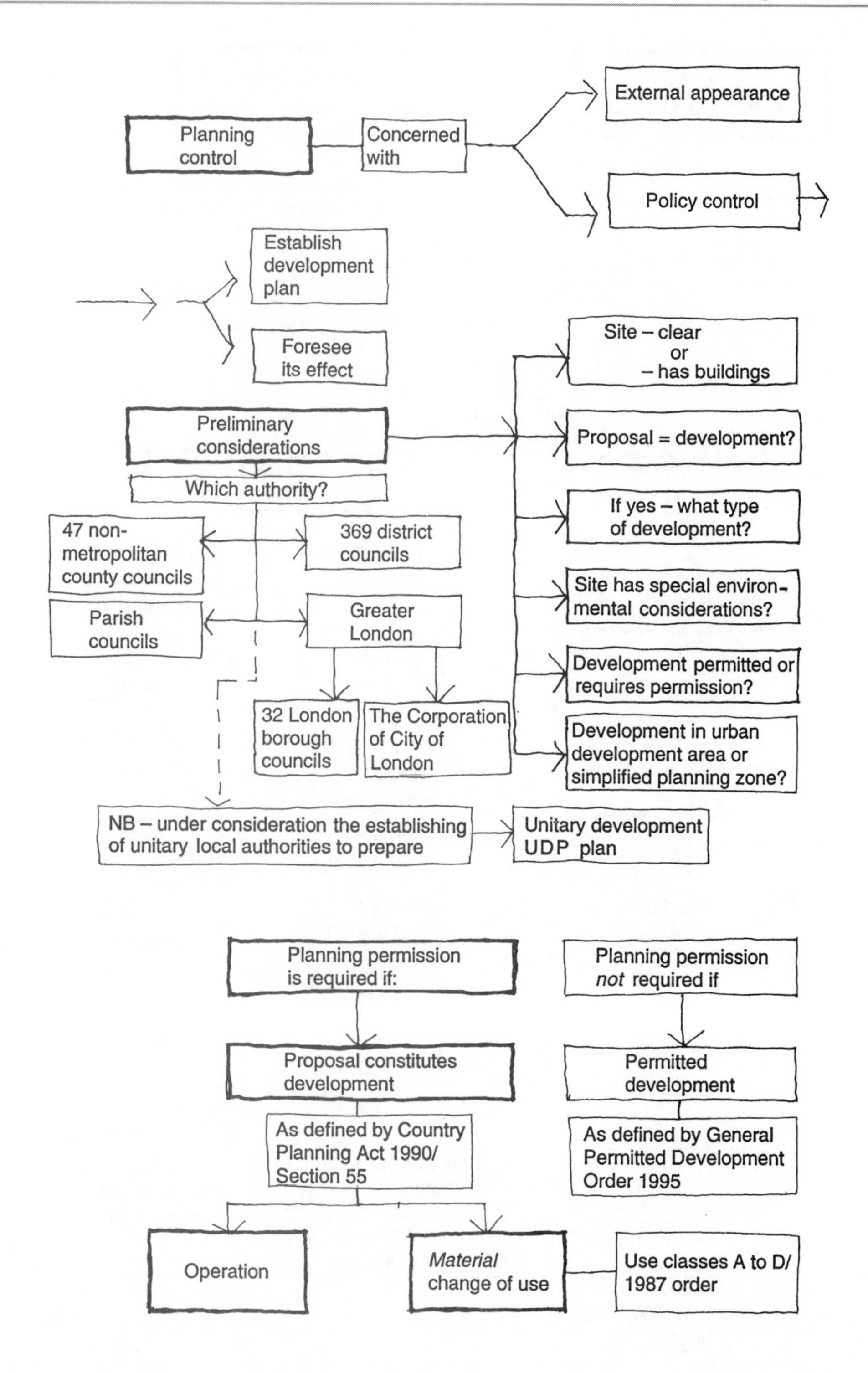
Planning control
Concerned with
External appearance
Policy control
Establish development plan
Foresee its effect
Site – clear or – has buildings
Preliminary considerations
Proposal = development?
If yes – what type of development?
Which authority?
47 non-metropolitan county councils
369 district councils
Site has special environmental considerations?
Parish councils
Greater London
Development permitted or requires permission?
32 London borough councils
The Corporation of City of London
Development in urban development area or simplified planning zone?
NB – under consideration the establishing of unitary local authorities to prepare
Unitary development UDP plan
Planning permission is required if:
Planning permission not required if
Proposal constitutes development
Permitted development
As defined by Country Planning Act 1990/ Section 55
As defined by General Permitted Development Order 1995
Operation
Material change of use
Use classes A to D/ 1987 order

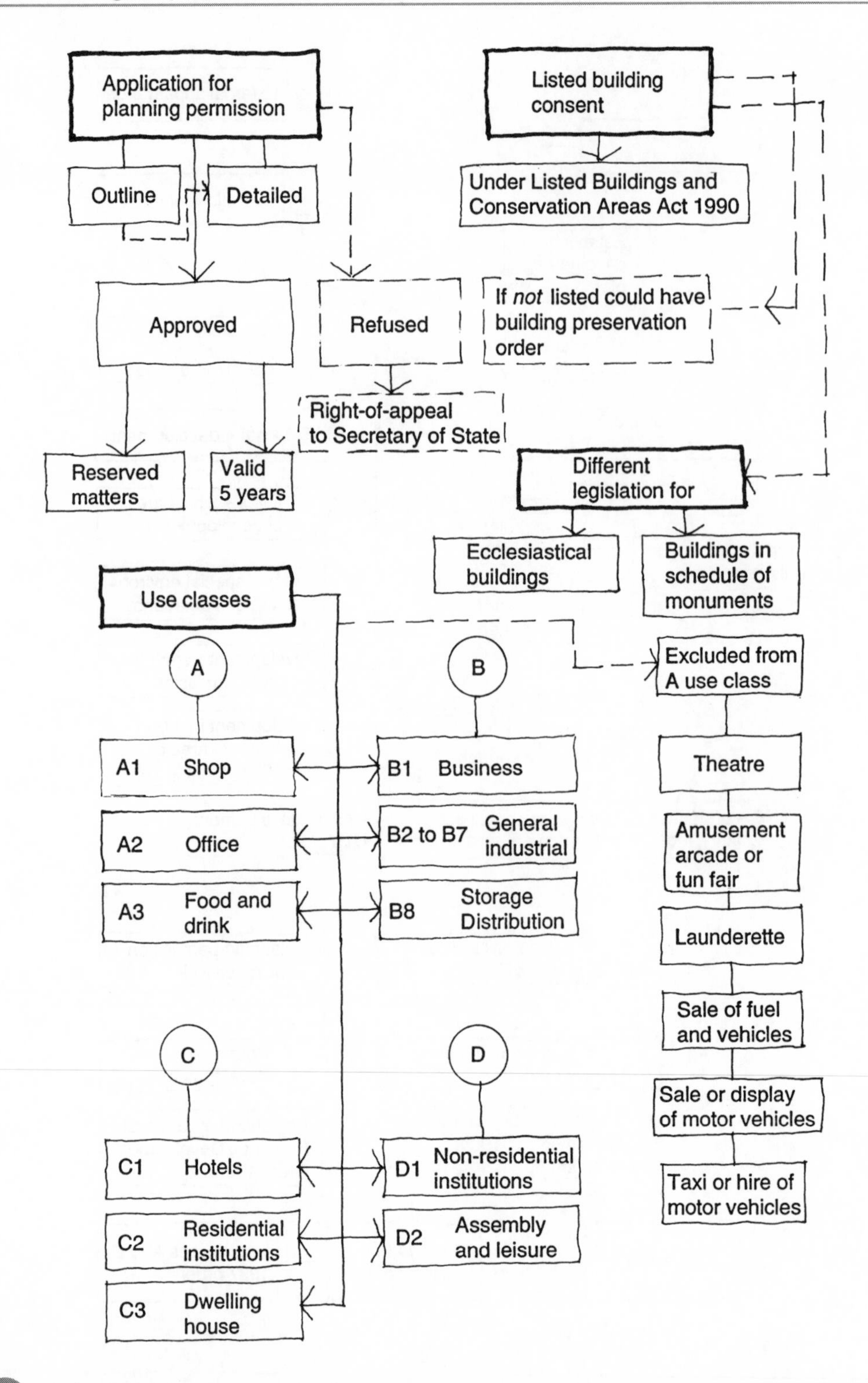

Application for planning permission
Outline
Detailed
Approved
Refused
Reserved matters
Valid 5 years
Right-of-appeal to Secretary of State
Listed building consent
Under Listed Buildings and Conservation Areas Act 1990
If *not* listed could have building preservation order
Different legislation for
Ecclesiastical buildings
Buildings in schedule of monuments
Use classes
A
B
A1 Shop
A2 Office
A3 Food and drink
B1 Business
B2 to B7 General industrial
B8 Storage Distribution
Excluded from A use class
Theatre
Amusement arcade or fun fair
Launderette
Sale of fuel and vehicles
Sale or display of motor vehicles
Taxi or hire of motor vehicles
C
D
C1 Hotels
C2 Residential institutions
C3 Dwelling house
D1 Non-residential institutions
D2 Assembly and leisure

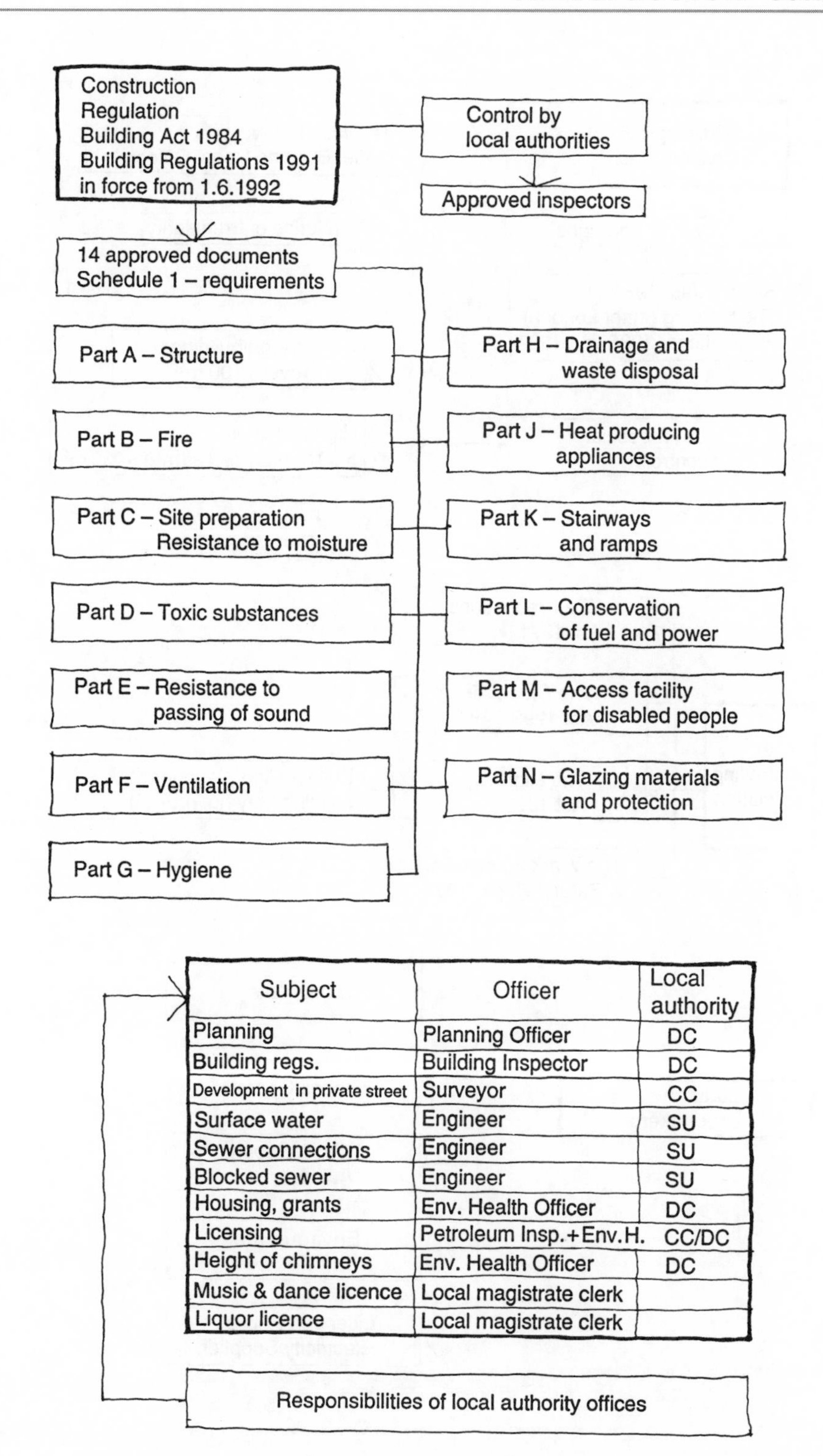

Subject	Officer	Local authority
Planning	Planning Officer	DC
Building regs.	Building Inspector	DC
Development in private street	Surveyor	CC
Surface water	Engineer	SU
Sewer connections	Engineer	SU
Blocked sewer	Engineer	SU
Housing, grants	Env. Health Officer	DC
Licensing	Petroleum Insp.+Env.H.	CC/DC
Height of chimneys	Env. Health Officer	DC
Music & dance licence	Local magistrate clerk	
Liquor licence	Local magistrate clerk	

Building control in London

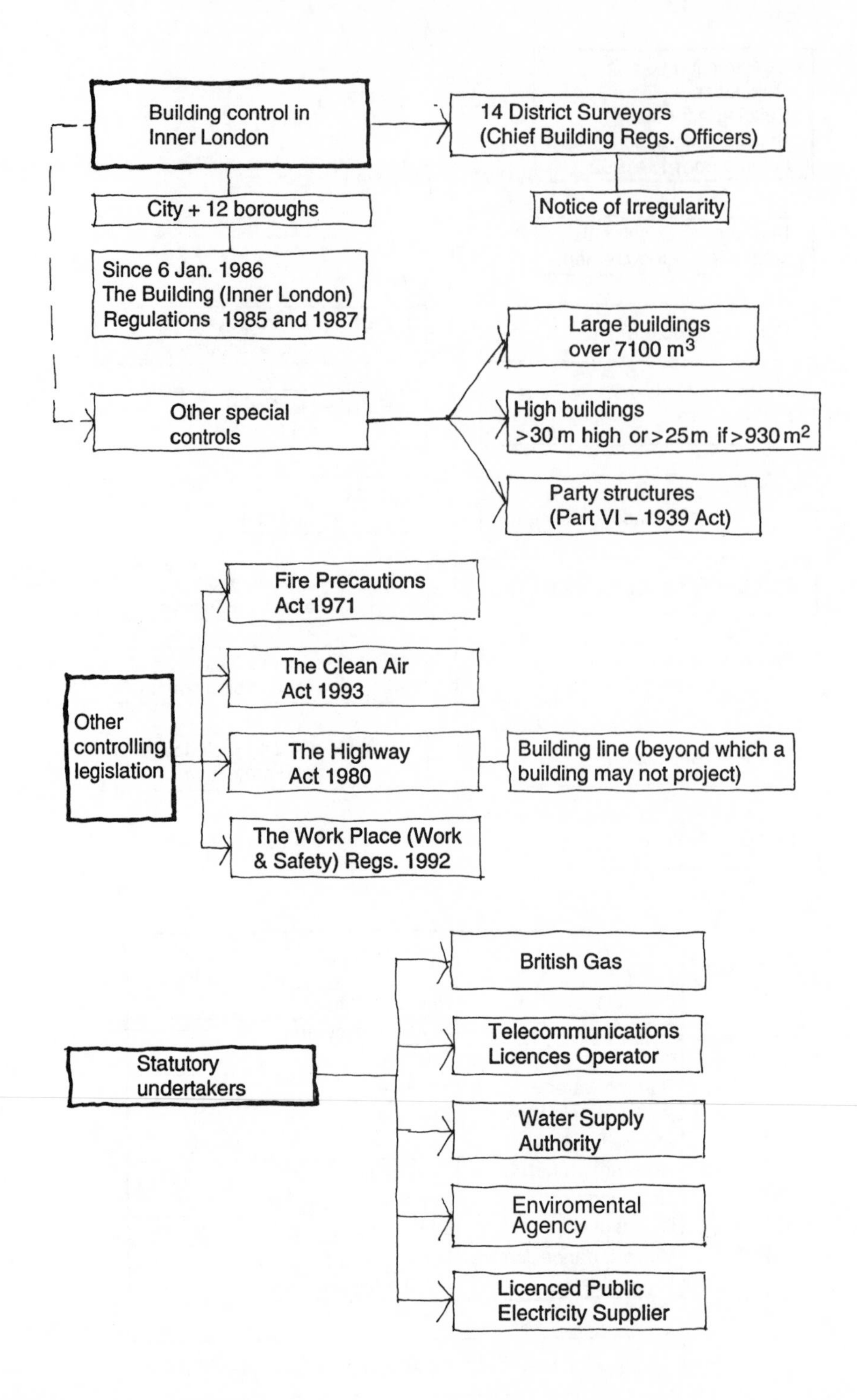

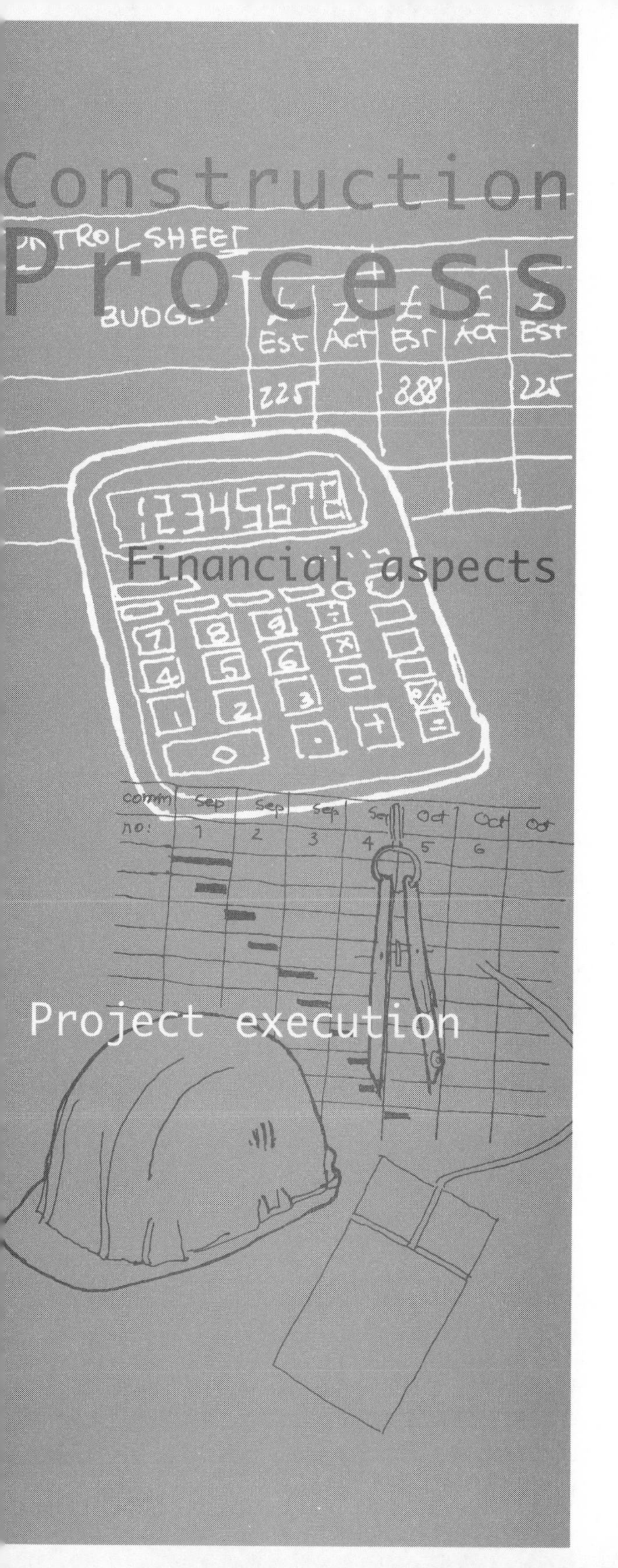

III. Construction Proces

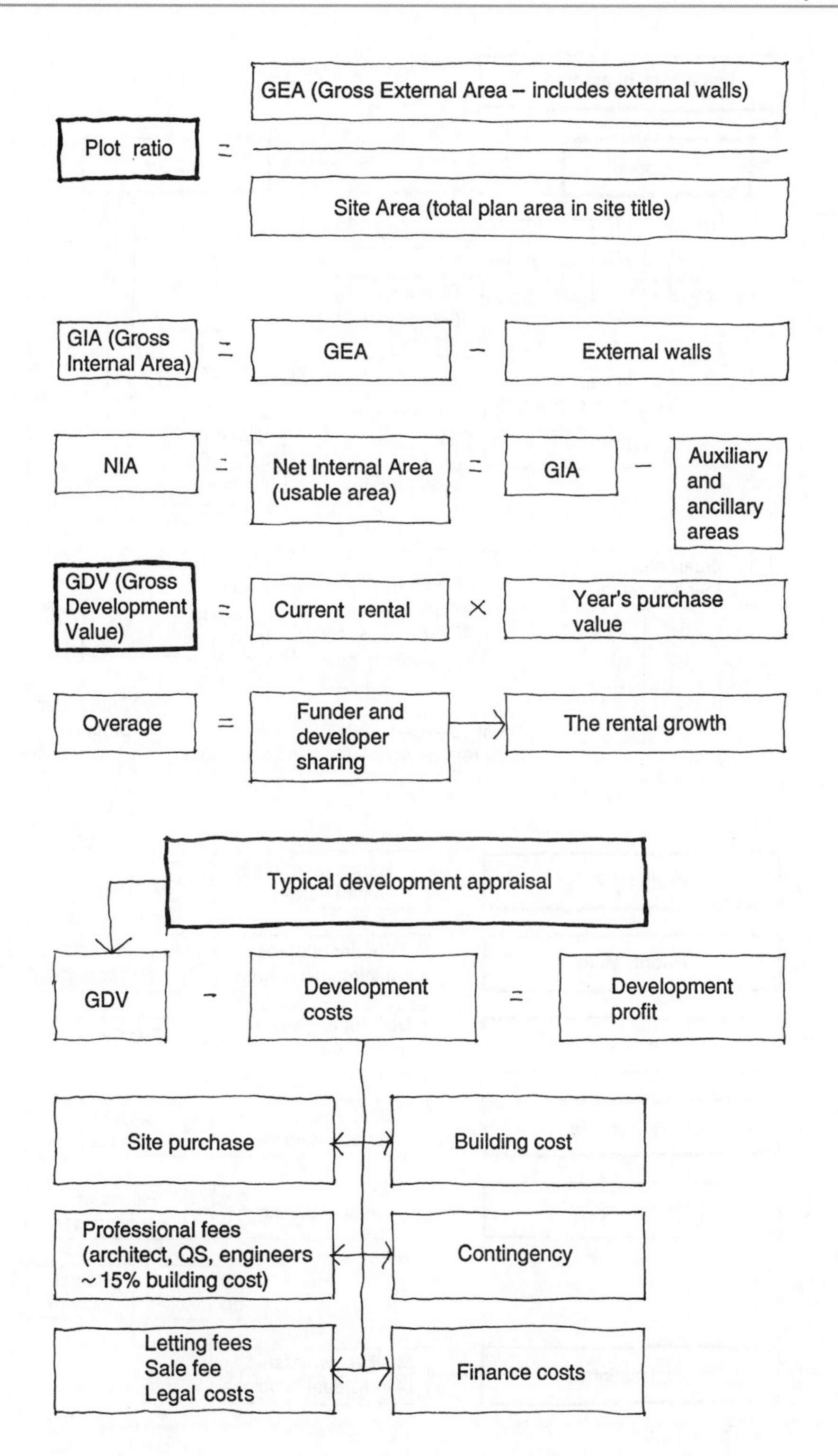

Plot ratio
=
GEA (Gross External Area – includes external walls)
Site Area (total plan area in site title)
GIA (Gross Internal Area)
=
GEA
–
External walls
NIA
=
Net Internal Area (usable area)
=
GIA
–
Auxiliary and ancillary areas
GDV (Gross Development Value)
=
Current rental
×
Year's purchase value
Overage
=
Funder and developer sharing
The rental growth
Typical development appraisal
GDV
–
Development costs
=
Development profit
Site purchase
Building cost
Professional fees (architect, QS, engineers ~15% building cost)
Contingency
Letting fees
Sale fee
Legal costs
Finance costs

Development funding methods

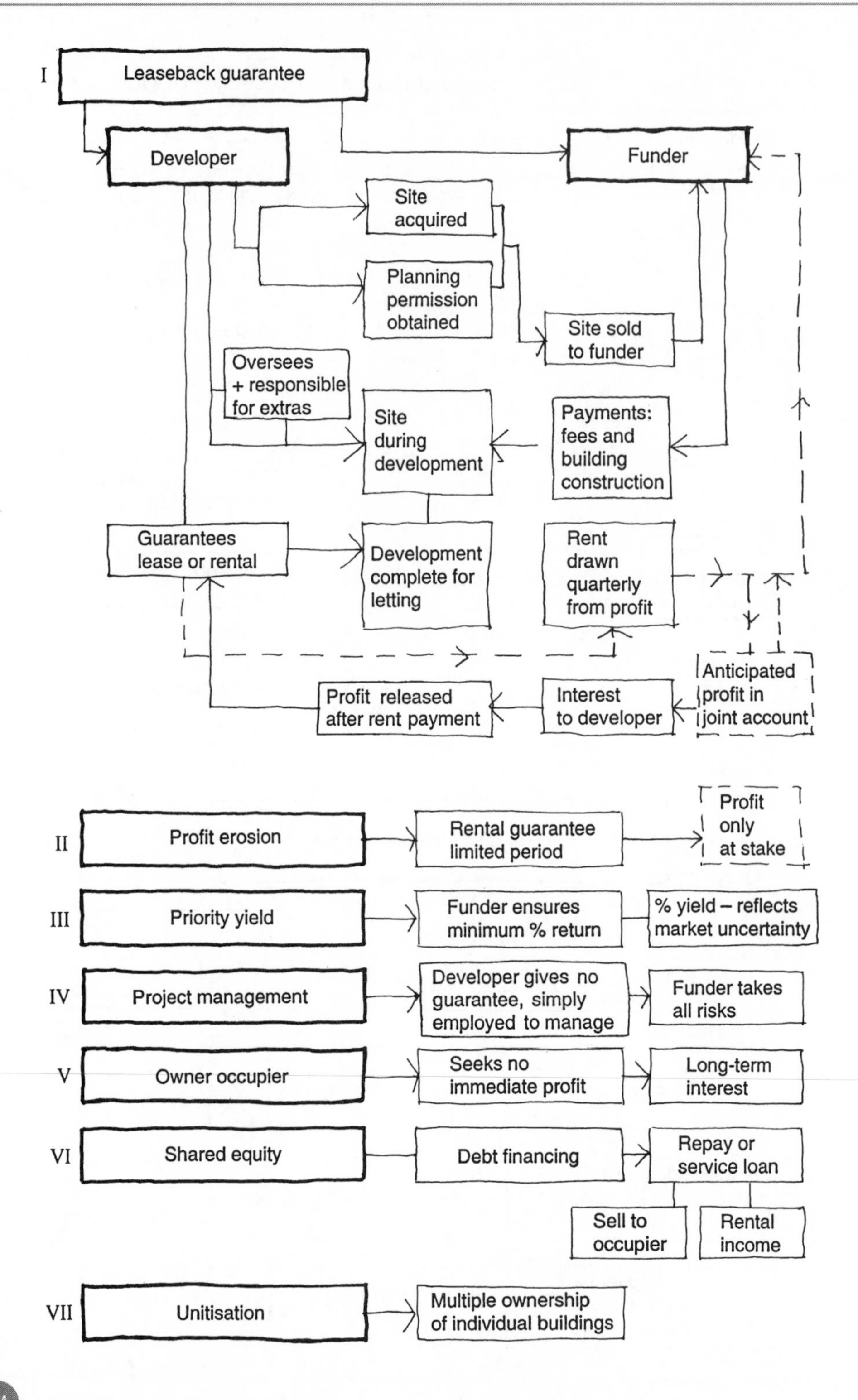

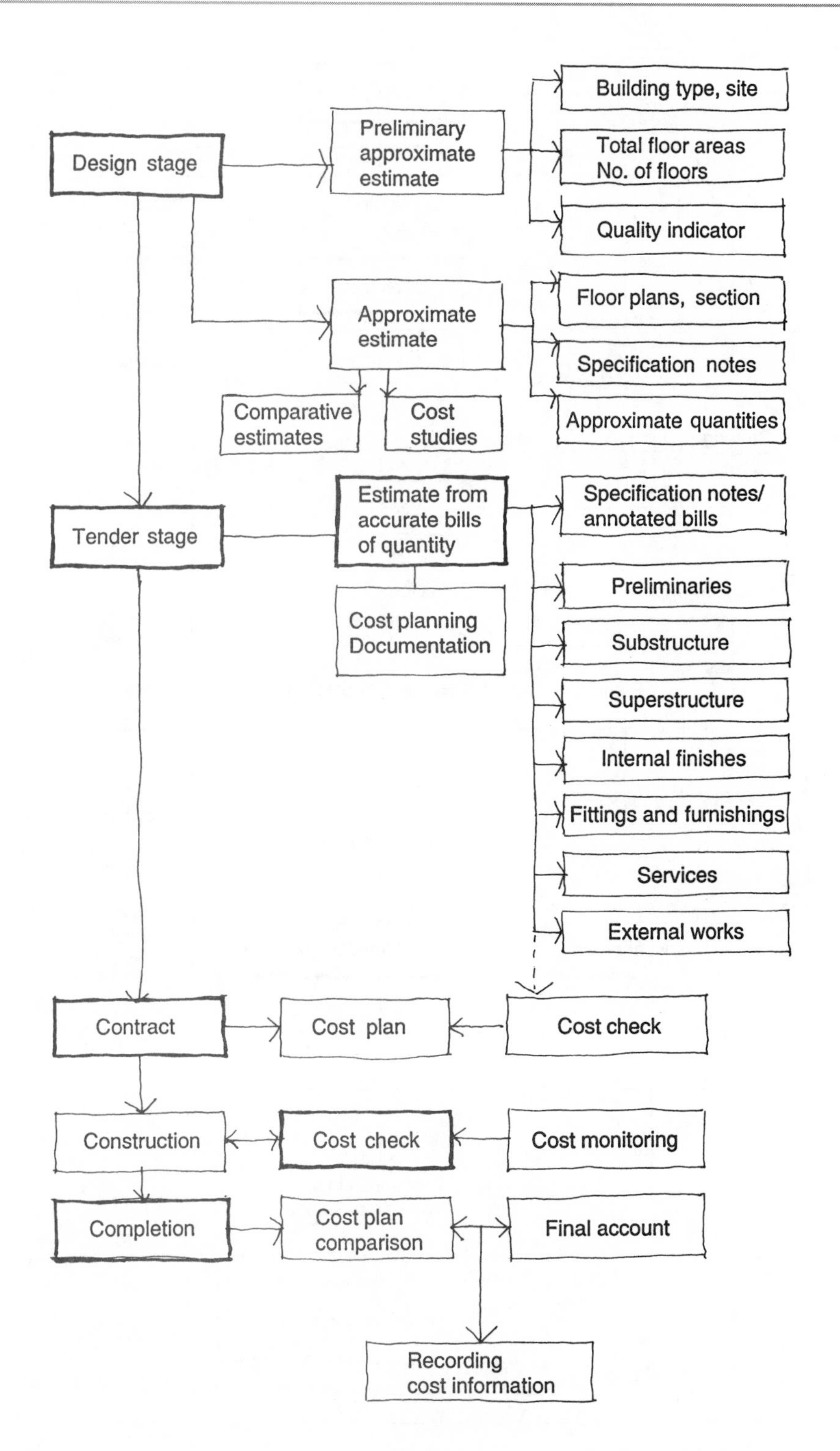
Design stage
Preliminary approximate estimate
Building type, site
Total floor areas
No. of floors
Quality indicator
Approximate estimate
Floor plans, section
Specification notes
Approximate quantities
Comparative estimates
Cost studies
Tender stage
Estimate from accurate bills of quantity
Specification notes/ annotated bills
Cost planning Documentation
Preliminaries
Substructure
Superstructure
Internal finishes
Fittings and furnishings
Services
External works
Contract
Cost plan
Cost check
Construction
Cost check
Cost monitoring
Completion
Cost plan comparison
Final account
Recording cost information

Project design

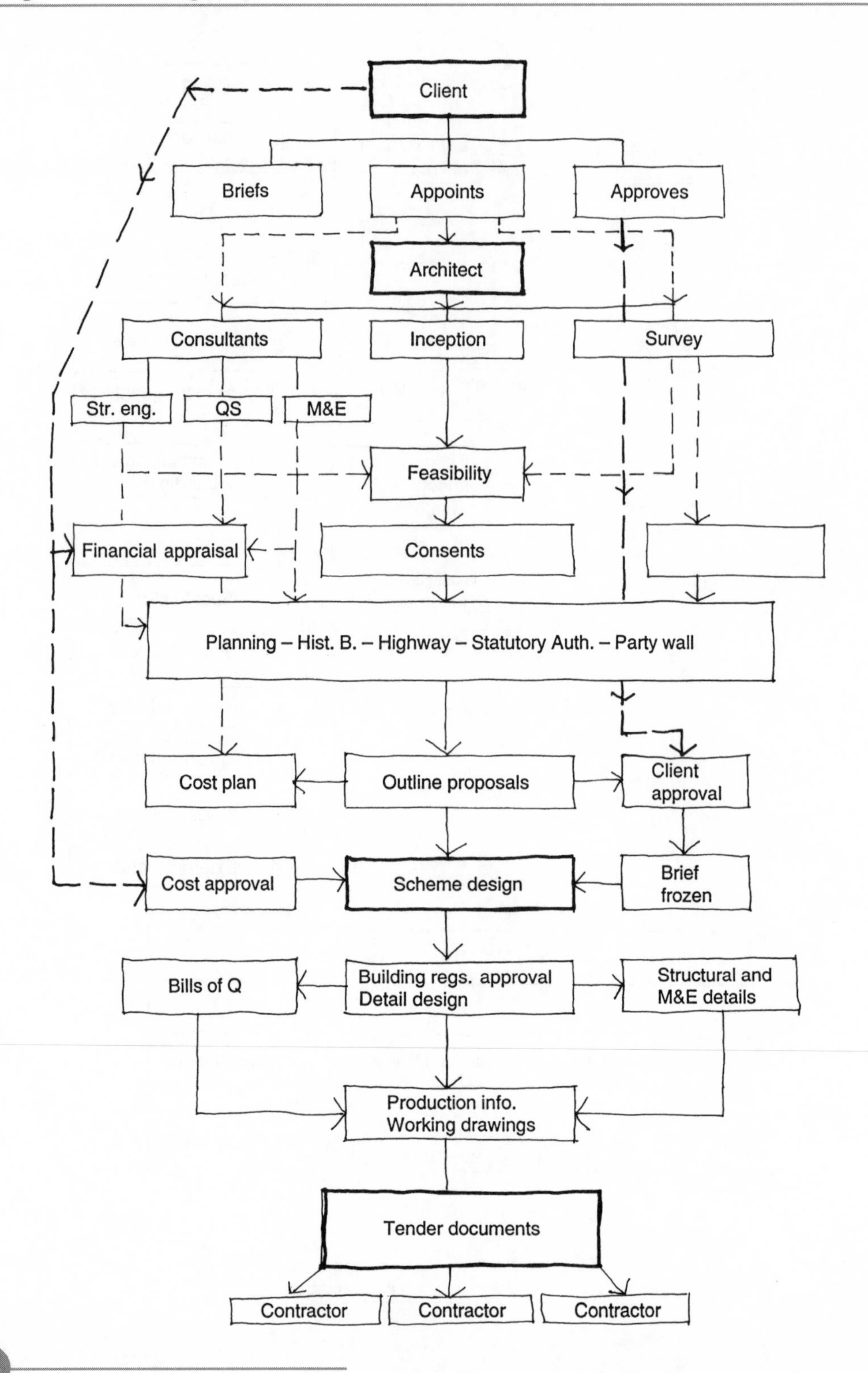

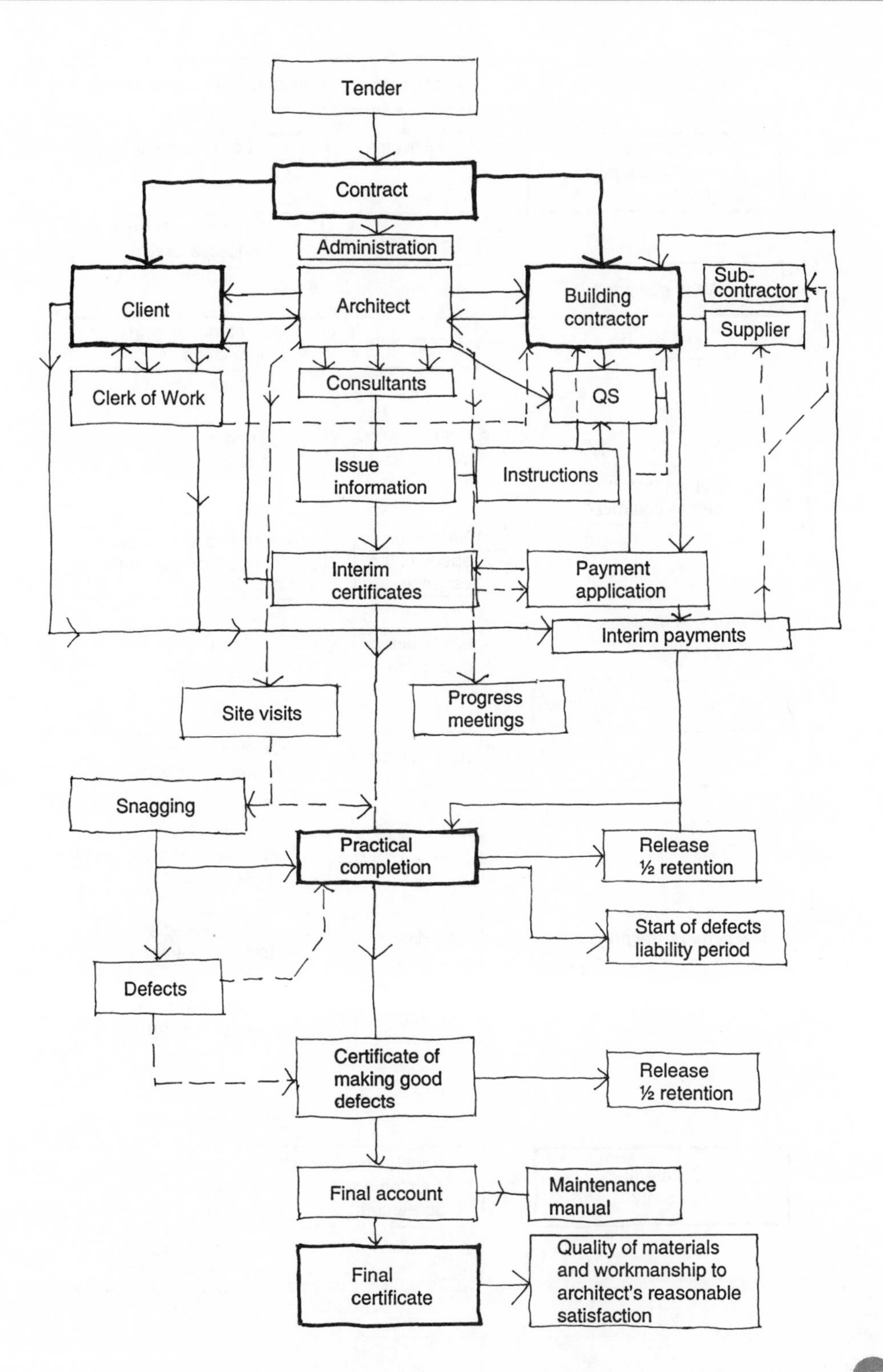
Tender
Contract
Administration
Client
Architect
Building contractor
Sub-contractor
Supplier
Clerk of Work
Consultants
QS
Issue information
Instructions
Interim certificates
Payment application
Interim payments
Site visits
Progress meetings
Snagging
Practical completion
Release ½ retention
Start of defects liability period
Defects
Certificate of making good defects
Release ½ retention
Final account
Maintenance manual
Final certificate
Quality of materials and workmanship to architect's reasonable satisfaction

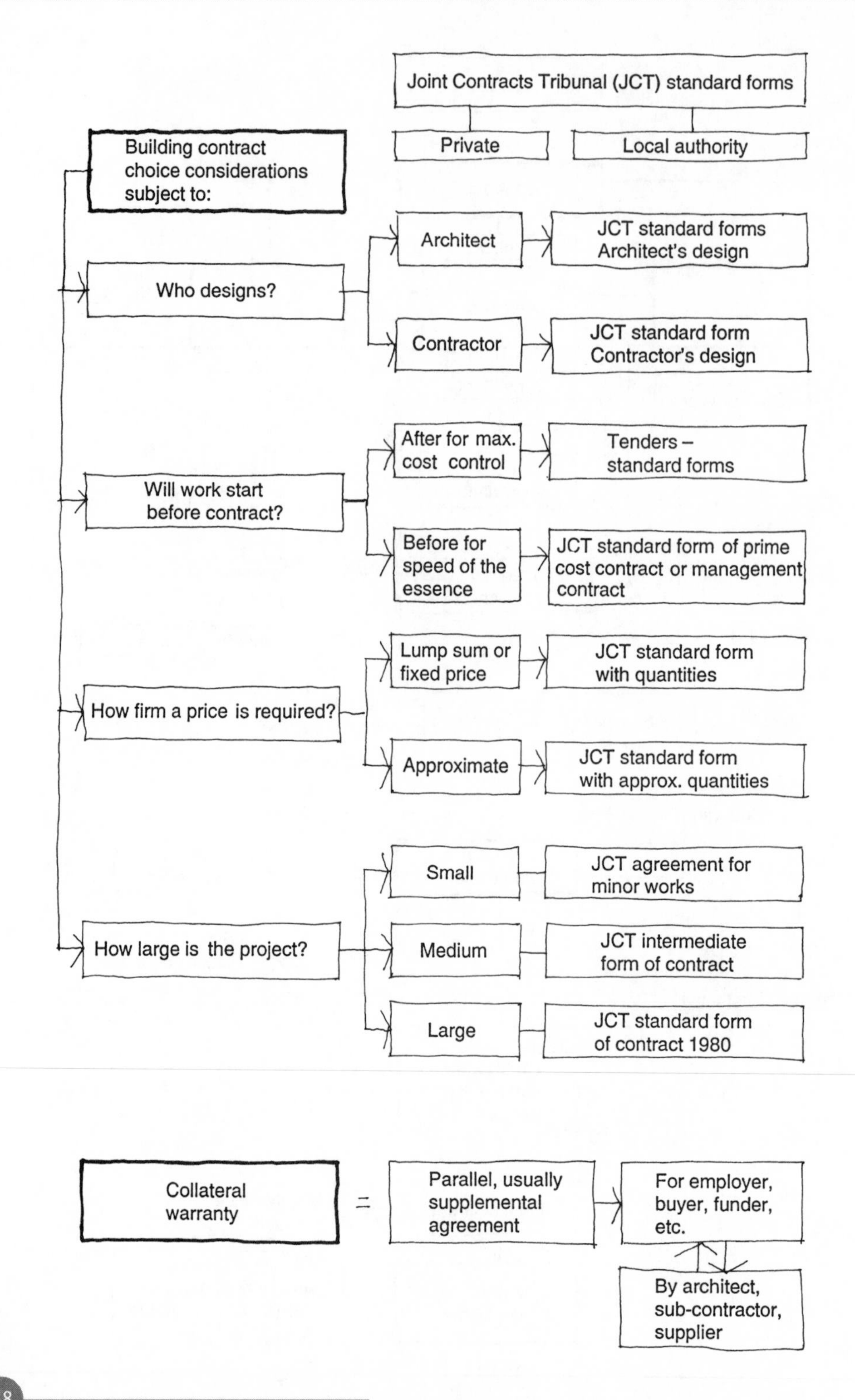

Joint Contracts Tribunal (JCT) standard forms
Private
Local authority
Building contract choice considerations subject to:
Who designs?
Architect
JCT standard forms Architect's design
Contractor
JCT standard form Contractor's design
Will work start before contract?
After for max. cost control
Tenders – standard forms
Before for speed of the essence
JCT standard form of prime cost contract or management contract
How firm a price is required?
Lump sum or fixed price
JCT standard form with quantities
Approximate
JCT standard form with approx. quantities
How large is the project?
Small
JCT agreement for minor works
Medium
JCT intermediate form of contract
Large
JCT standard form of contract 1980
Collateral warranty
=
Parallel, usually supplemental agreement
For employer, buyer, funder, etc.
By architect, sub-contractor, supplier

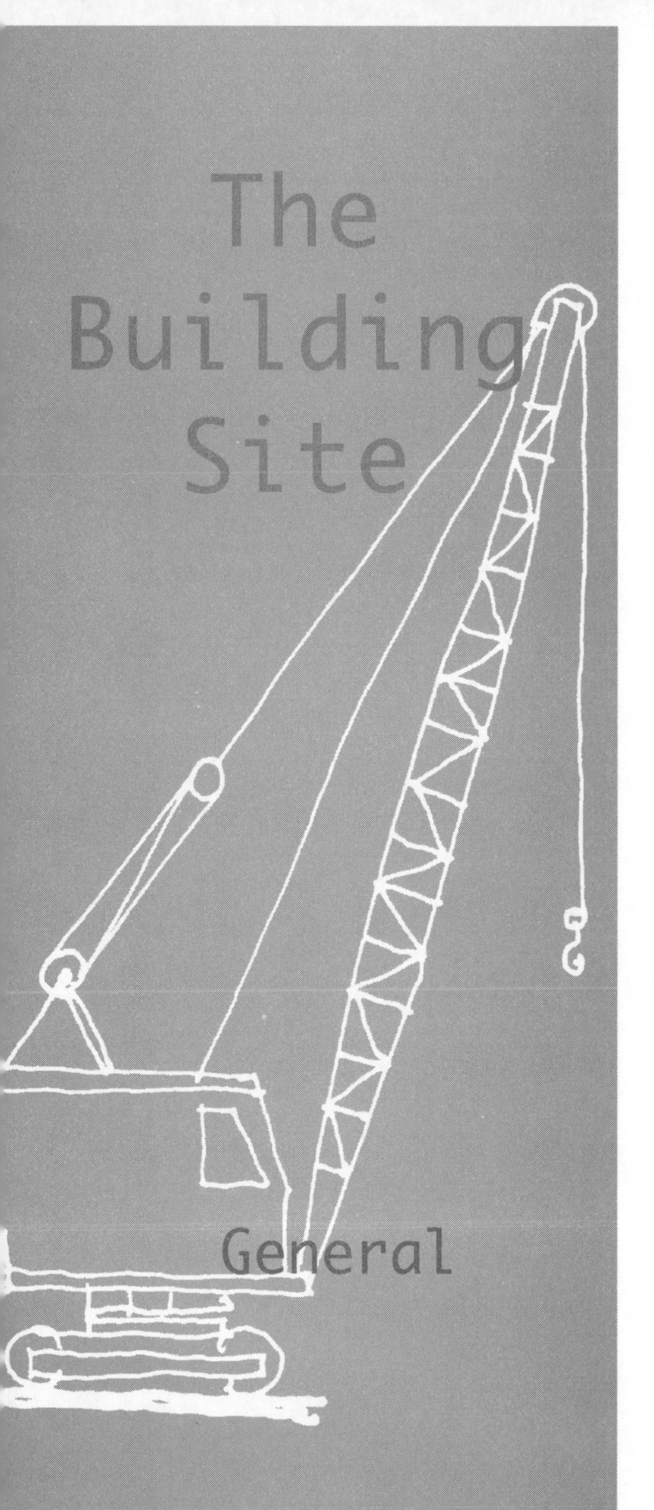

IV. The Building Site

General

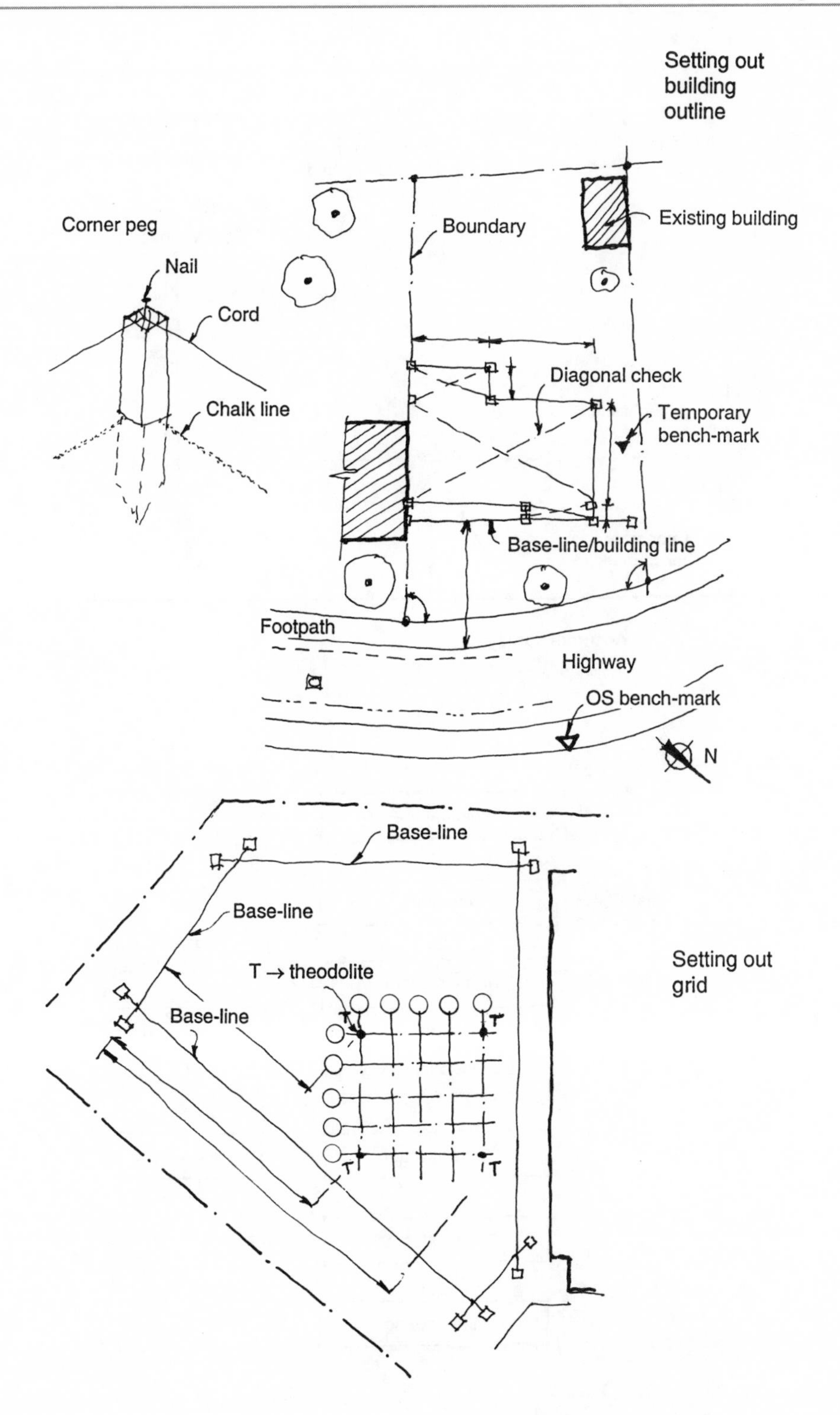
Setting out building outline
Corner peg
Nail
Cord
Chalk line
Boundary
Existing building
Diagonal check
Temporary bench-mark
Base-line/building line
Footpath
Highway
OS bench-mark
N
Base-line
Base-line
T → theodolite
Base-line
T
T
T
T
Setting out grid

Accurate survey

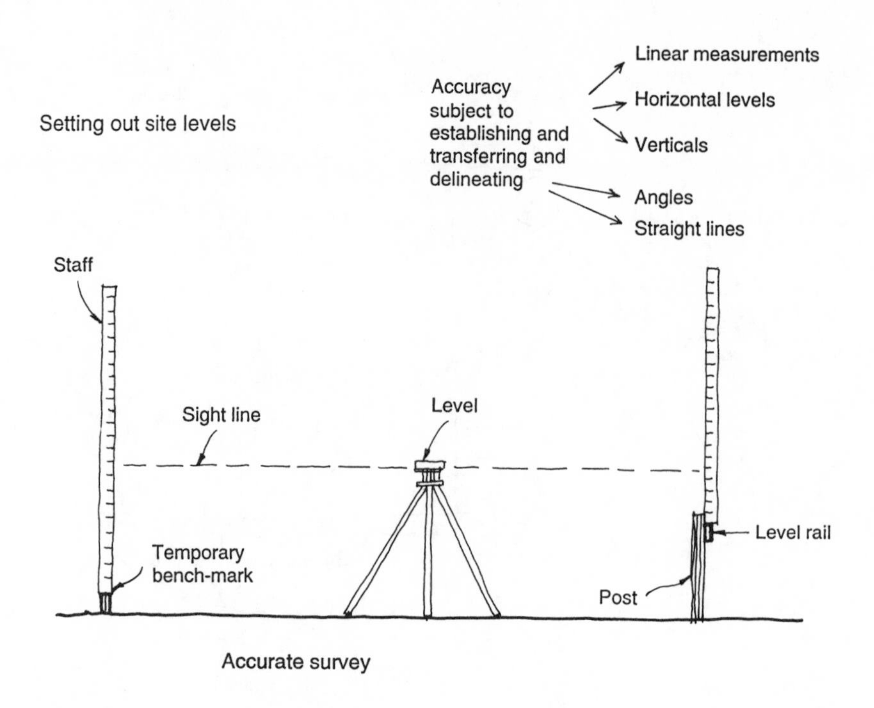

Accurate survey

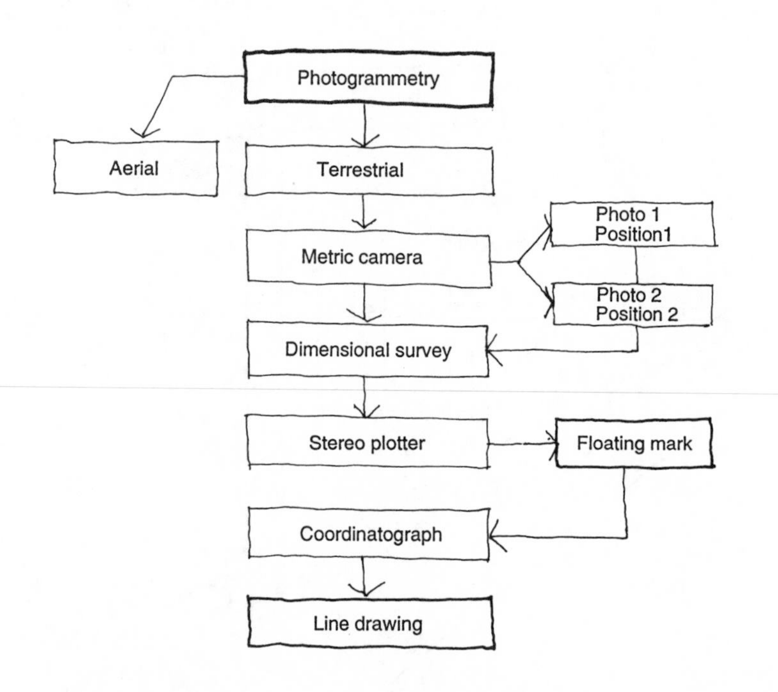

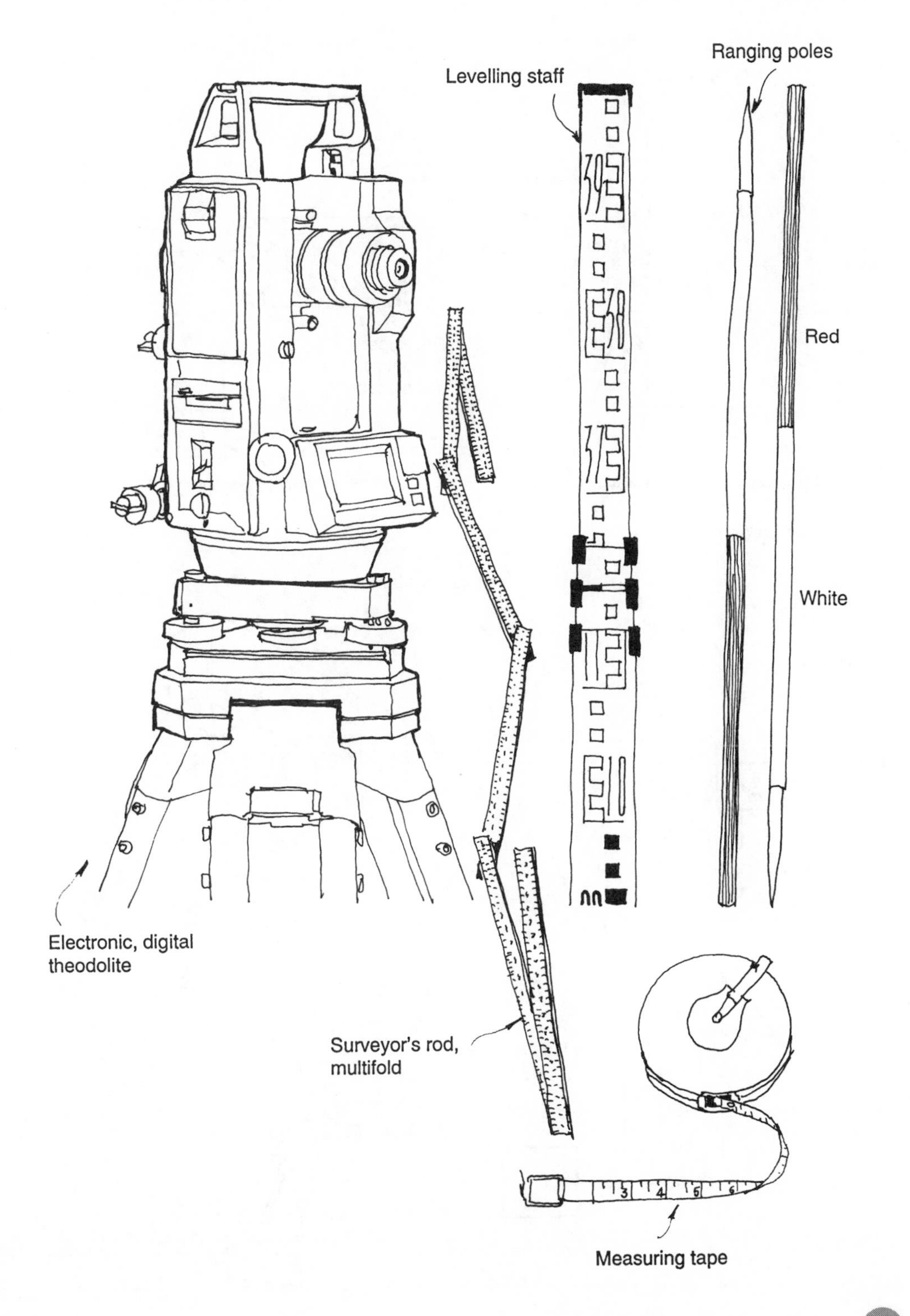
Ranging poles
Levelling staff
Red
White
Electronic, digital theodolite
Surveyor's rod, multifold
Measuring tape

Non-destructive survey: method 1

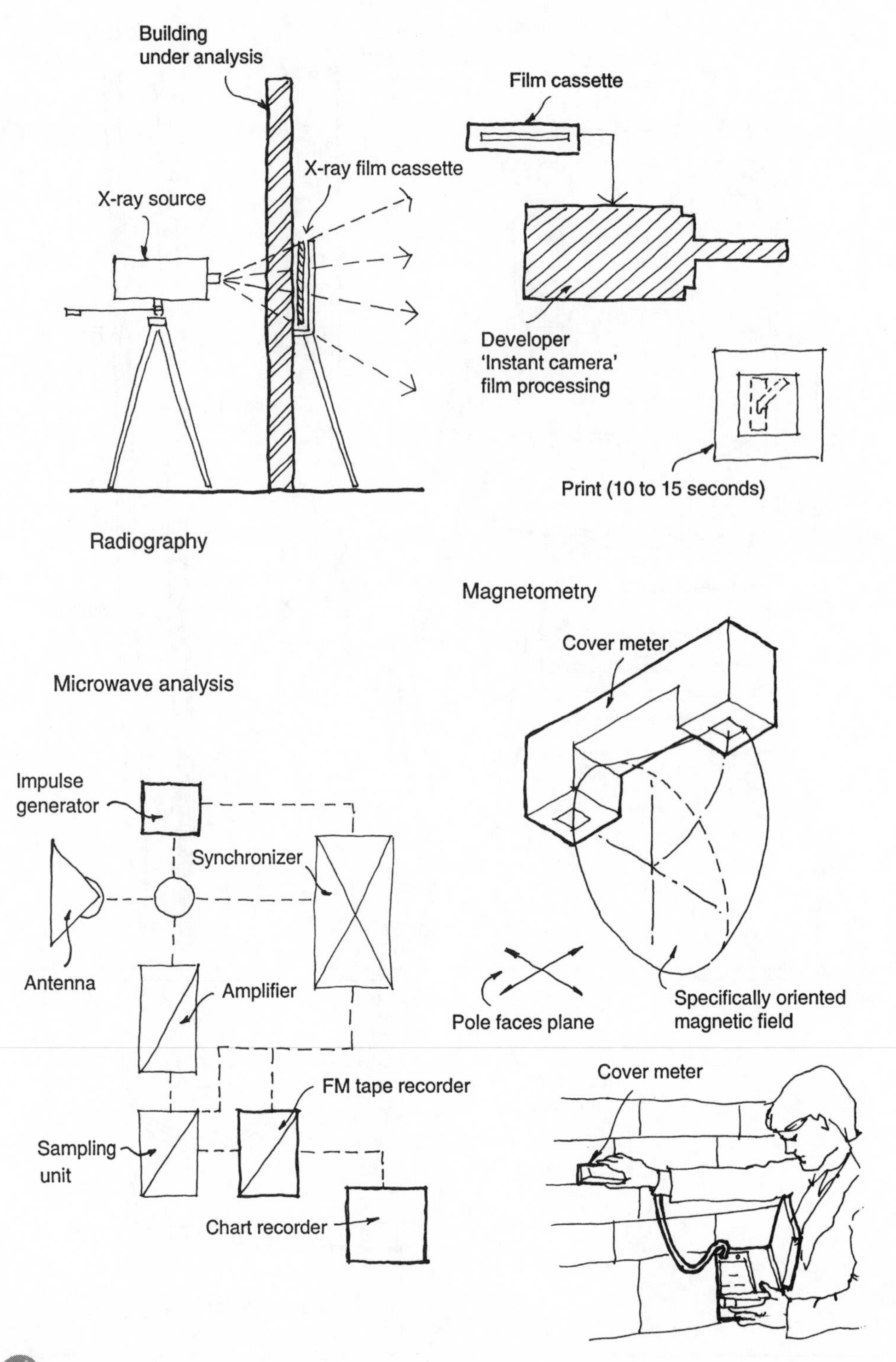

LVDT (Linear Variable Differential Transducers)

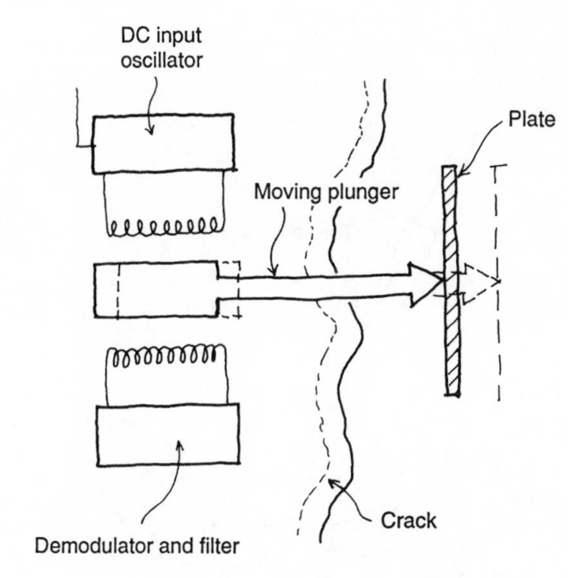

Transducer monitoring

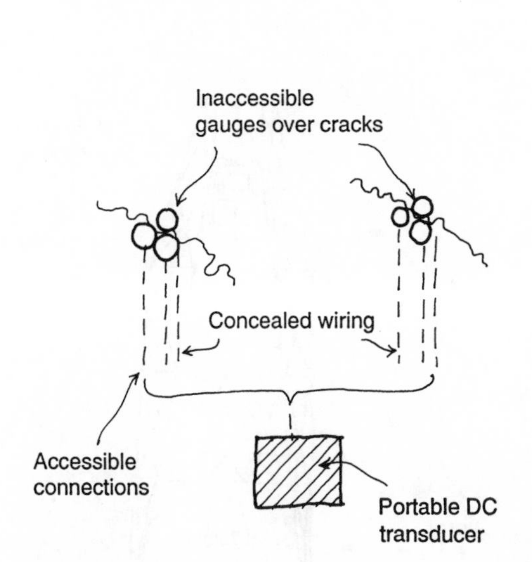

Portable read-out unit

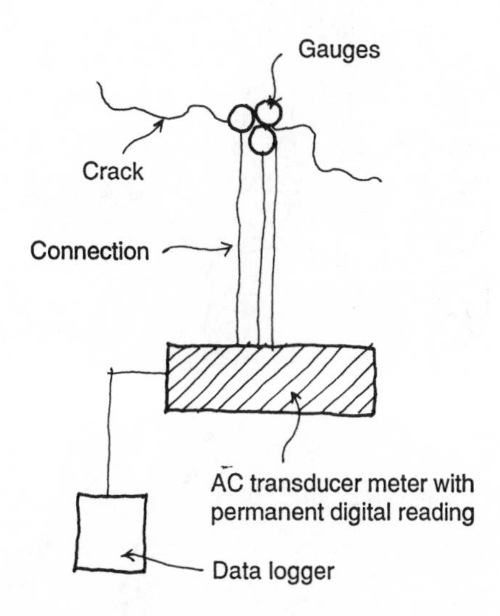

Monitoring with permanent transducer meter

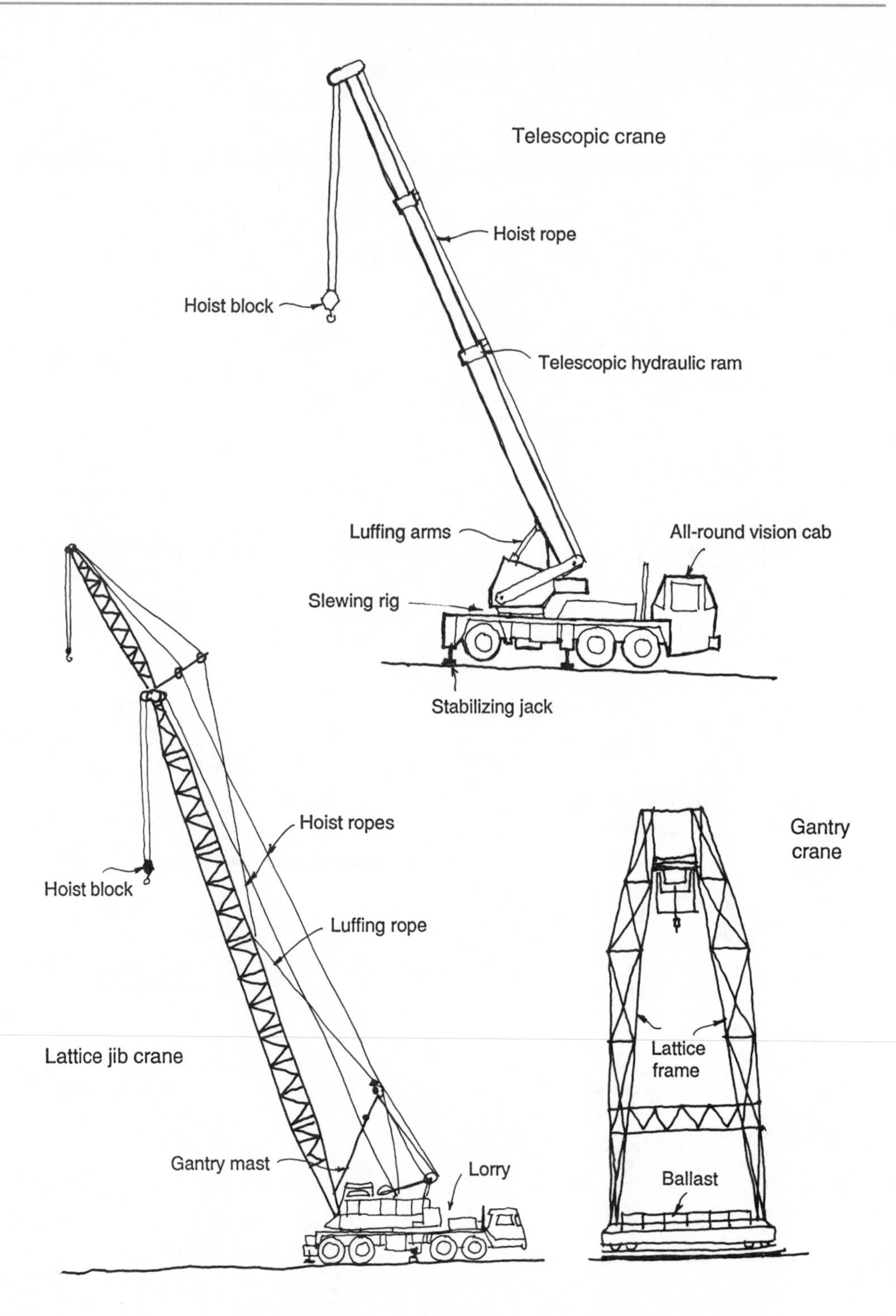

Telescopic crane
Hoist rope
Hoist block
Telescopic hydraulic ram
Luffing arms
All-round vision cab
Slewing rig
Stabilizing jack
Hoist ropes
Hoist block
Luffing rope
Gantry crane
Lattice jib crane
Lattice frame
Gantry mast
Lorry
Ballast

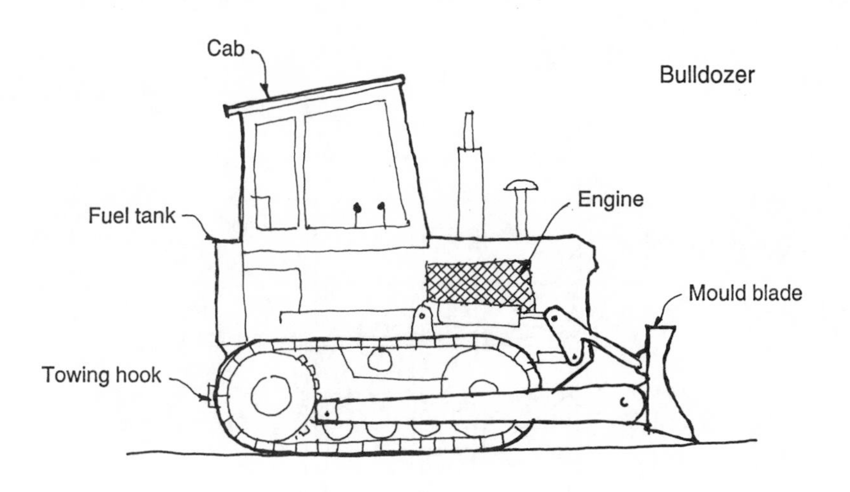
Cab
Bulldozer
Fuel tank
Engine
Mould blade
Towing hook

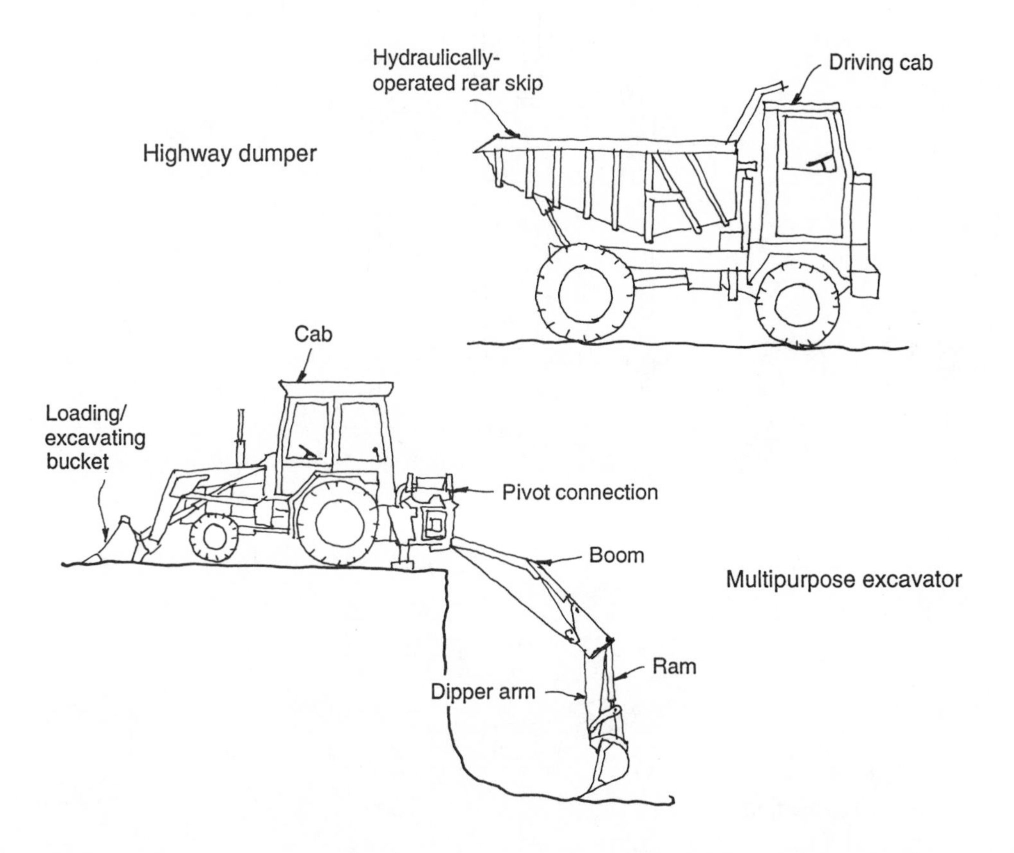
Hydraulically-
operated rear skip
Driving cab
Highway dumper
Cab
Loading/
excavating
bucket
Pivot connection
Boom
Multipurpose excavator
Ram
Dipper arm

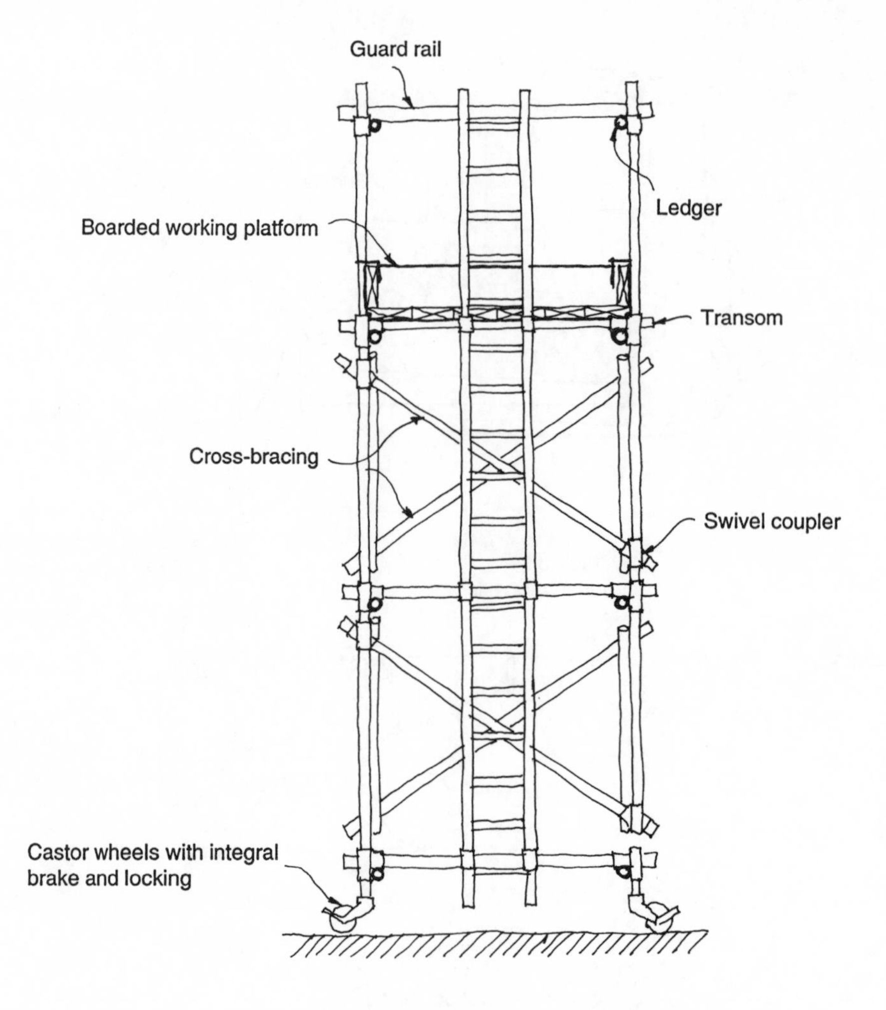
Scaffolding tower
Guard rail
Ledger
Boarded working platform
Transom
Cross-bracing
Swivel coupler
Castor wheels with integral brake and locking

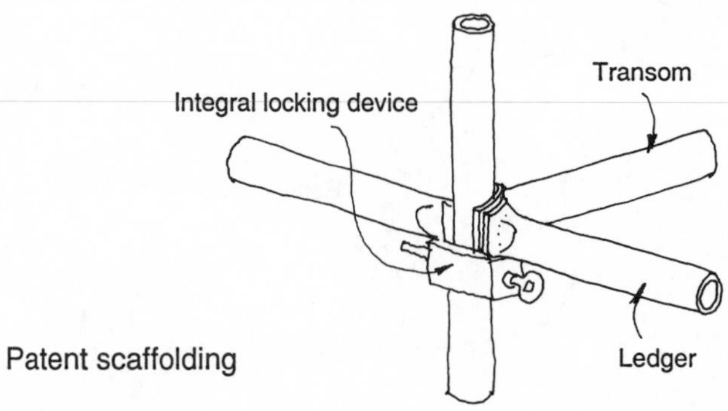
Transom
Integral locking device
Patent scaffolding
Ledger

Independent scaffold
(truss-out)

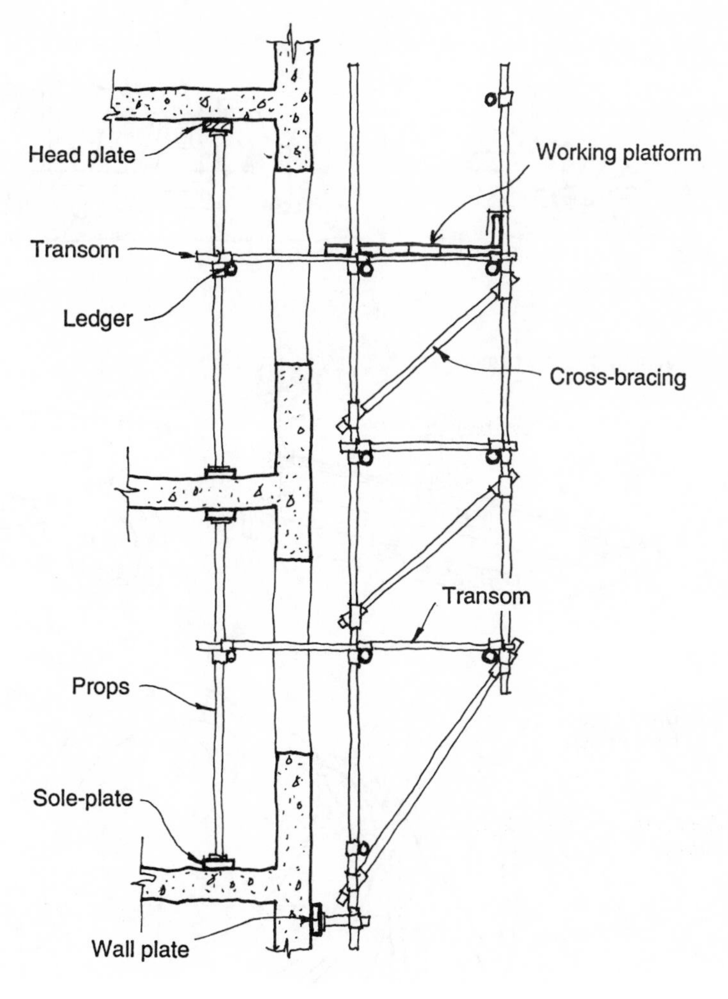

Scaffolding board

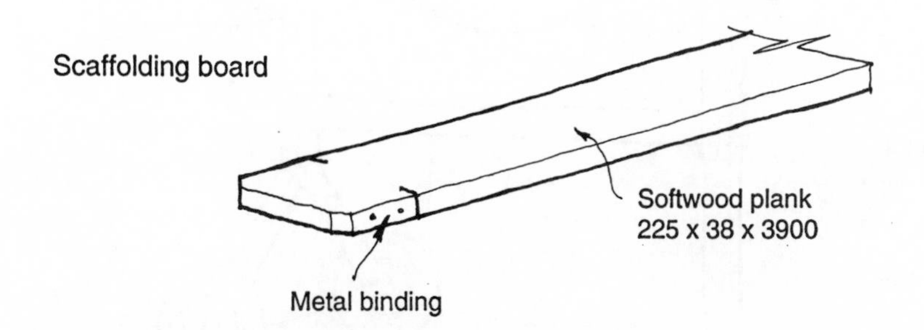

Concrete plant

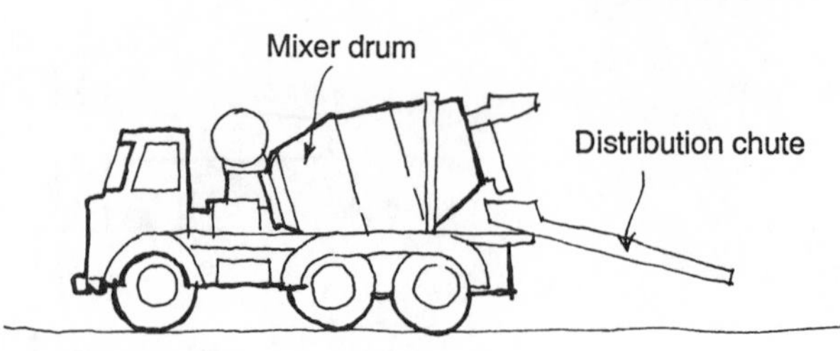

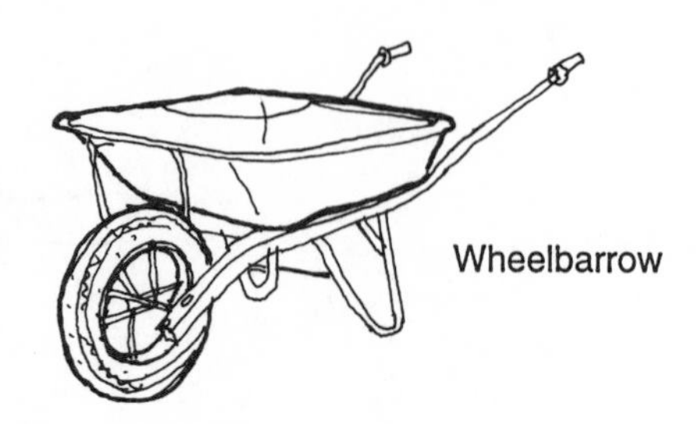

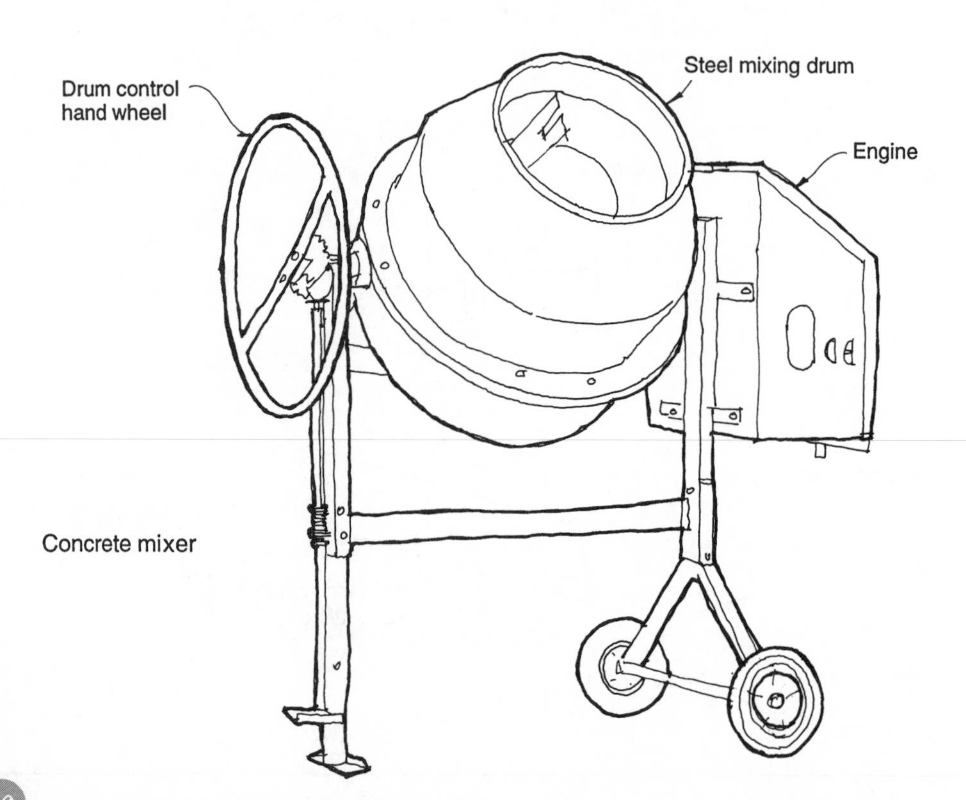

Generator set

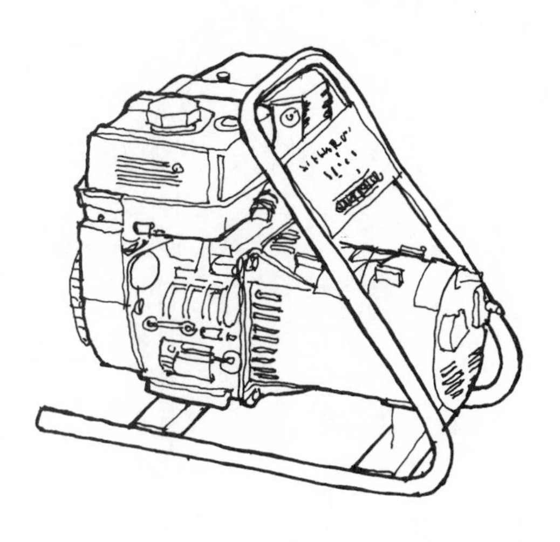

Rammer

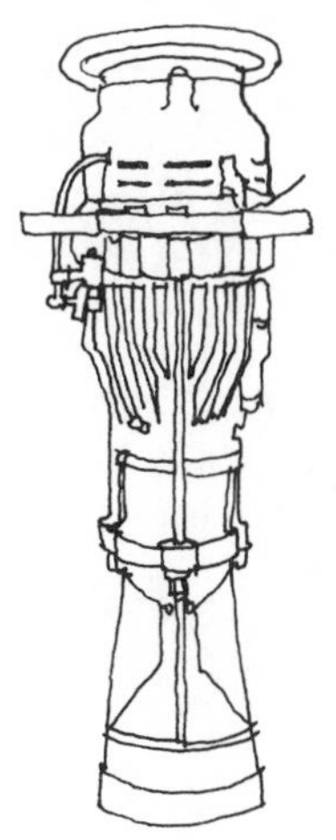

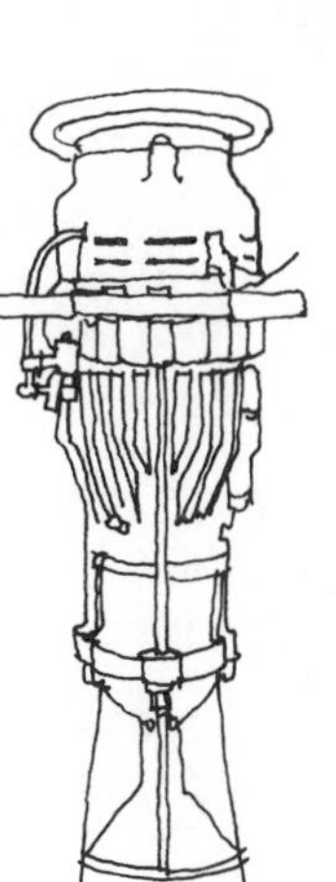

Compressor

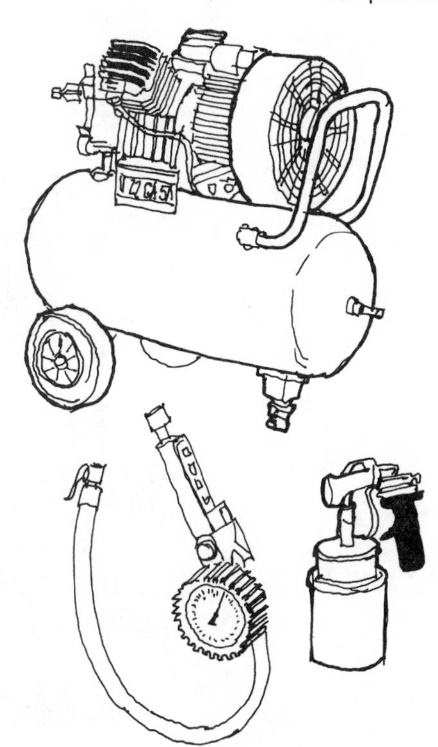

Pick

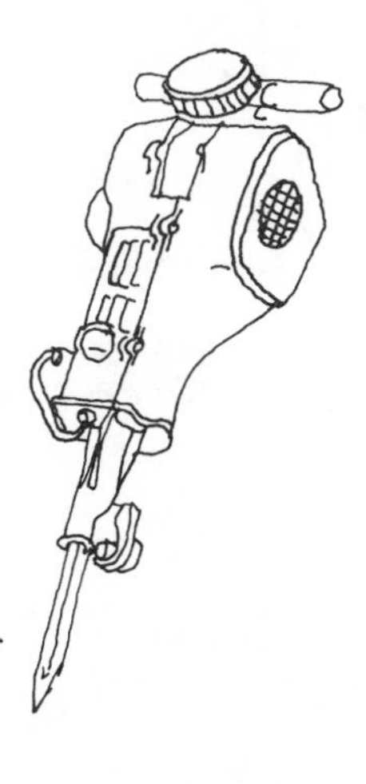

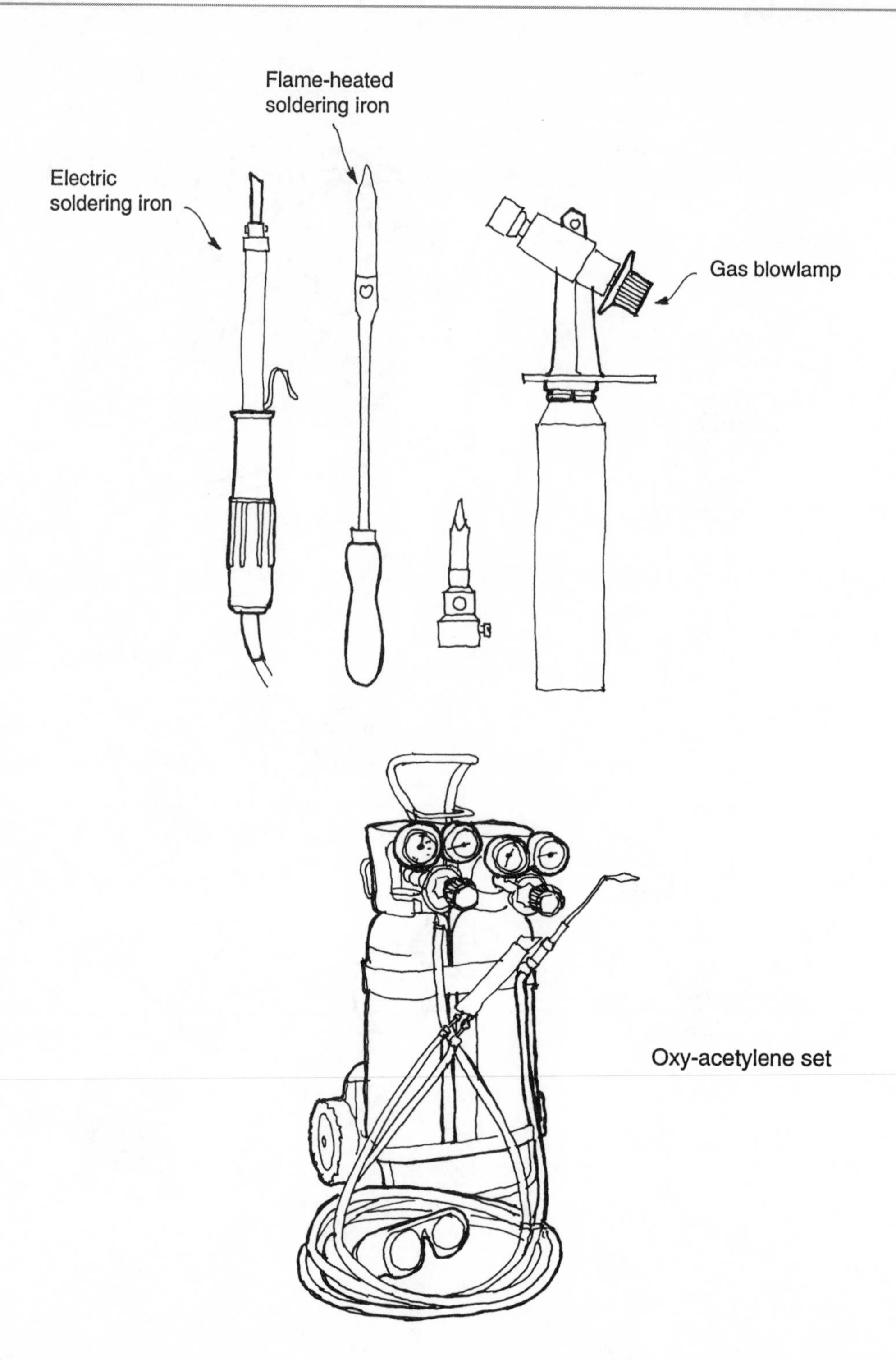
Flame-heated
soldering iron
Electric
soldering iron
Gas blowlamp
Oxy-acetylene set

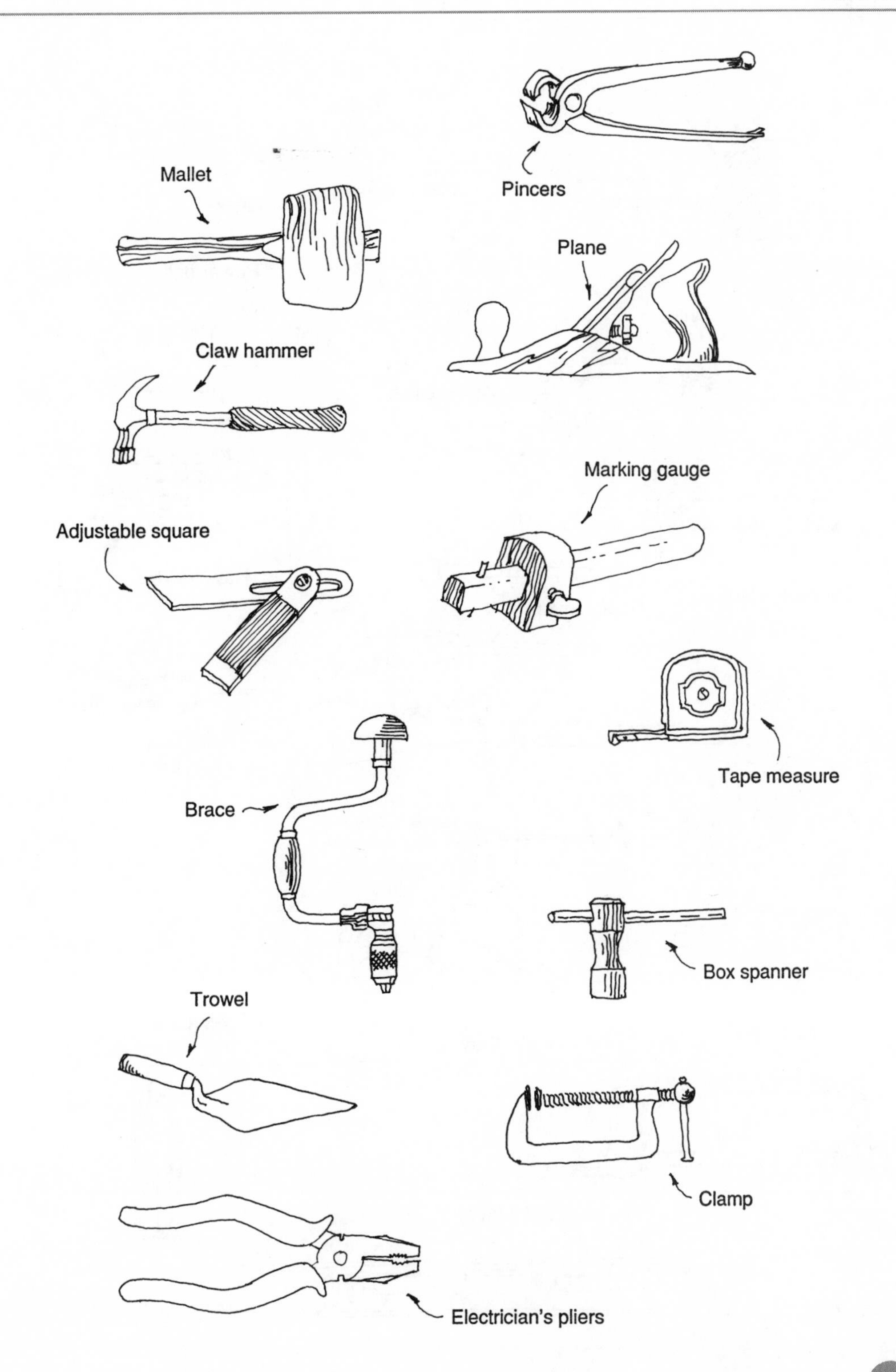
Pincers
Mallet
Plane
Claw hammer
Marking gauge
Adjustable square
Tape measure
Brace
Box spanner
Trowel
Clamp
Electrician's pliers

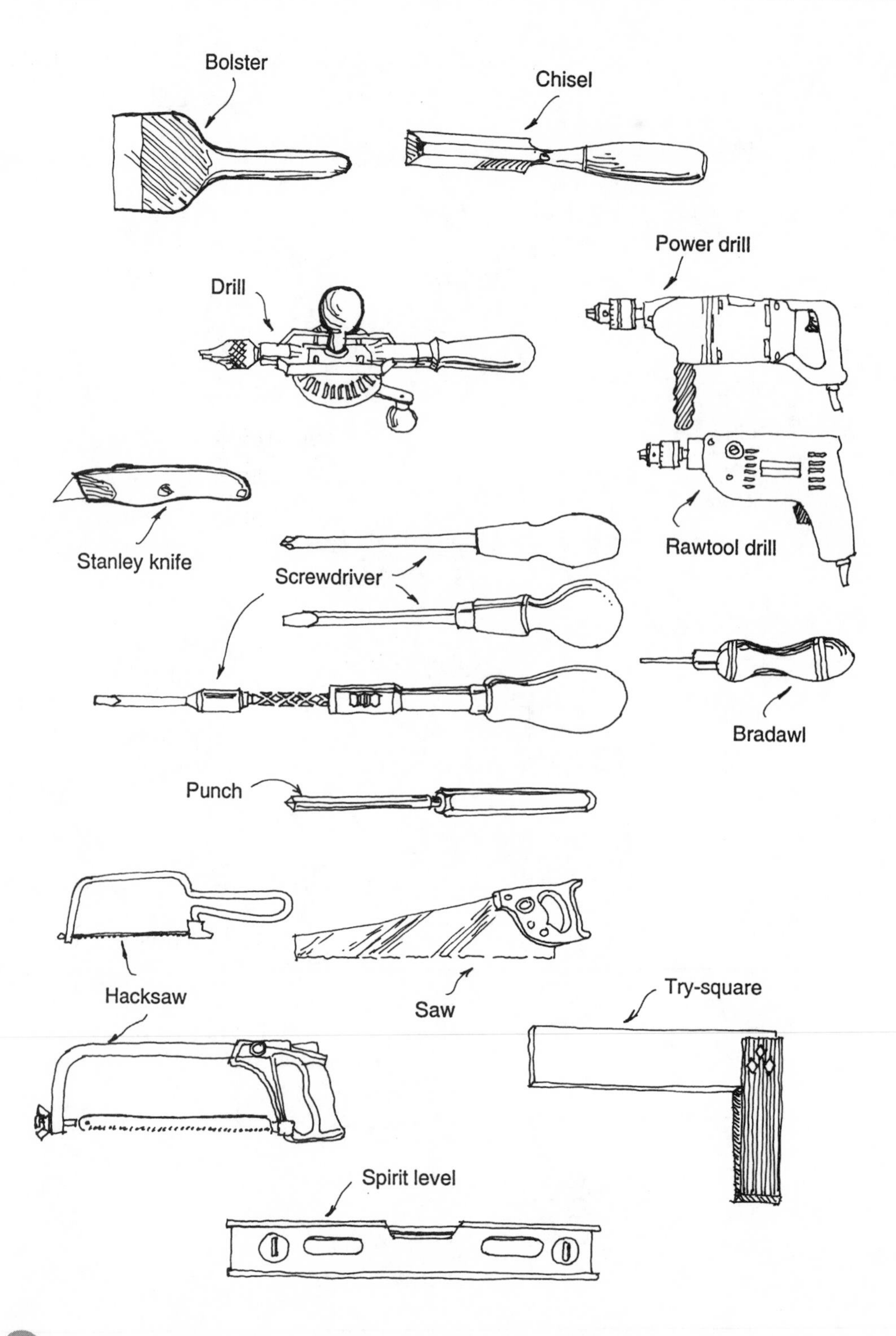
Bolster
Chisel
Power drill
Drill
Stanley knife
Rawtool drill
Screwdriver
Bradawl
Punch
Hacksaw
Saw
Try-square
Spirit level

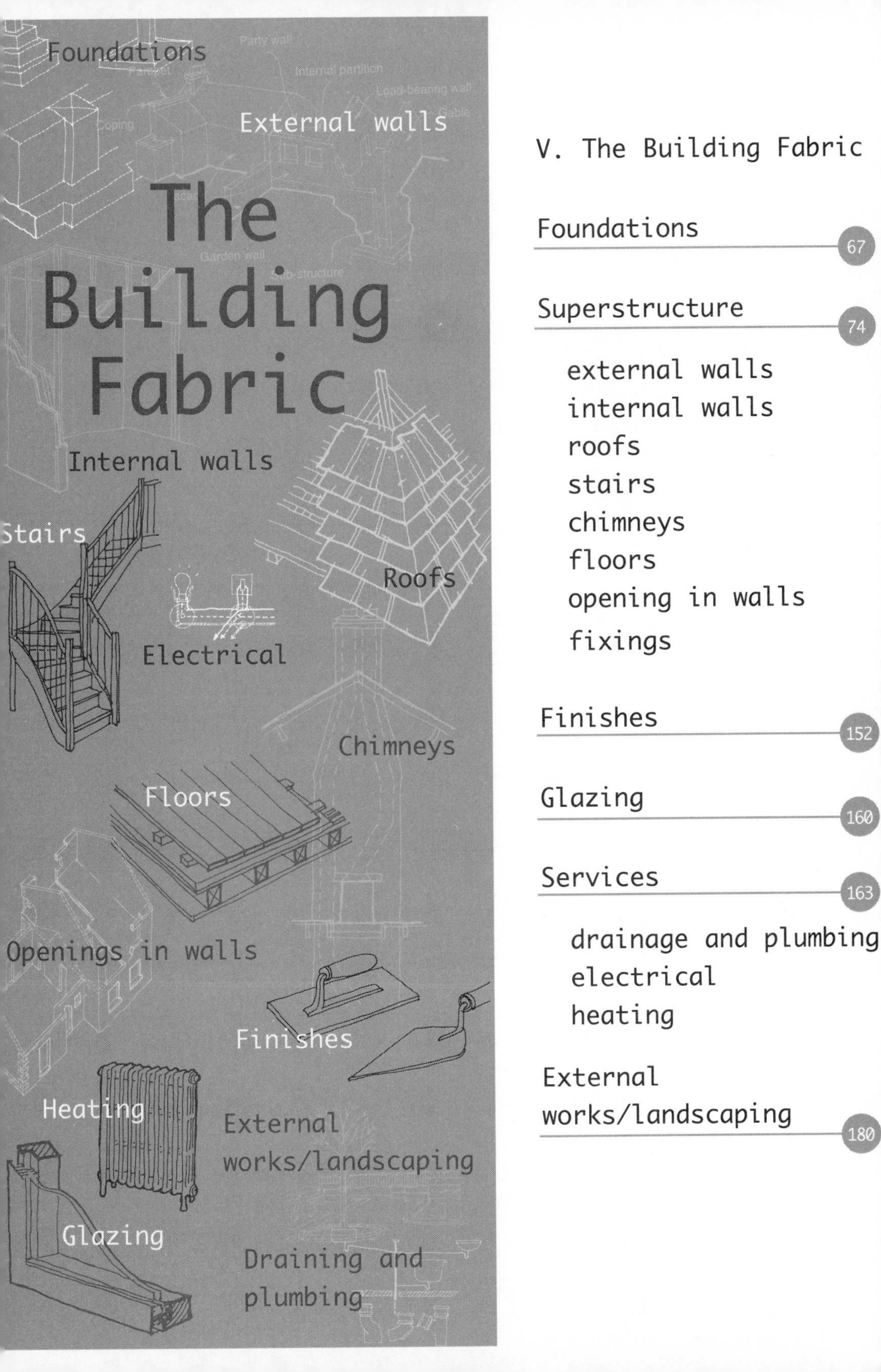

V. The Building Fabric

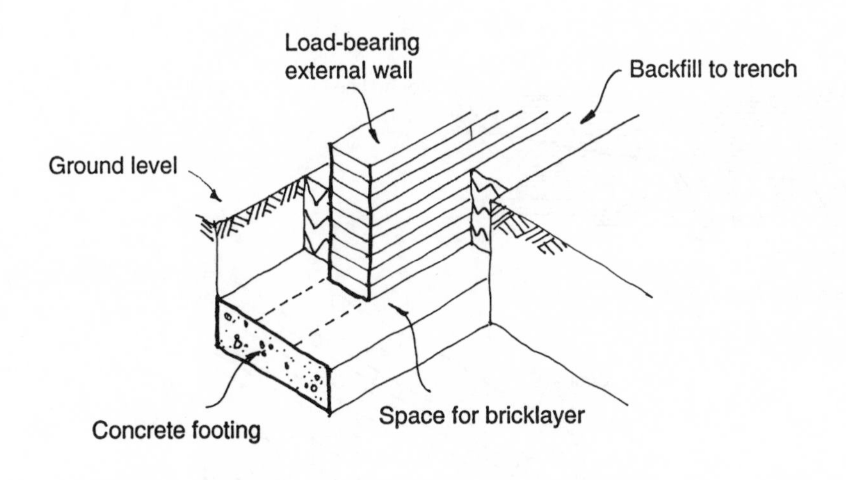

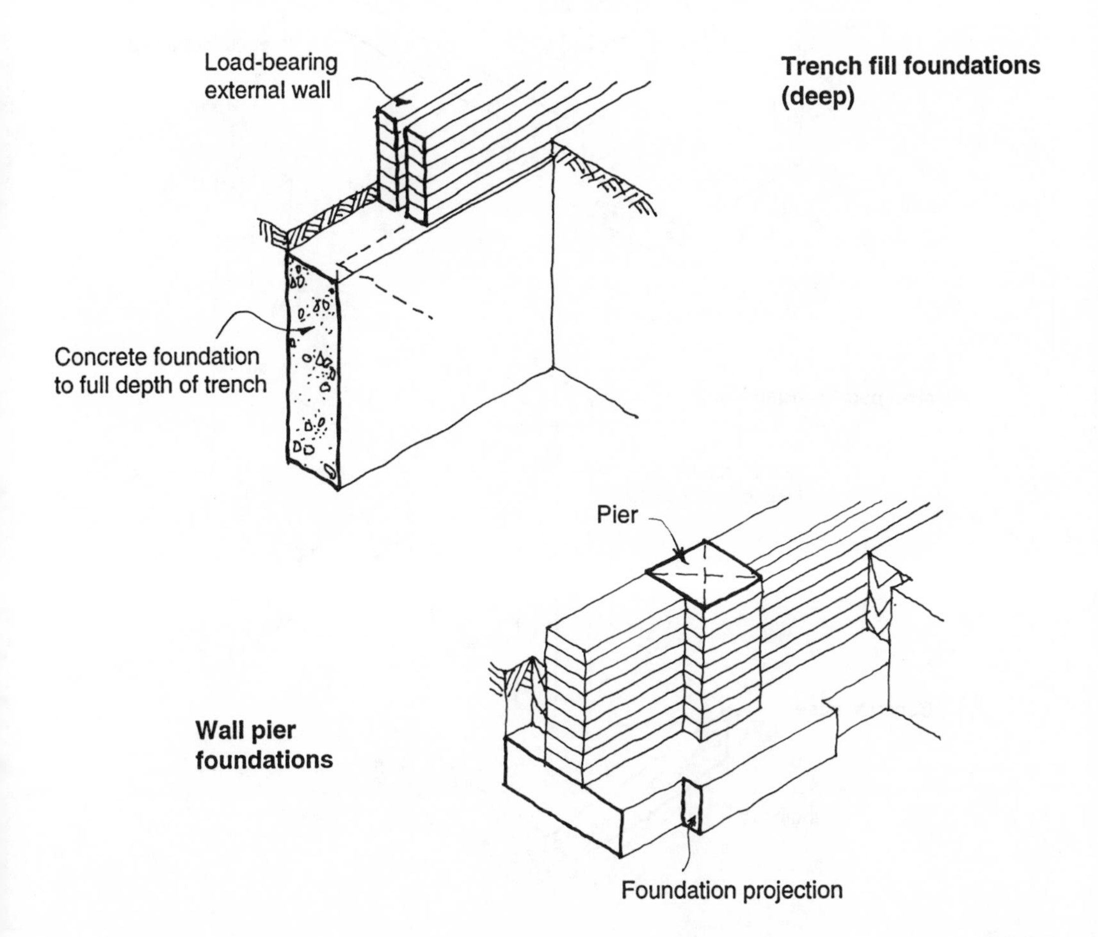

Trench fill foundations (deep)

Wall pier foundations

Stepped foundations

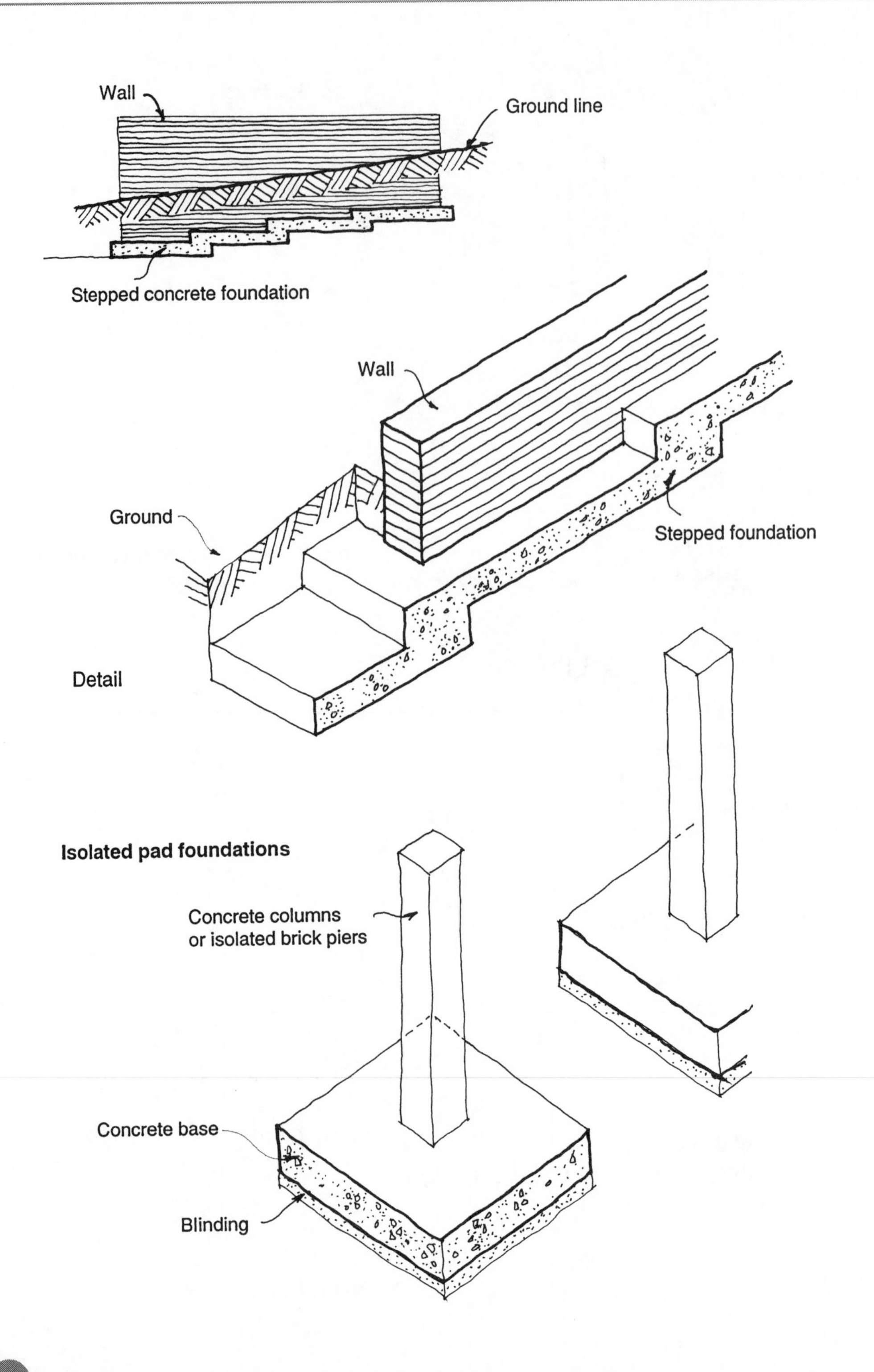

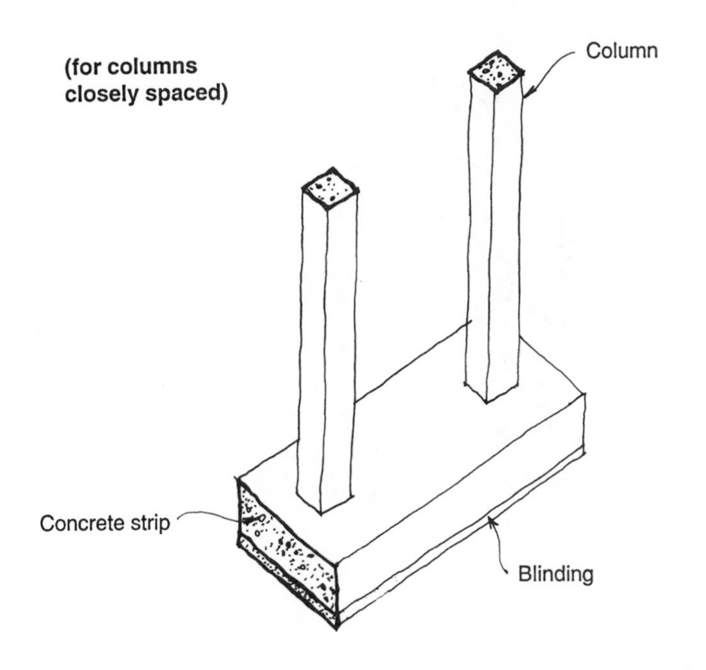

Raft foundations

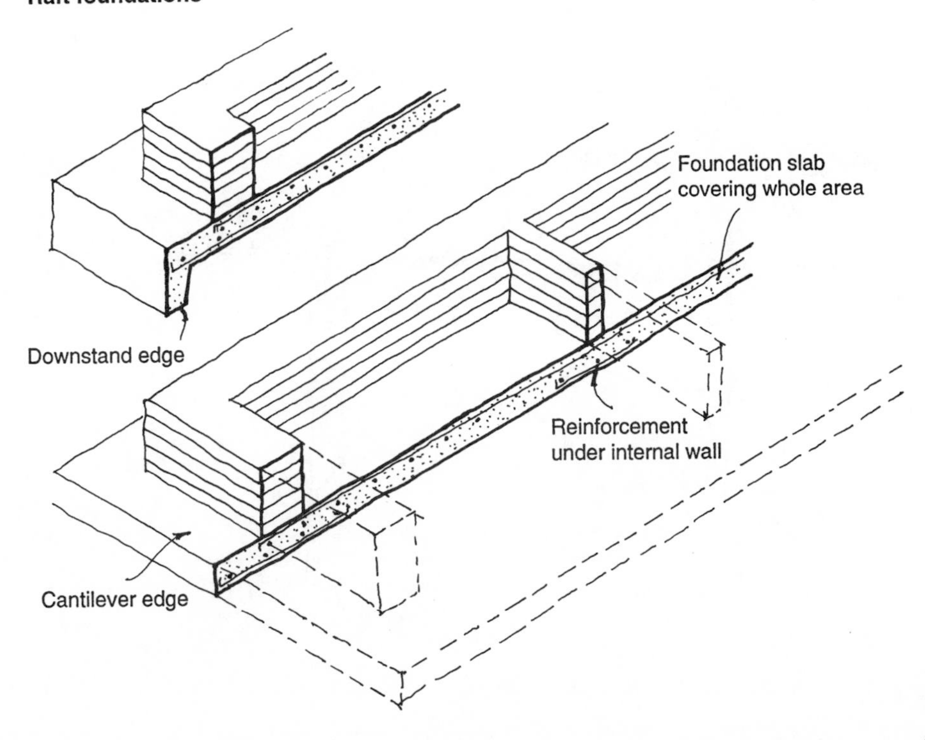

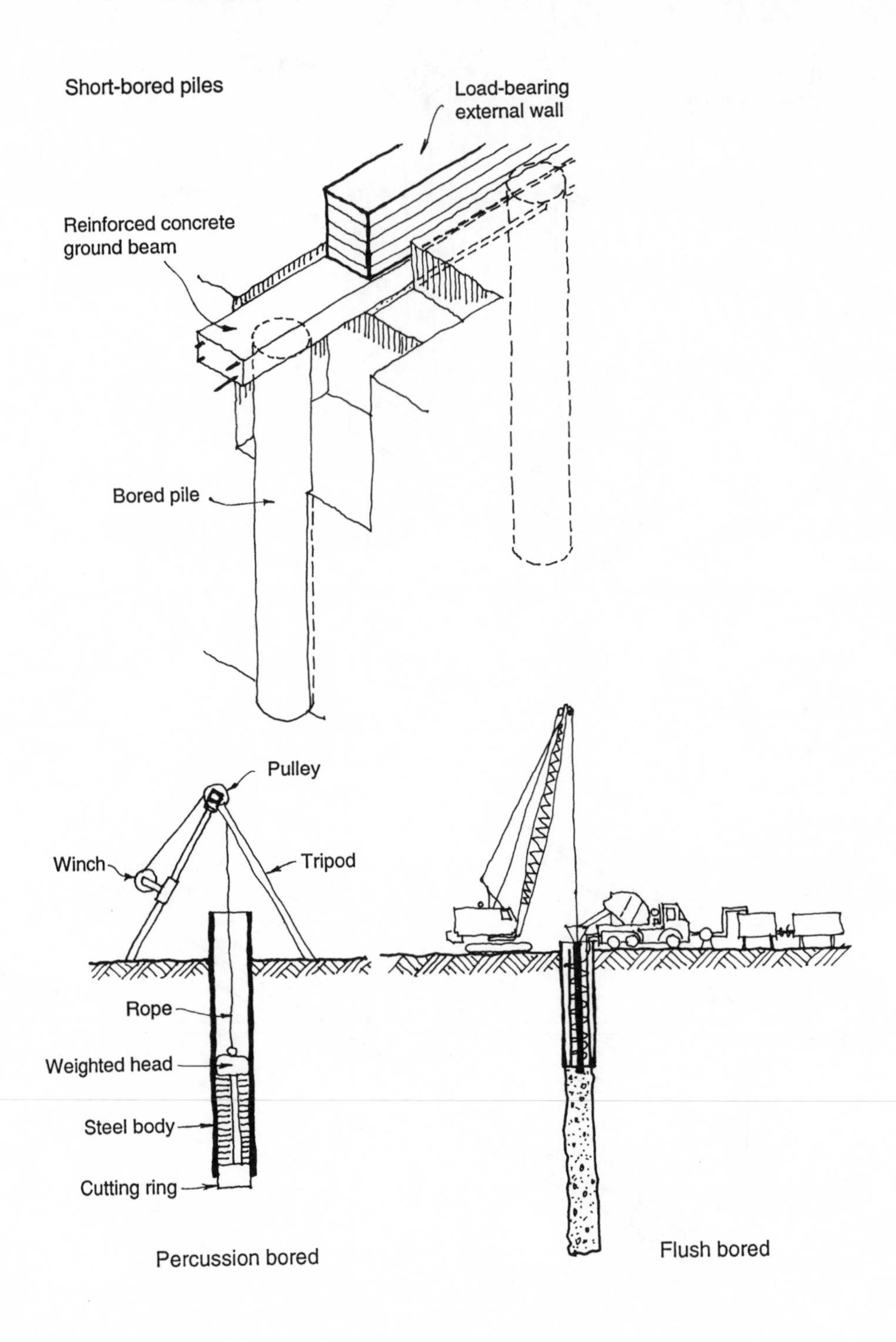
Short-bored piles
Load-bearing external wall
Reinforced concrete ground beam
Bored pile
Pulley
Winch
Tripod
Rope
Weighted head
Steel body
Cutting ring
Percussion bored
Flush bored

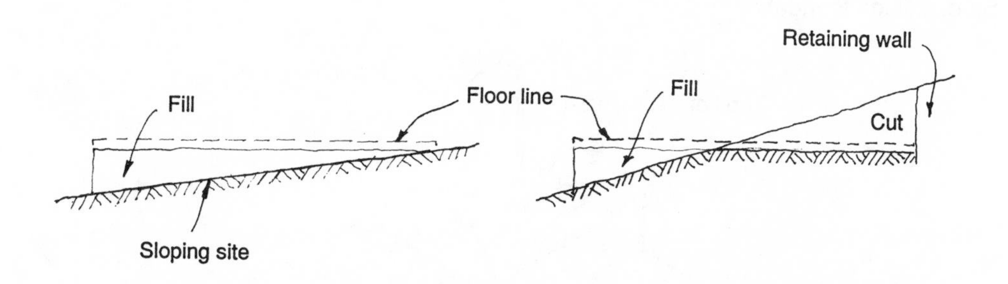
Fill
Floor line
Sloping site
Fill
Retaining wall
Cut

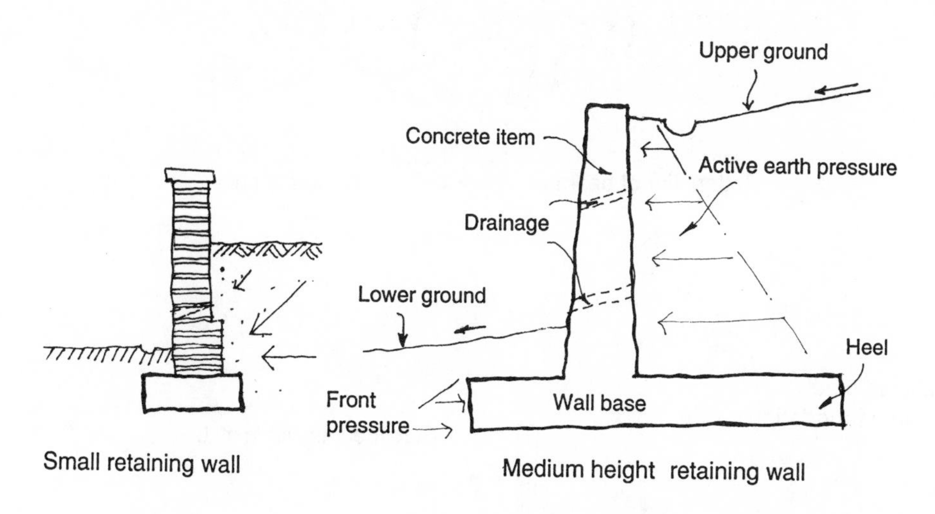
Upper ground
Concrete item
Active earth pressure
Drainage
Lower ground
Heel
Front pressure
Wall base
Small retaining wall
Medium height retaining wall

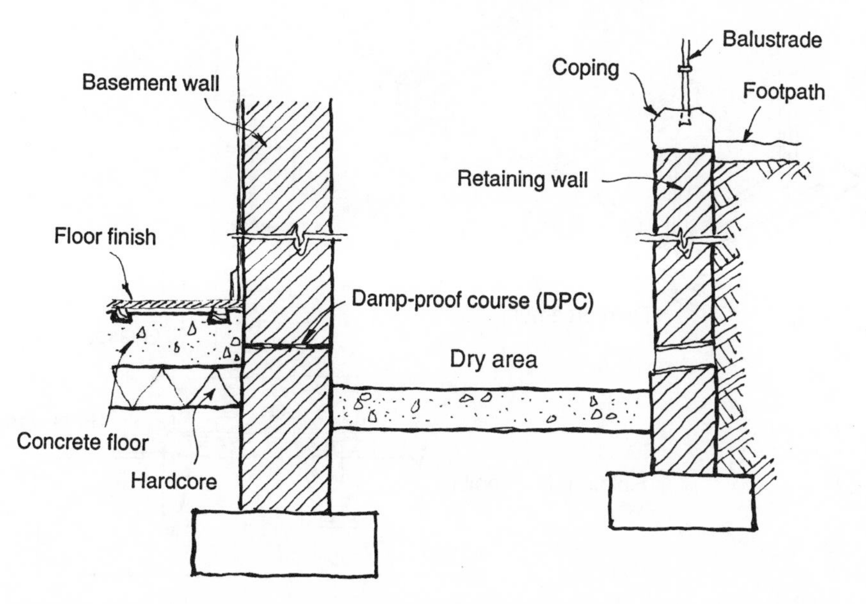
Basement wall
Balustrade
Coping
Footpath
Retaining wall
Floor finish
Damp-proof course (DPC)
Dry area
Concrete floor
Hardcore

Steel grillage foundation

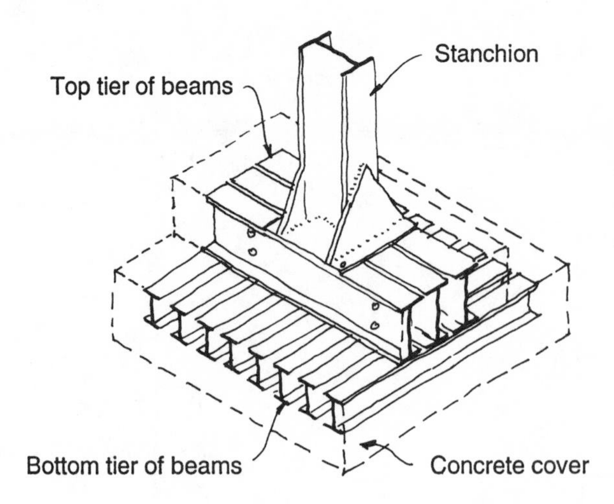

Multiple steel column foundation

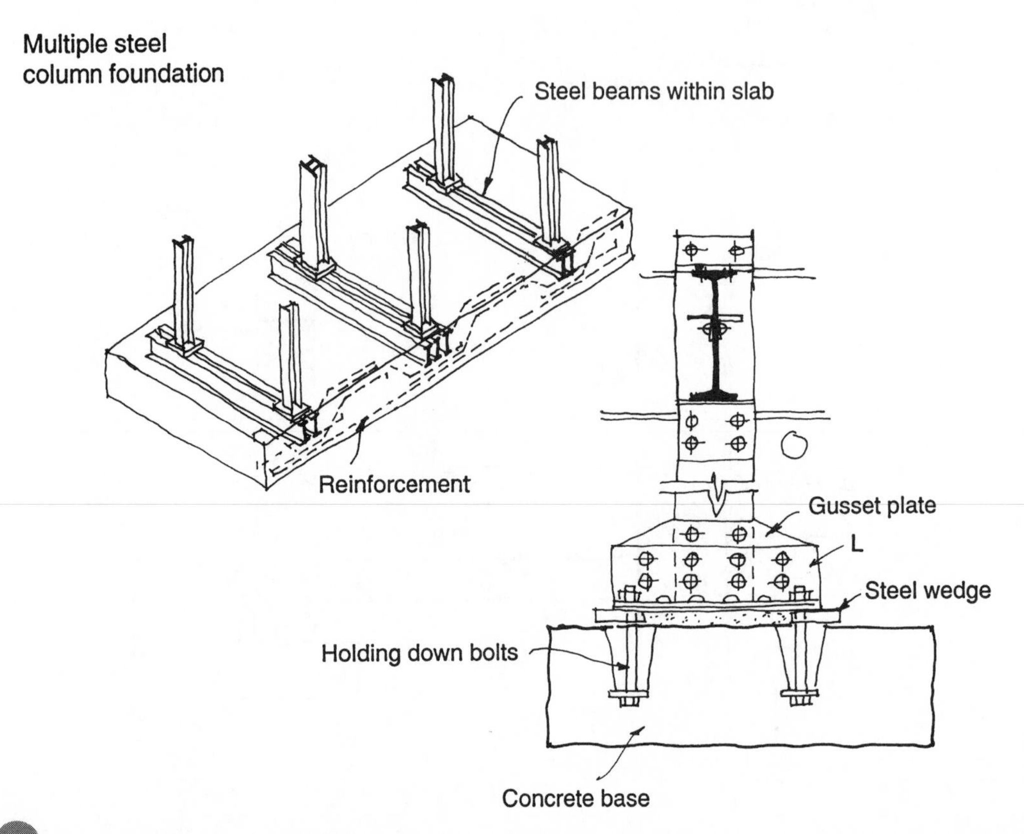

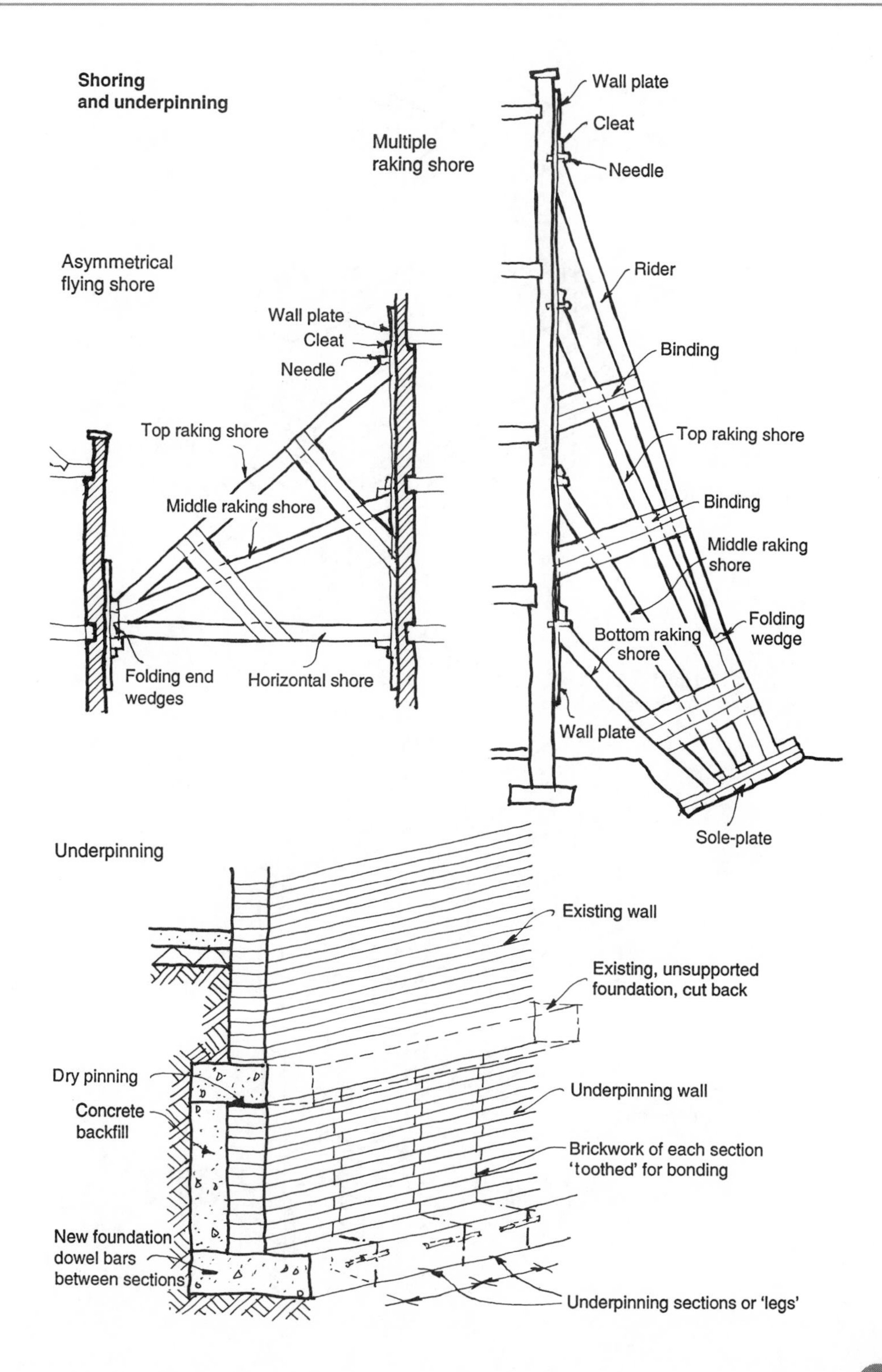
Shoring
and underpinning
Multiple
raking shore
Wall plate
Cleat
Needle
Rider
Binding
Top raking shore
Binding
Middle raking
shore
Folding
wedge
Bottom raking
shore
Wall plate
Sole-plate
Asymmetrical
flying shore
Wall plate
Cleat
Needle
Top raking shore
Middle raking shore
Folding end
wedges
Horizontal shore
Underpinning
Existing wall
Existing, unsupported
foundation, cut back
Dry pinning
Concrete
backfill
Underpinning wall
Brickwork of each section
'toothed' for bonding
New foundation
dowel bars
between sections
Underpinning sections or 'legs'

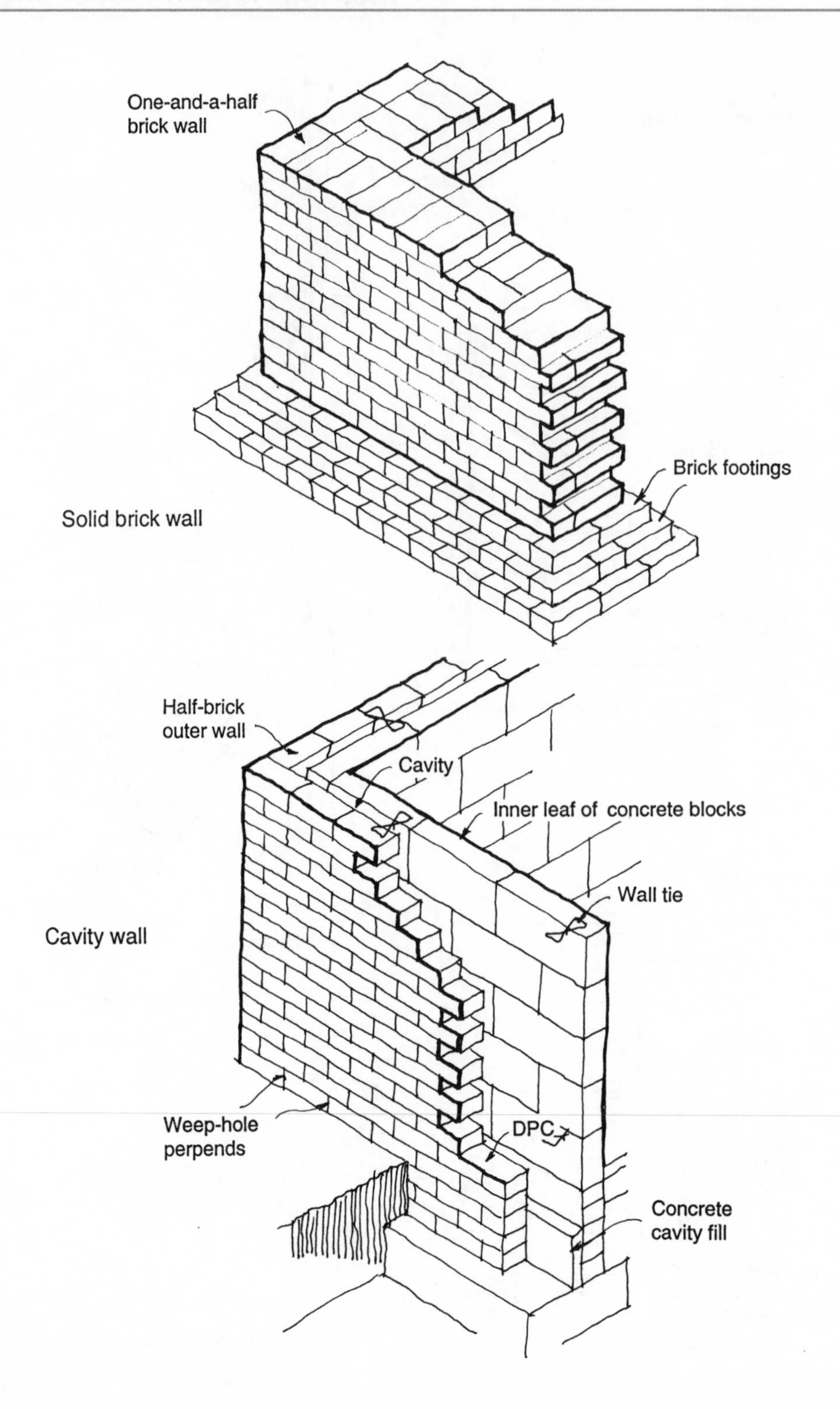
One-and-a-half
brick wall
Brick footings
Solid brick wall
Half-brick
outer wall
Cavity
Inner leaf of concrete blocks
Wall tie
Cavity wall
Weep-hole
perpends
DPC
Concrete
cavity fill

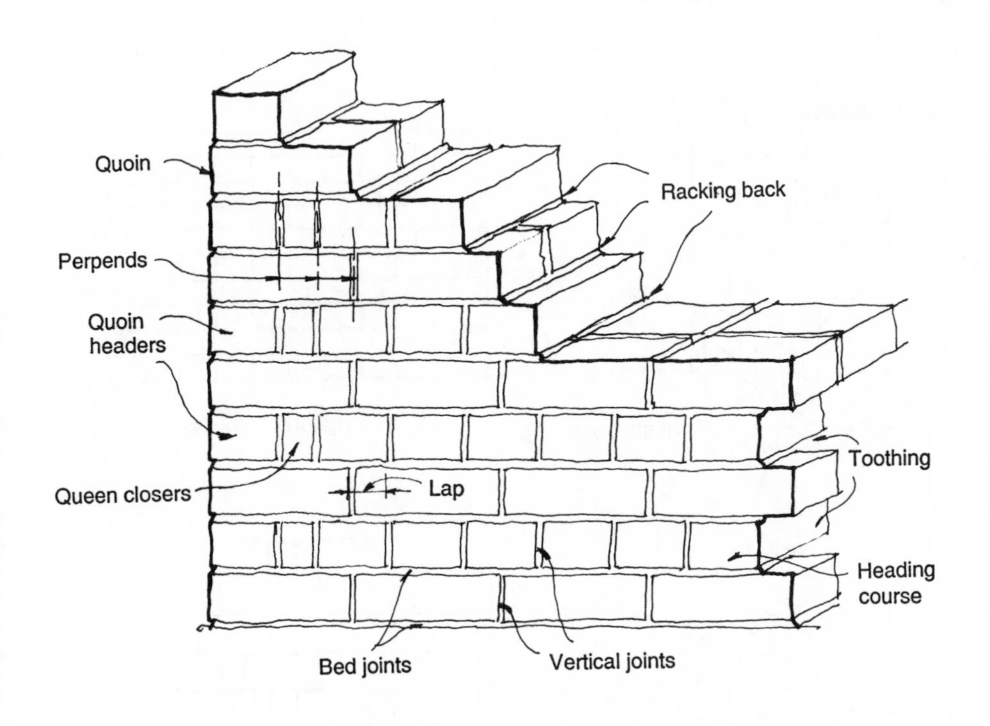

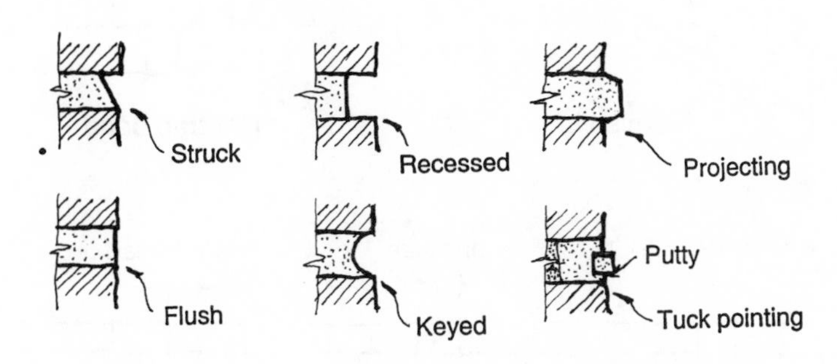

Type of joints

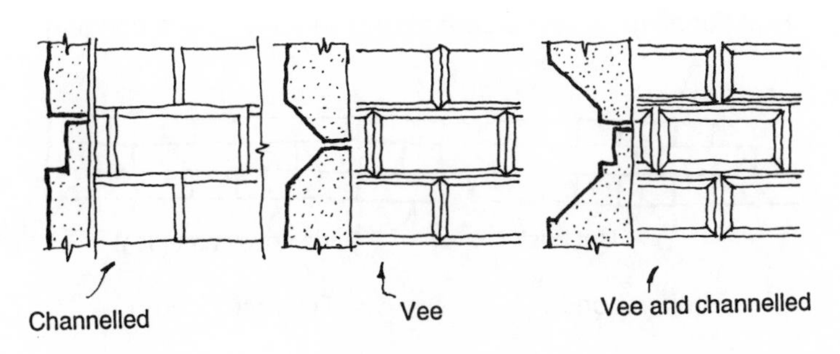

Rusticated joints

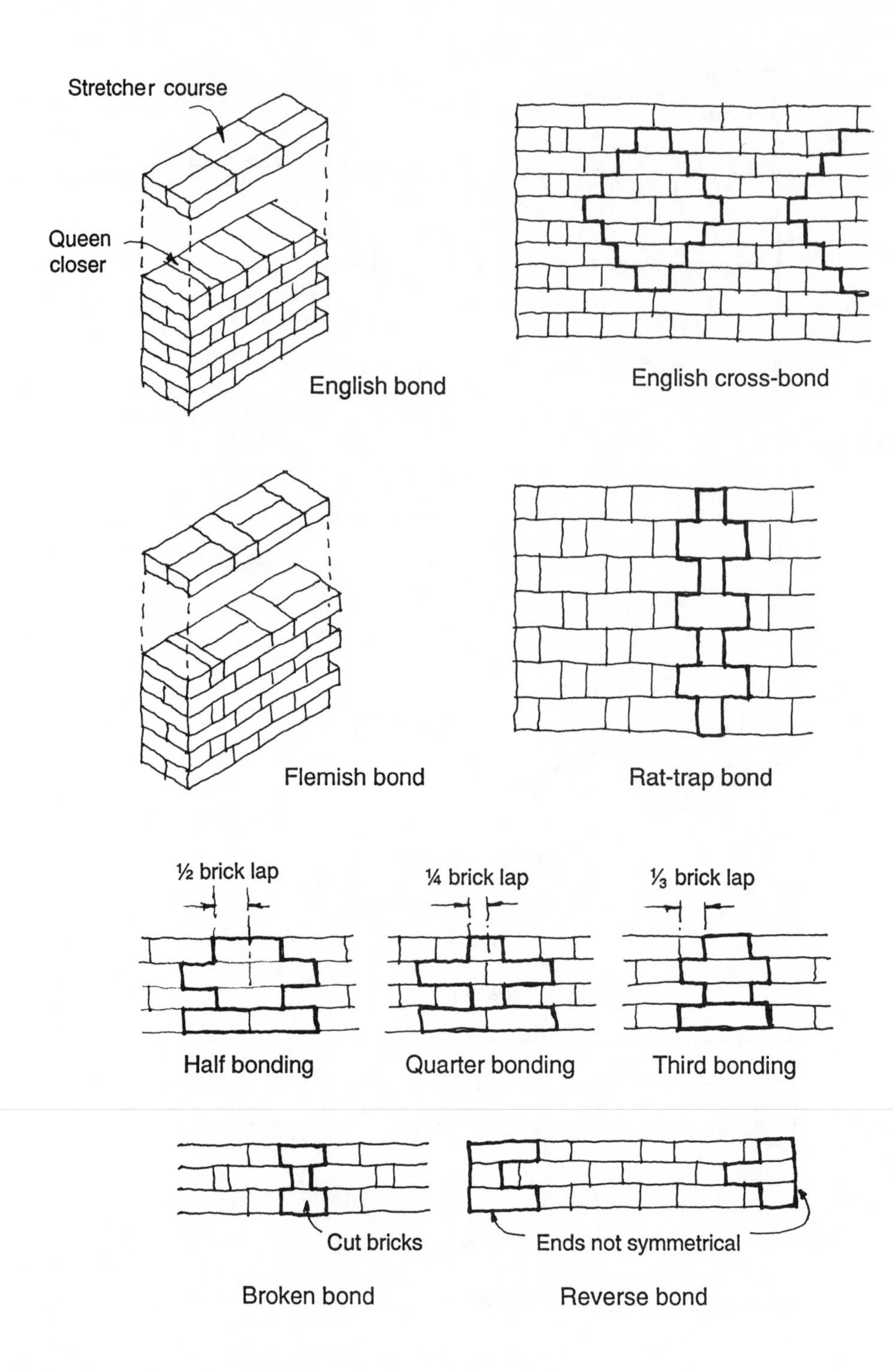
Stretcher course
Queen closer
English bond
English cross-bond
Flemish bond
Rat-trap bond
½ brick lap
¼ brick lap
⅓ brick lap
Half bonding
Quarter bonding
Third bonding
Cut bricks
Ends not symmetrical
Broken bond
Reverse bond

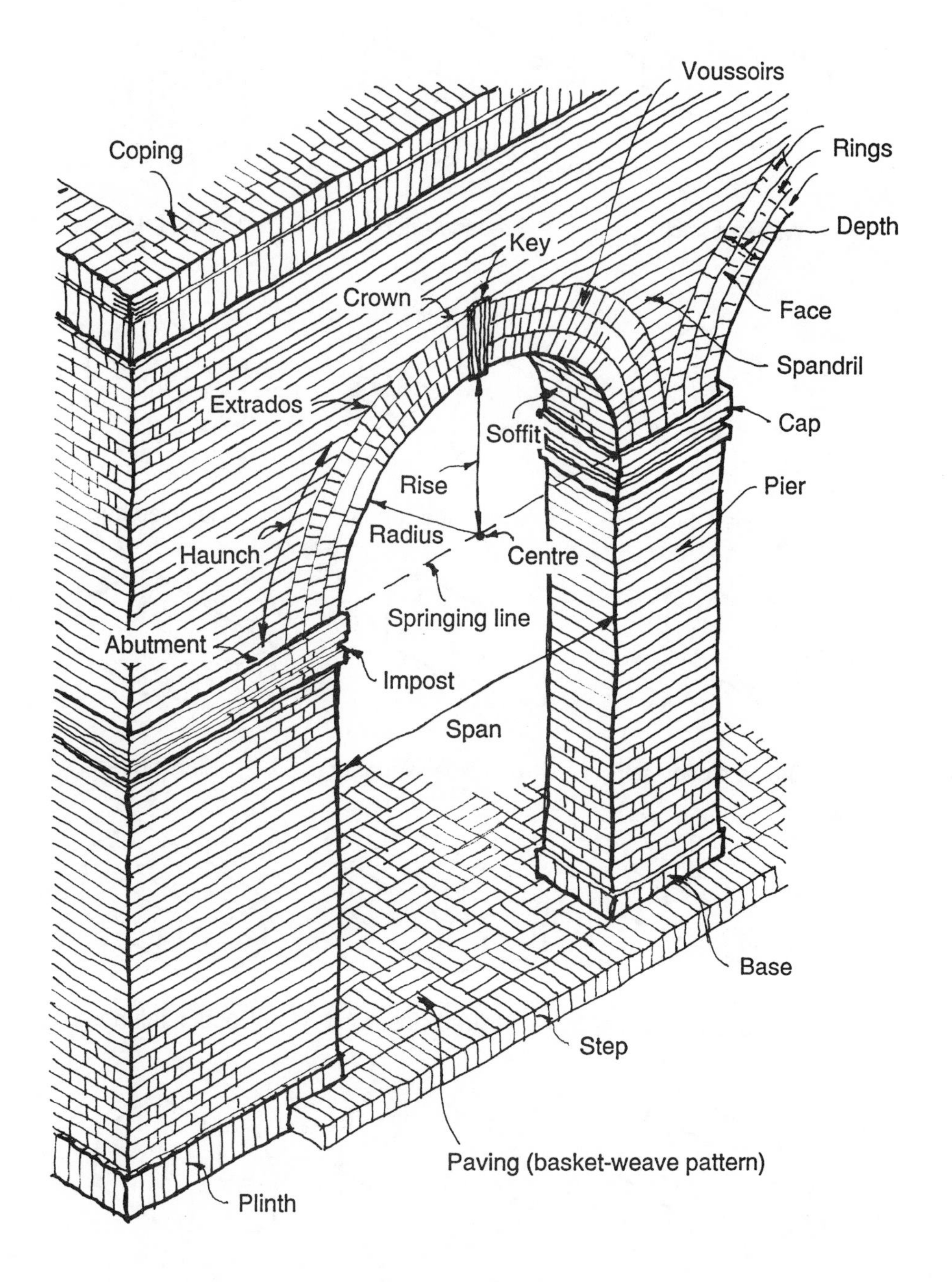
Voussoirs
Coping
Rings
Depth
Key
Crown
Face
Spandril
Extrados
Soffit
Cap
Rise
Pier
Radius
Centre
Haunch
Springing line
Abutment
Impost
Span
Base
Step
Paving (basket-weave pattern)
Plinth

Brick types

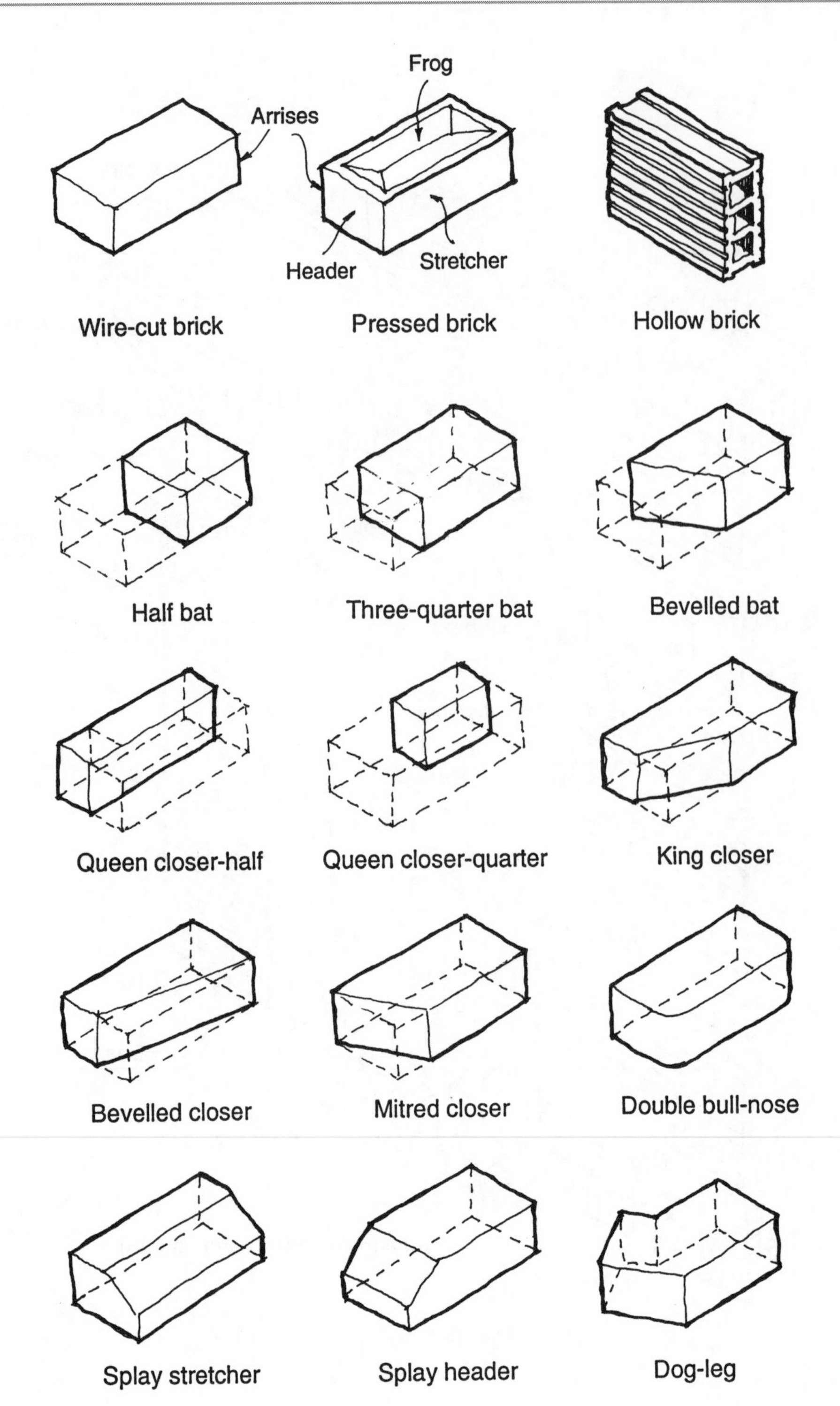
Frog
Arrises
Header
Stretcher
Wire-cut brick
Pressed brick
Hollow brick
Half bat
Three-quarter bat
Bevelled bat
Queen closer-half
Queen closer-quarter
King closer
Bevelled closer
Mitred closer
Double bull-nose
Splay stretcher
Splay header
Dog-leg

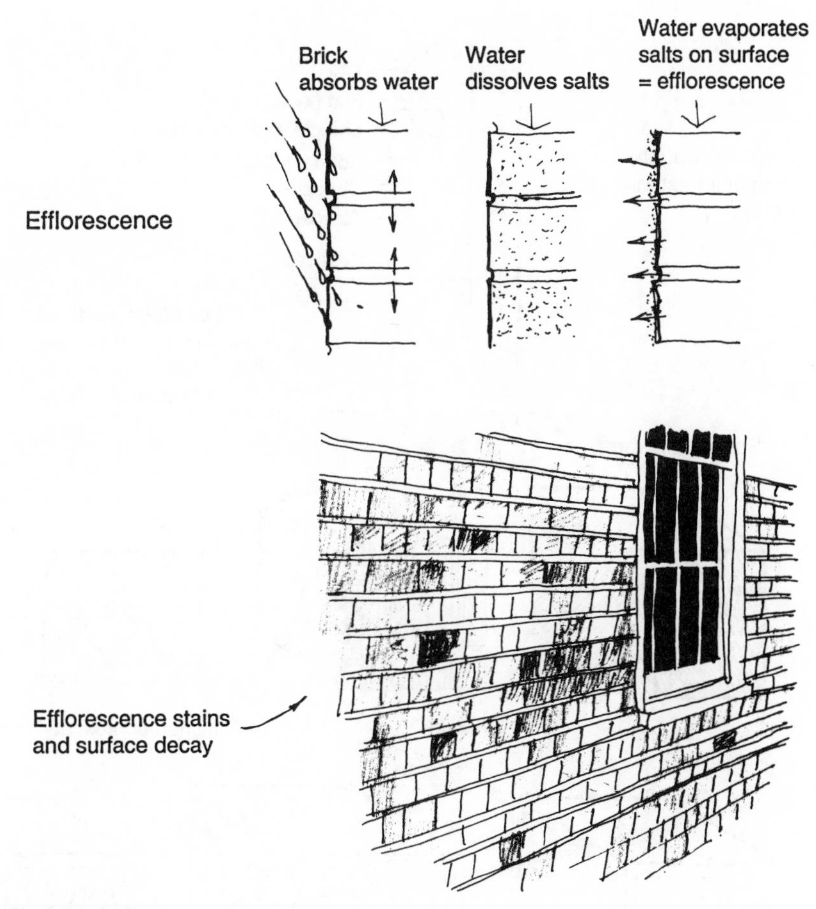

Condensation

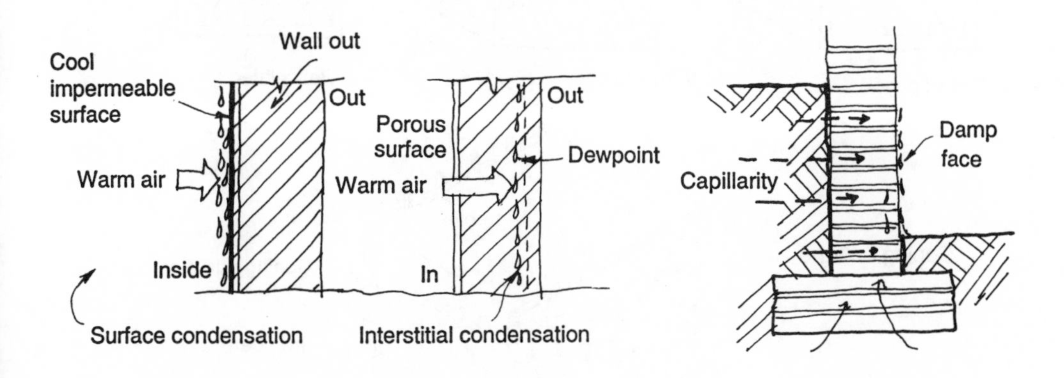

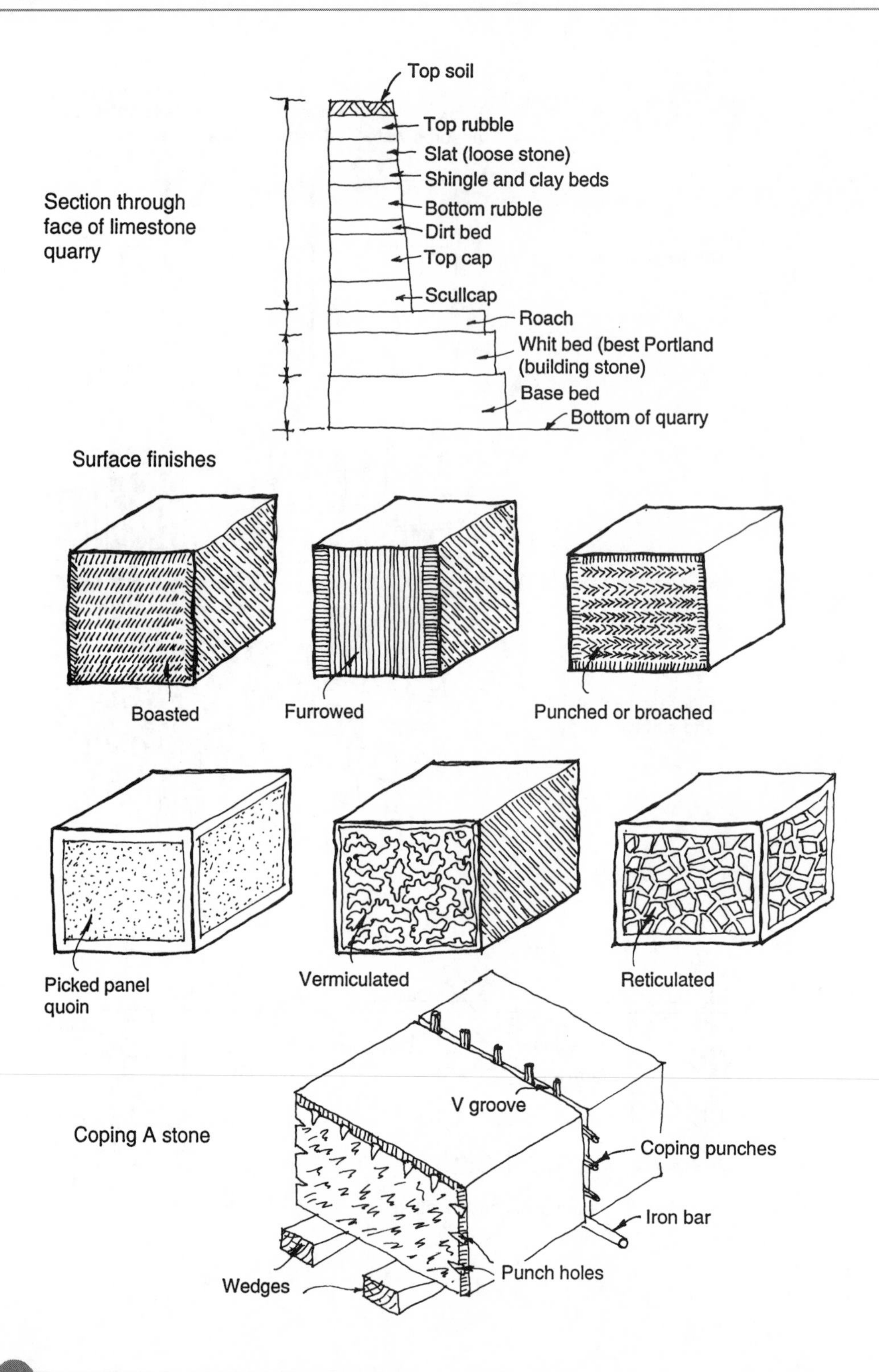
Section through face of limestone quarry
Top soil
Top rubble
Slat (loose stone)
Shingle and clay beds
Bottom rubble
Dirt bed
Top cap
Scullcap
Roach
Whit bed (best Portland (building stone)
Base bed
Bottom of quarry
Surface finishes
Boasted
Furrowed
Punched or broached
Picked panel quoin
Vermiculated
Reticulated
Coping A stone
V groove
Coping punches
Iron bar
Punch holes
Wedges

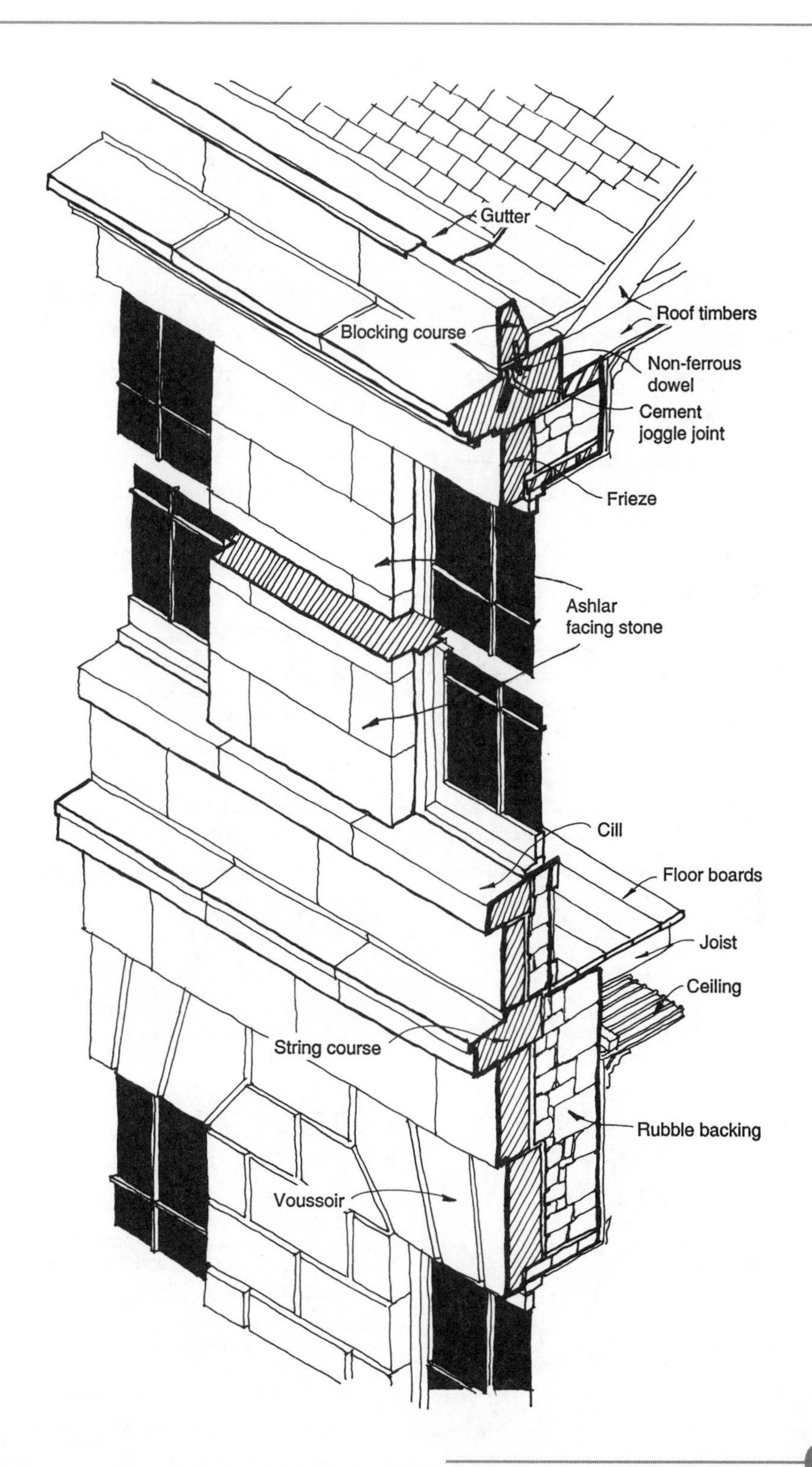
Gutter
Roof timbers
Blocking course
Non-ferrous dowel
Cement joggle joint
Frieze
Ashlar facing stone
Cill
Floor boards
Joist
Ceiling
String course
Rubble backing
Voussoir

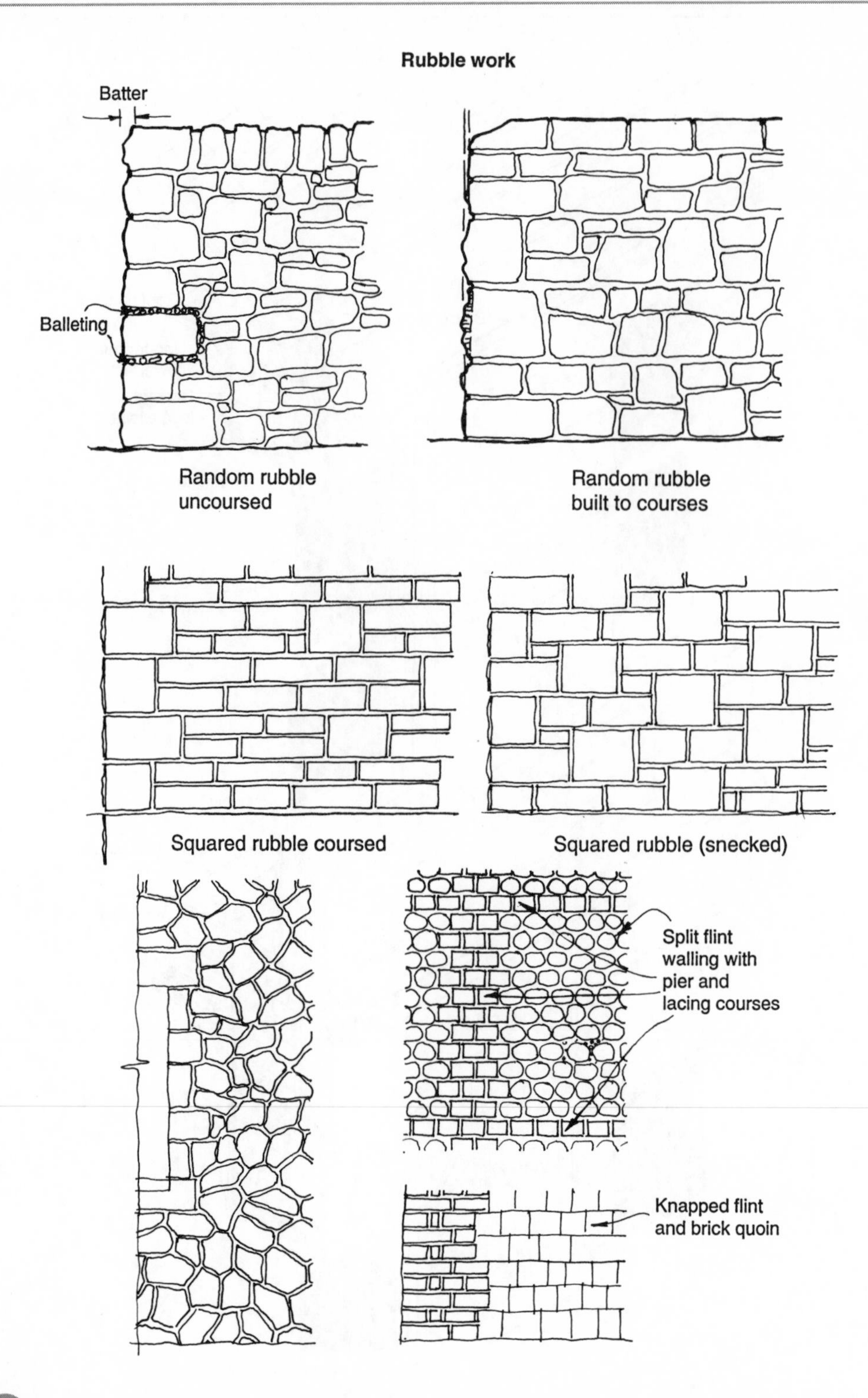
Rubble work
Batter
Balleting
Random rubble uncoursed
Random rubble built to courses
Squared rubble coursed
Squared rubble (snecked)
Split flint walling with pier and lacing courses
Knapped flint and brick quoin

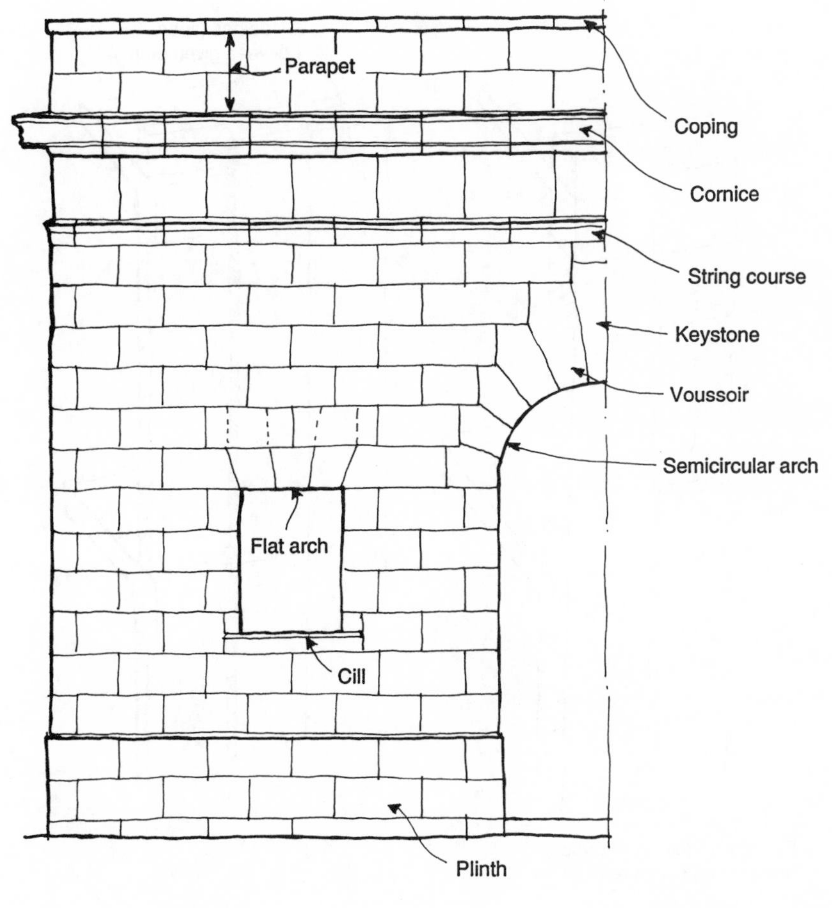
Parapet
Coping
Cornice
String course
Keystone
Voussoir
Semicircular arch
Flat arch
Cill
Plinth

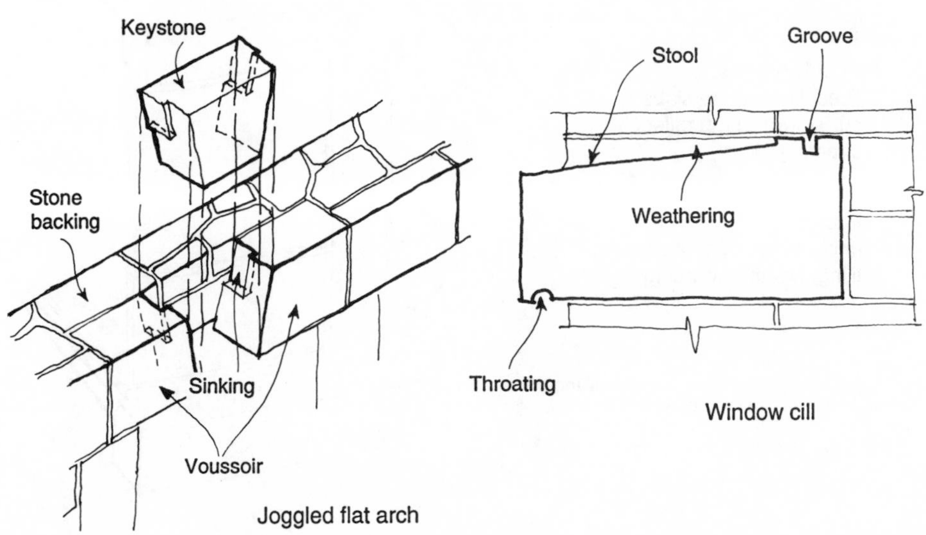
Keystone
Stone backing
Sinking
Voussoir
Joggled flat arch
Groove
Stool
Weathering
Throating
Window cill

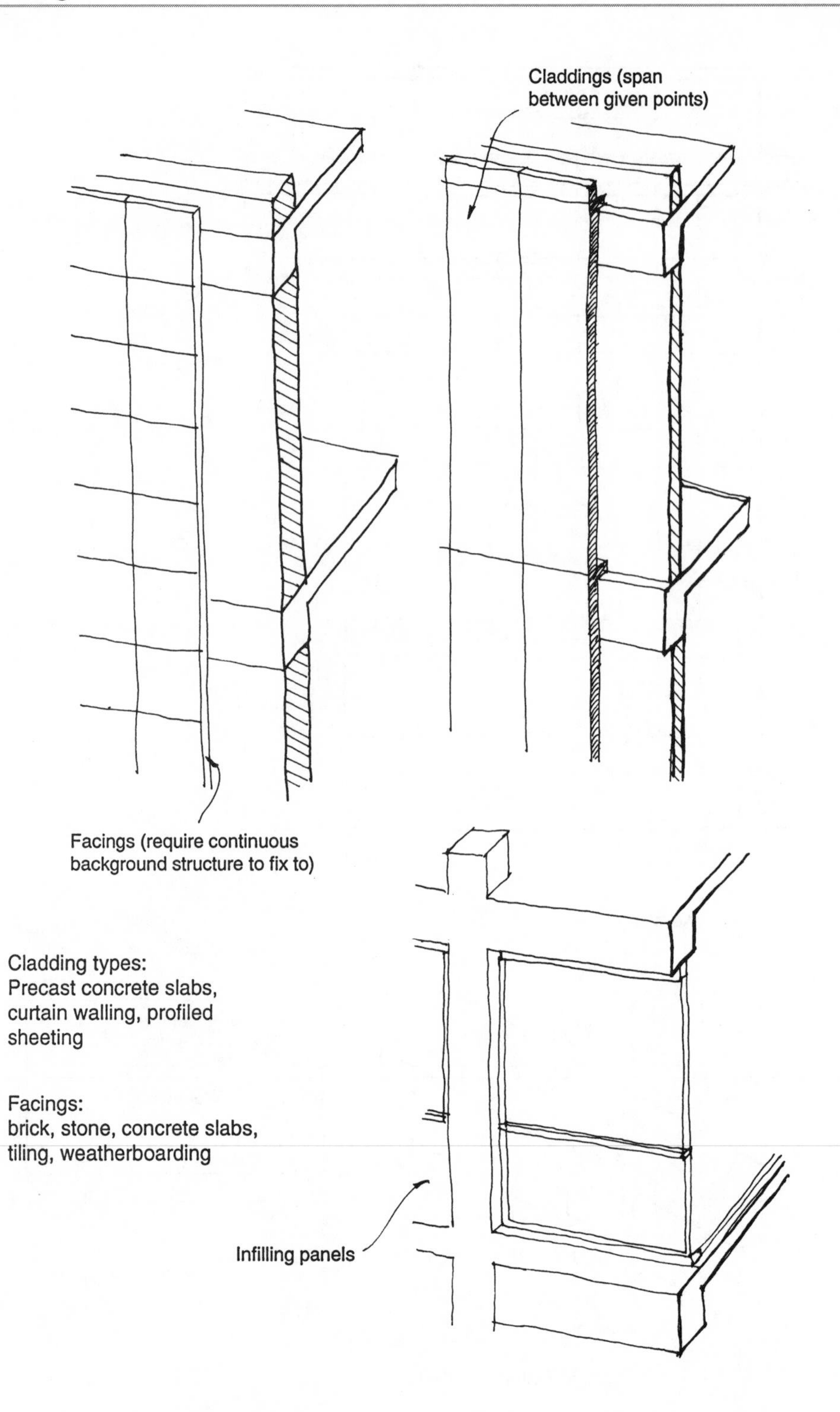
Claddings (span between given points)
Facings (require continuous background structure to fix to)
Cladding types:
Precast concrete slabs, curtain walling, profiled sheeting
Facings:
brick, stone, concrete slabs, tiling, weatherboarding
Infilling panels

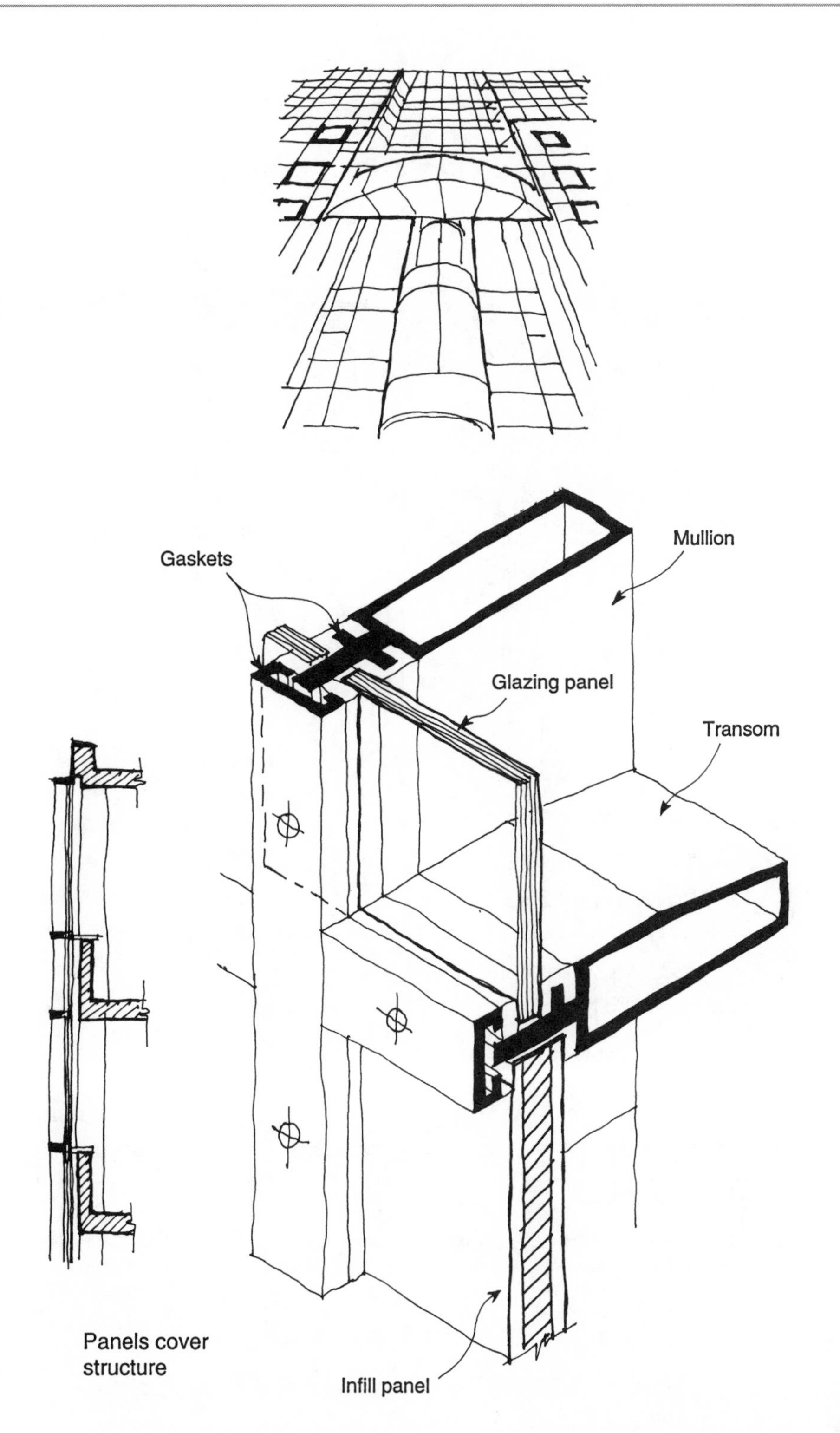
Gaskets
Mullion
Glazing panel
Transom
Panels cover structure
Infill panel

(glass fibre reinforced cement)

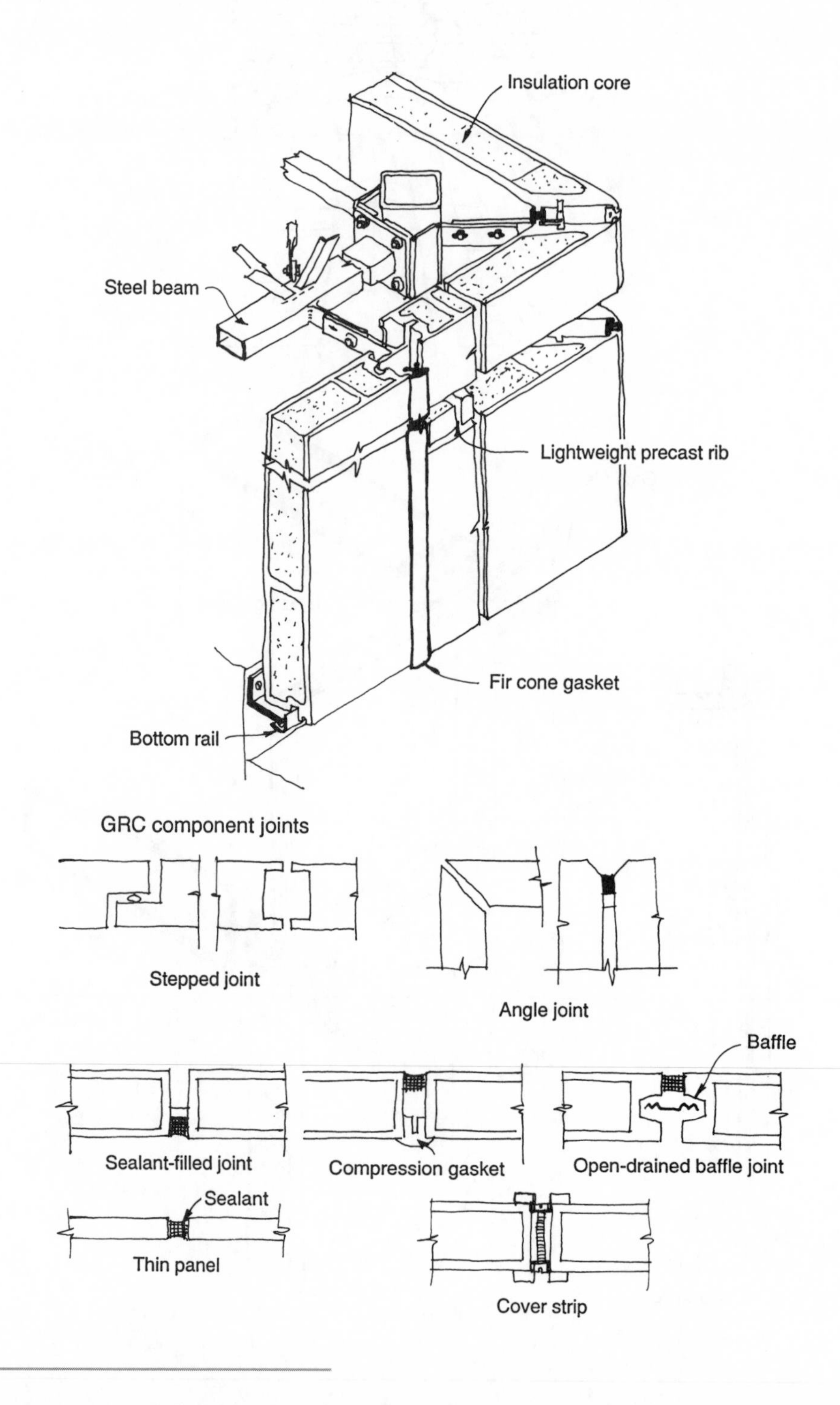

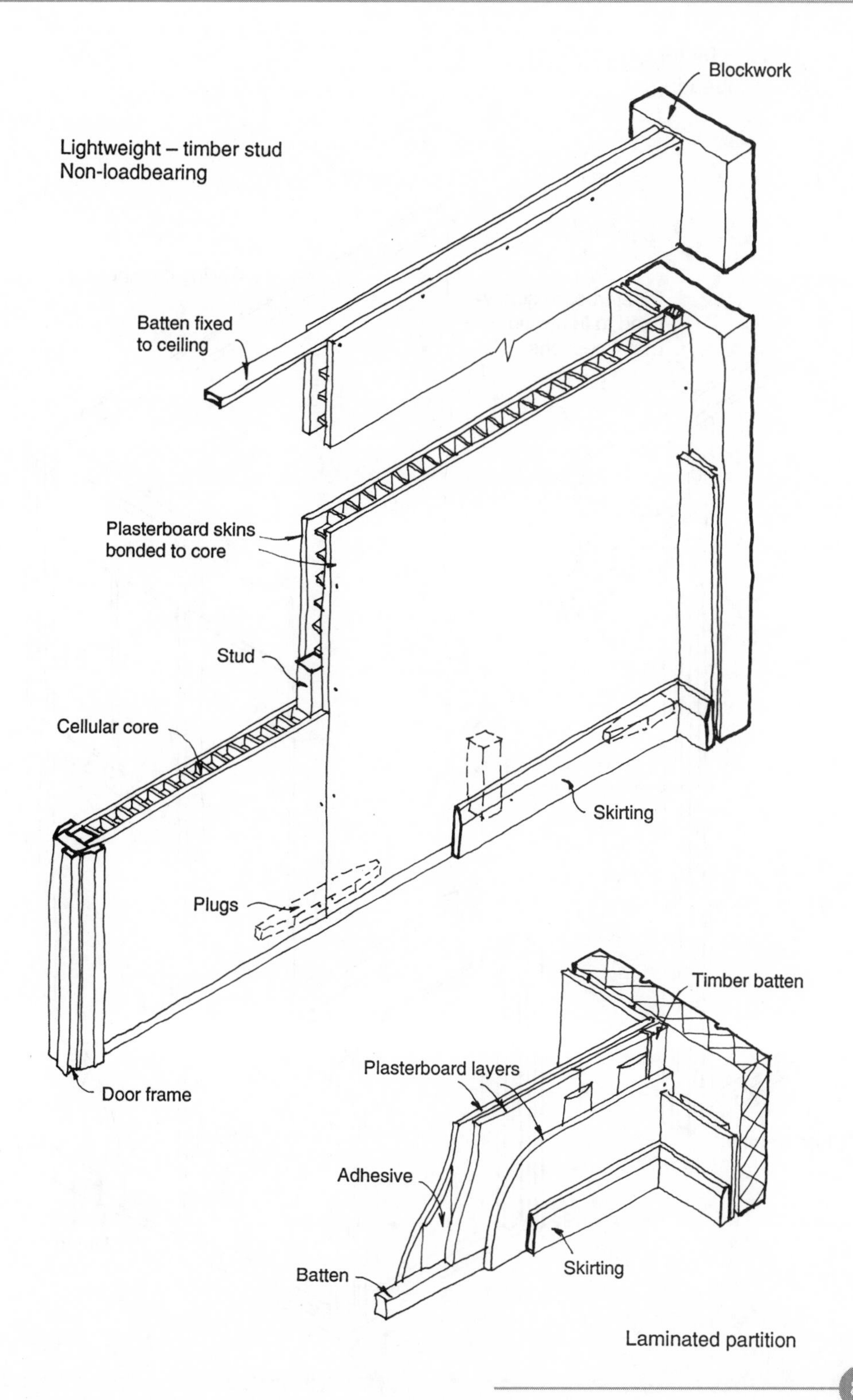
Blockwork
Lightweight – timber stud
Non-loadbearing
Batten fixed
to ceiling
Plasterboard skins
bonded to core
Stud
Cellular core
Skirting
Plugs
Door frame
Timber batten
Plasterboard layers
Adhesive
Skirting
Batten
Laminated partition

Internal – Metal stud
Non-loadbearing

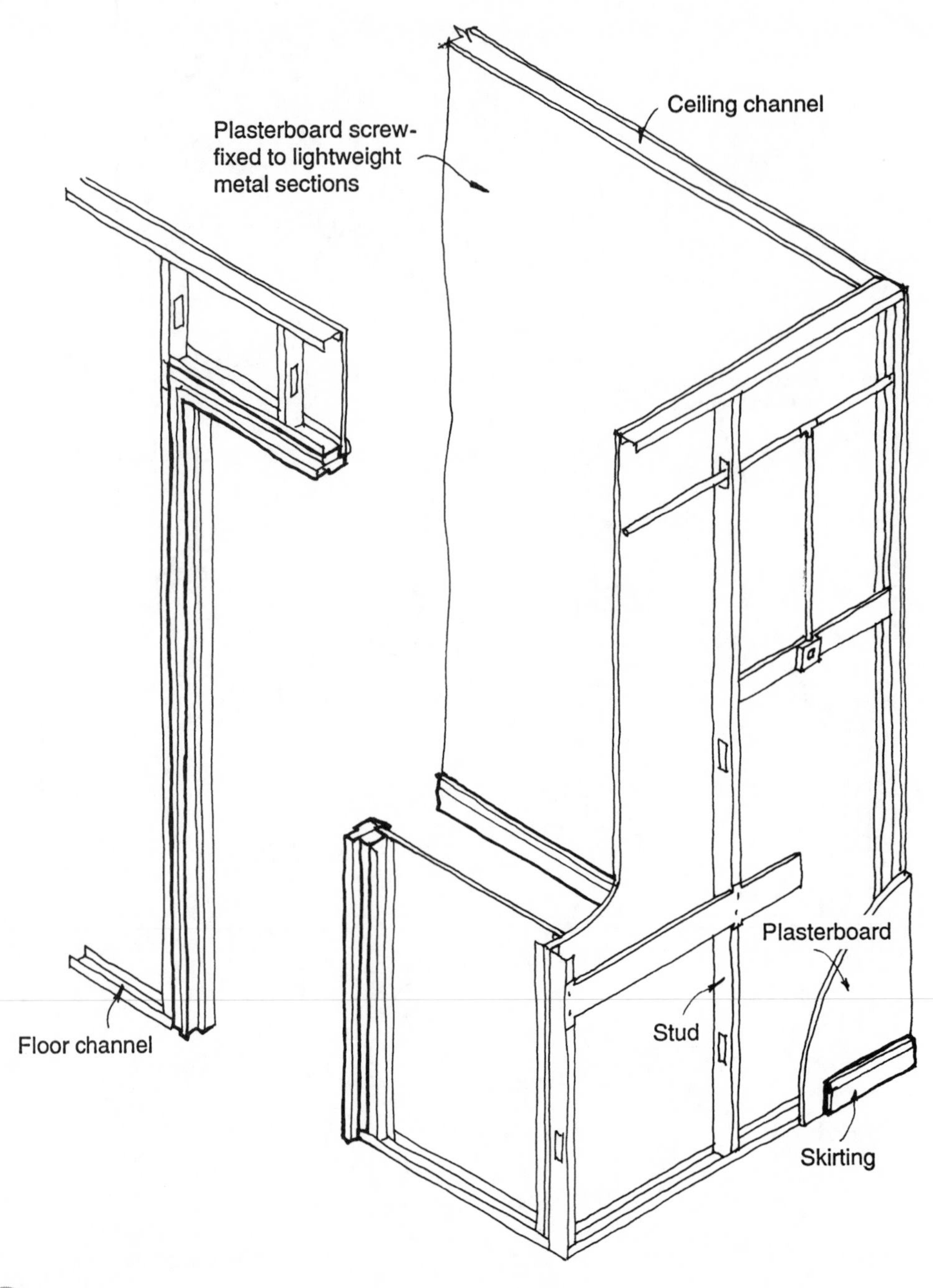

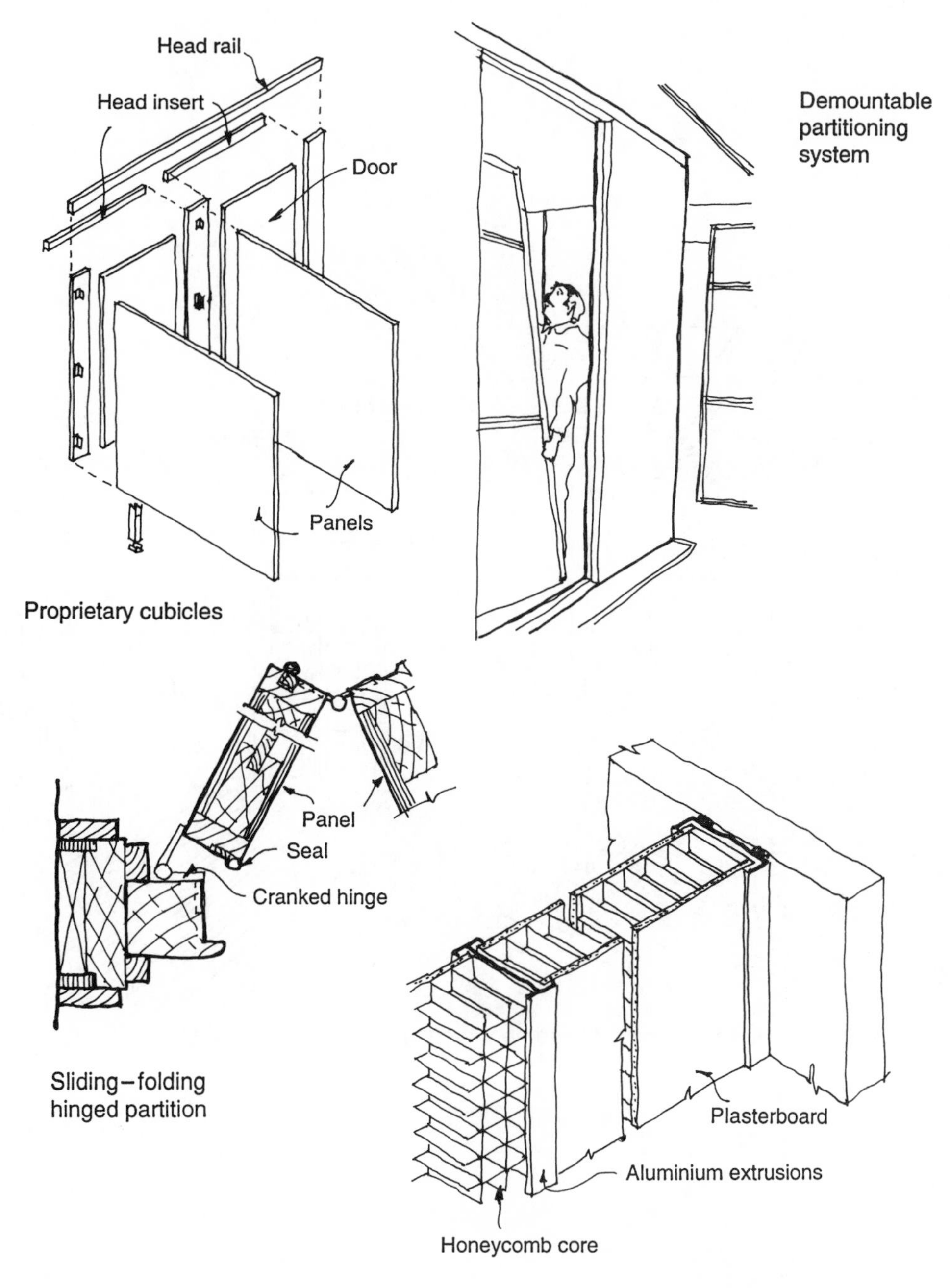

Frame and panel partition system

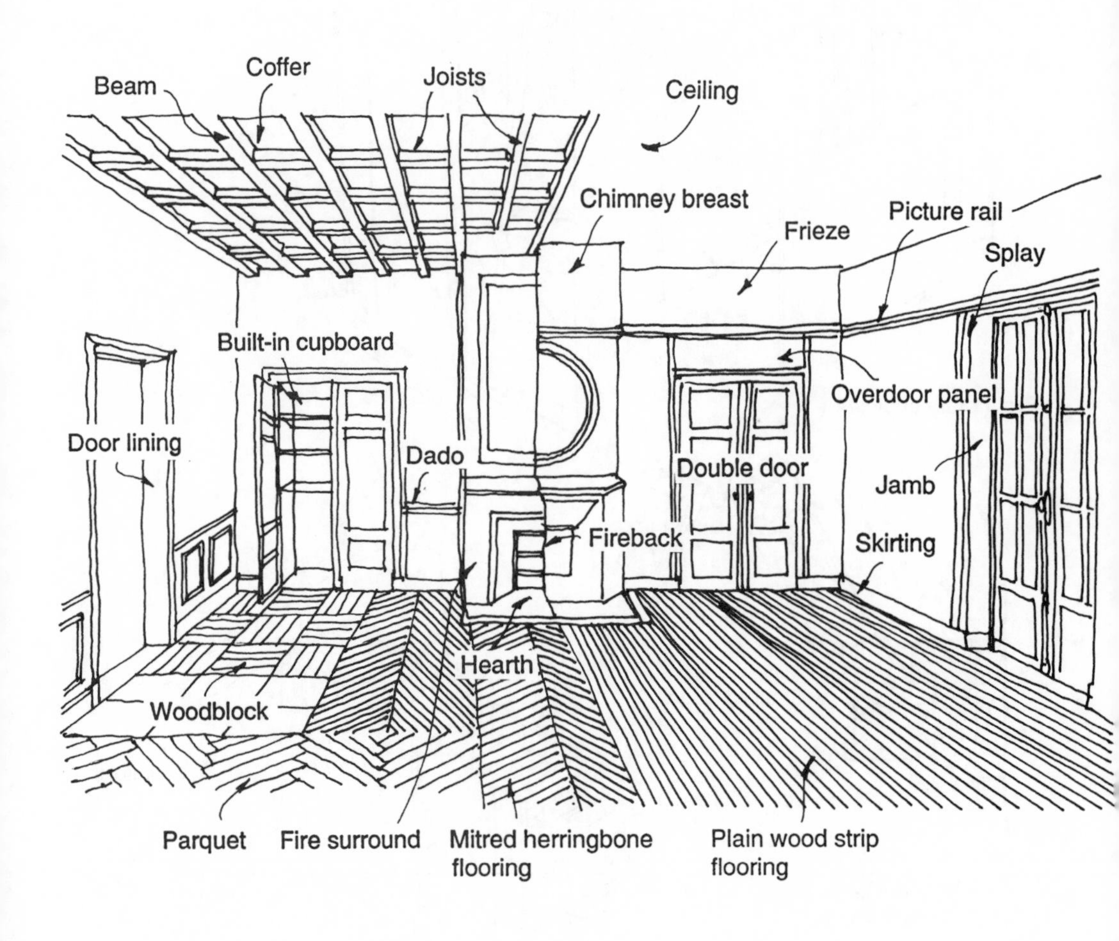

Beam
Coffer
Joists
Ceiling
Chimney breast
Frieze
Picture rail
Splay
Built-in cupboard
Overdoor panel
Door lining
Dado
Double door
Jamb
Fireback
Skirting
Hearth
Woodblock
Parquet
Fire surround
Mitred herringbone flooring
Plain wood strip flooring

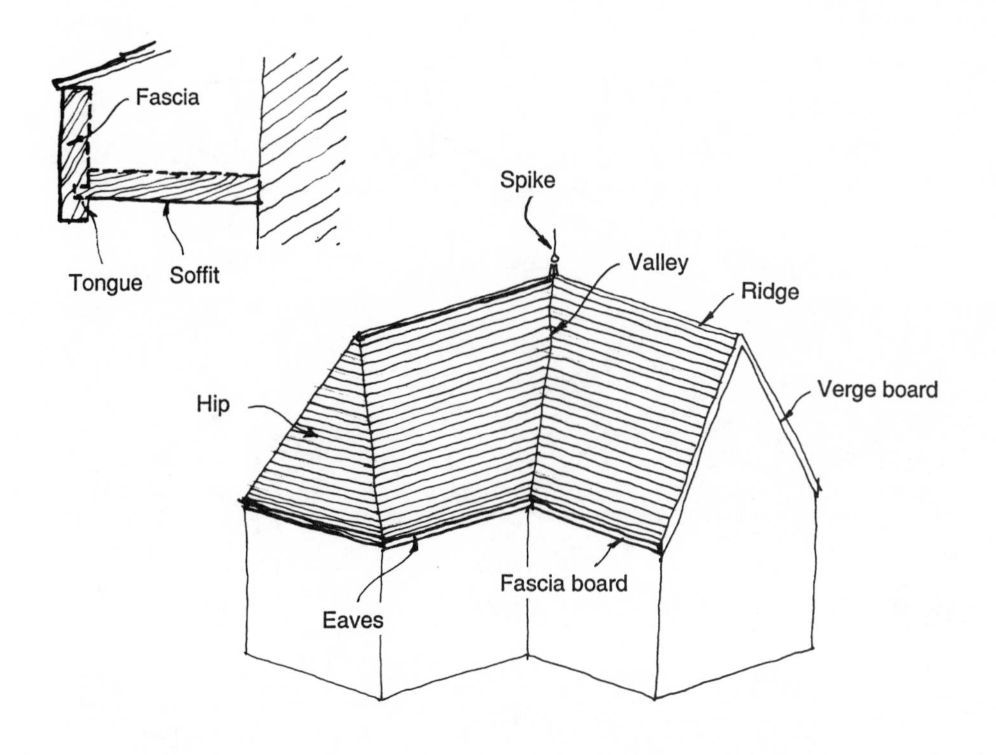

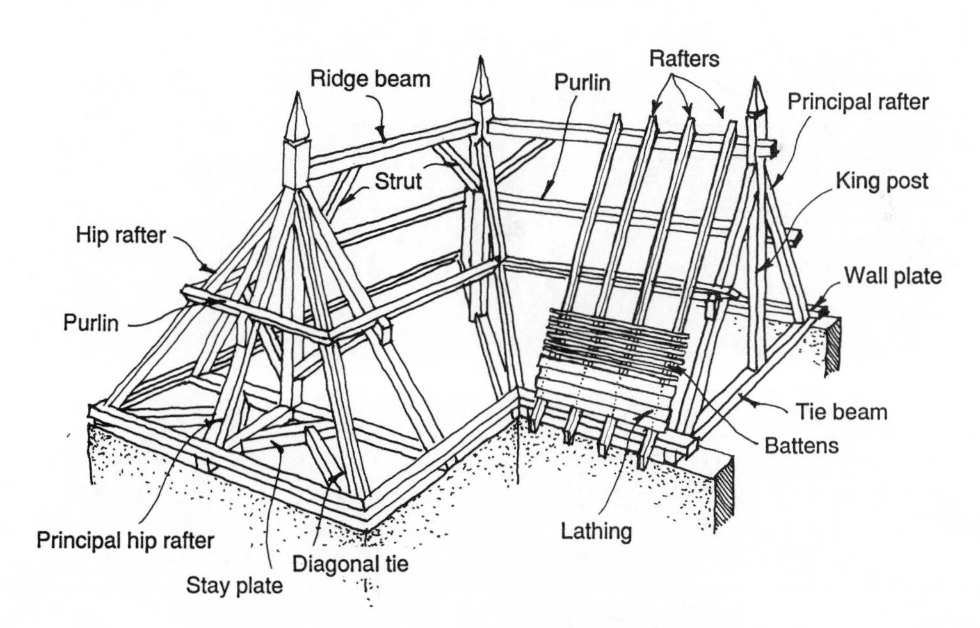

Traditional timber roof construction

Types of pitched roof

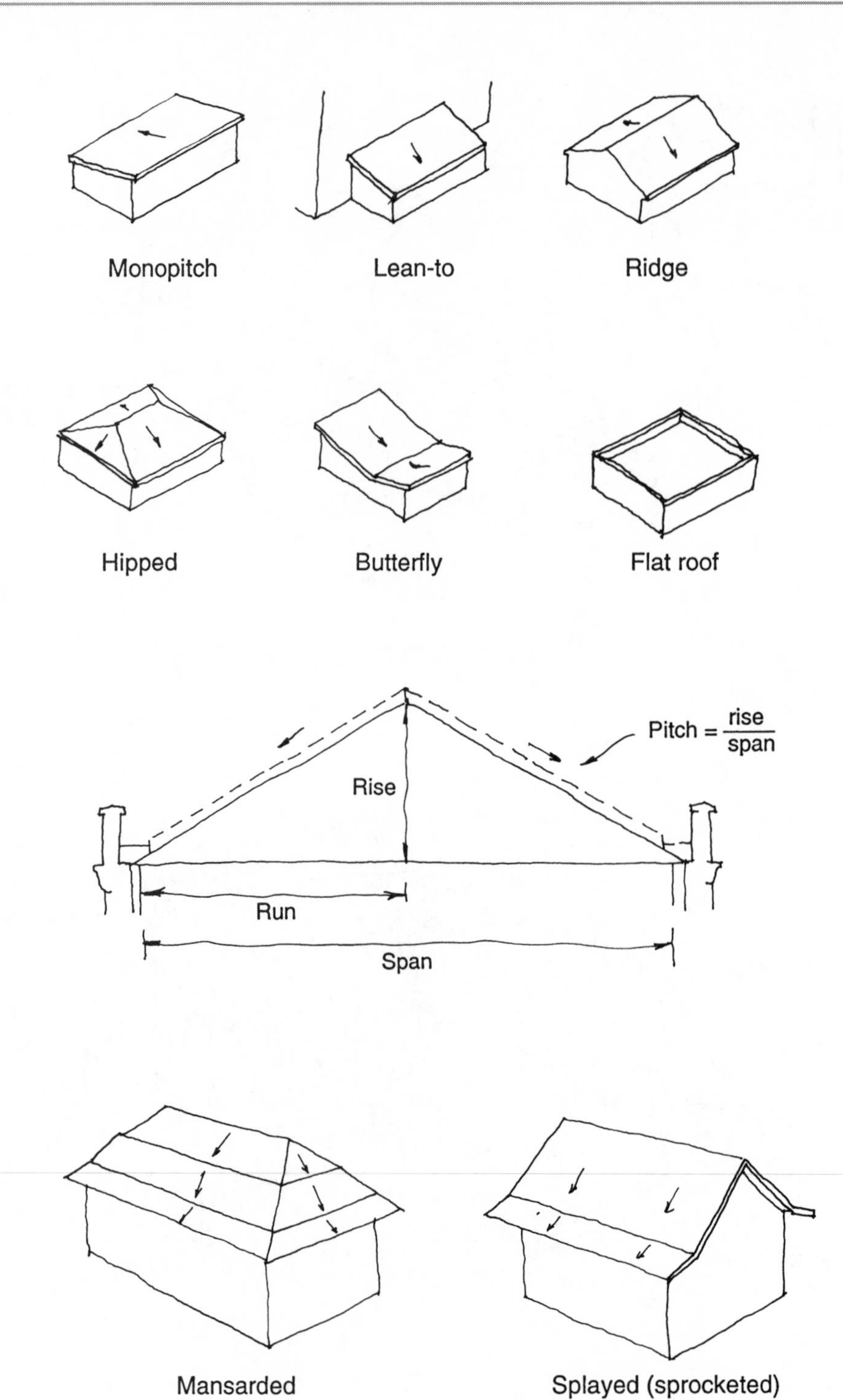

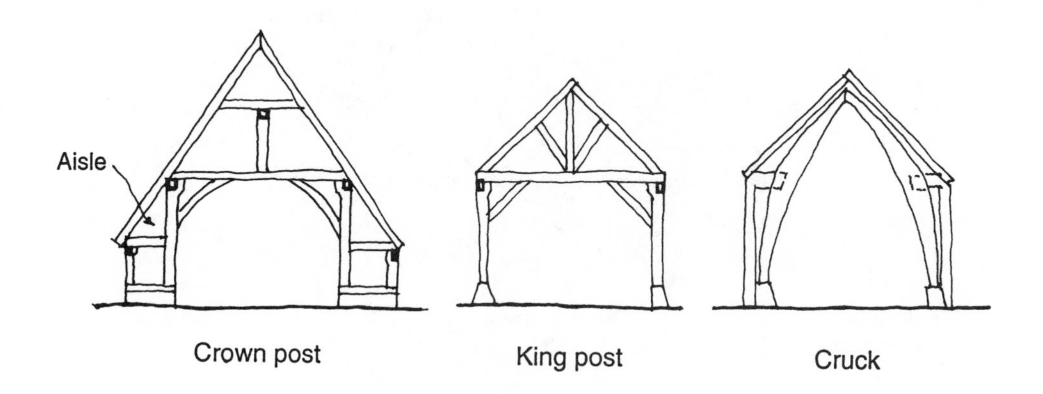
Aisle
Crown post
King post
Cruck

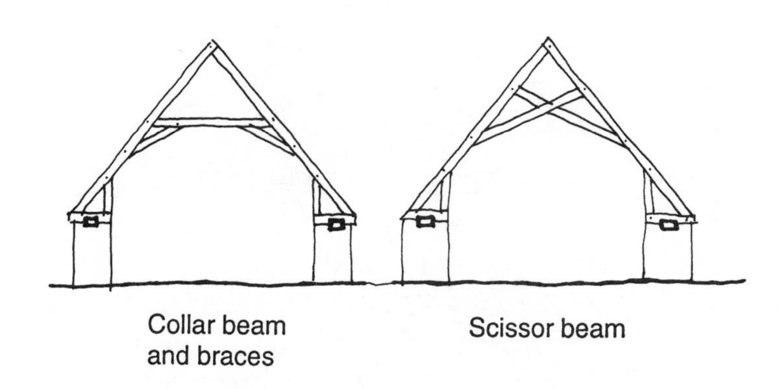
Collar beam and braces
Scissor beam

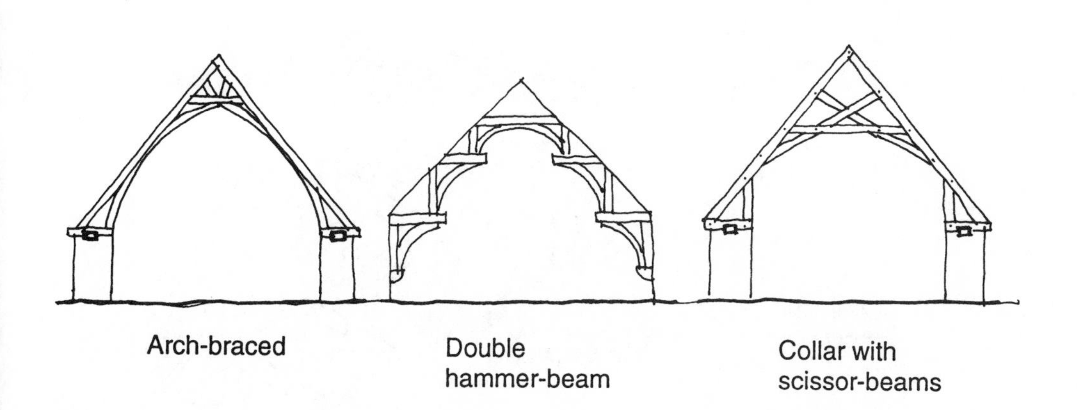
Arch-braced
Double hammer-beam
Collar with scissor-beams

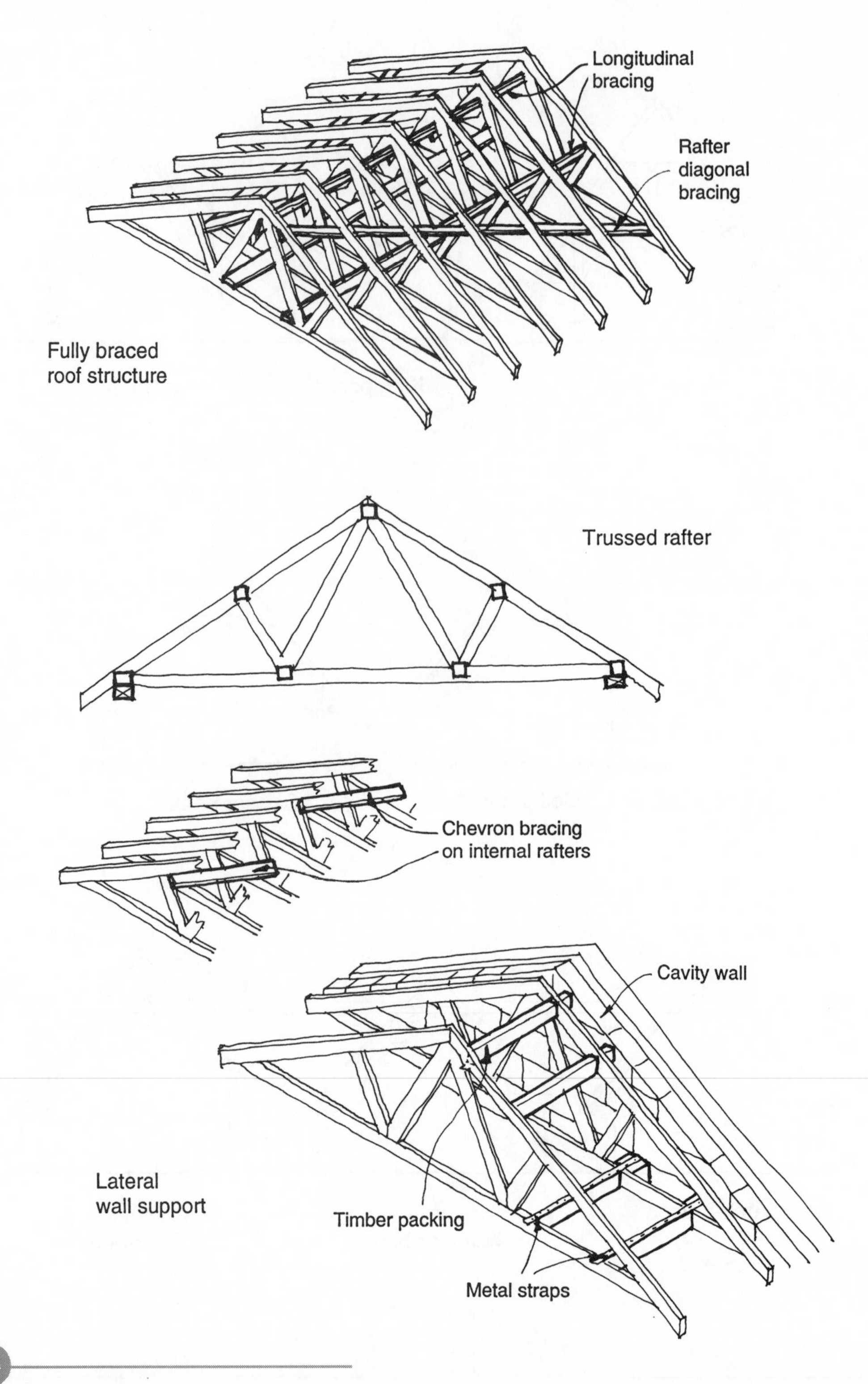
Longitudinal bracing
Rafter diagonal bracing
Fully braced roof structure
Trussed rafter
Chevron bracing on internal rafters
Cavity wall
Lateral wall support
Timber packing
Metal straps

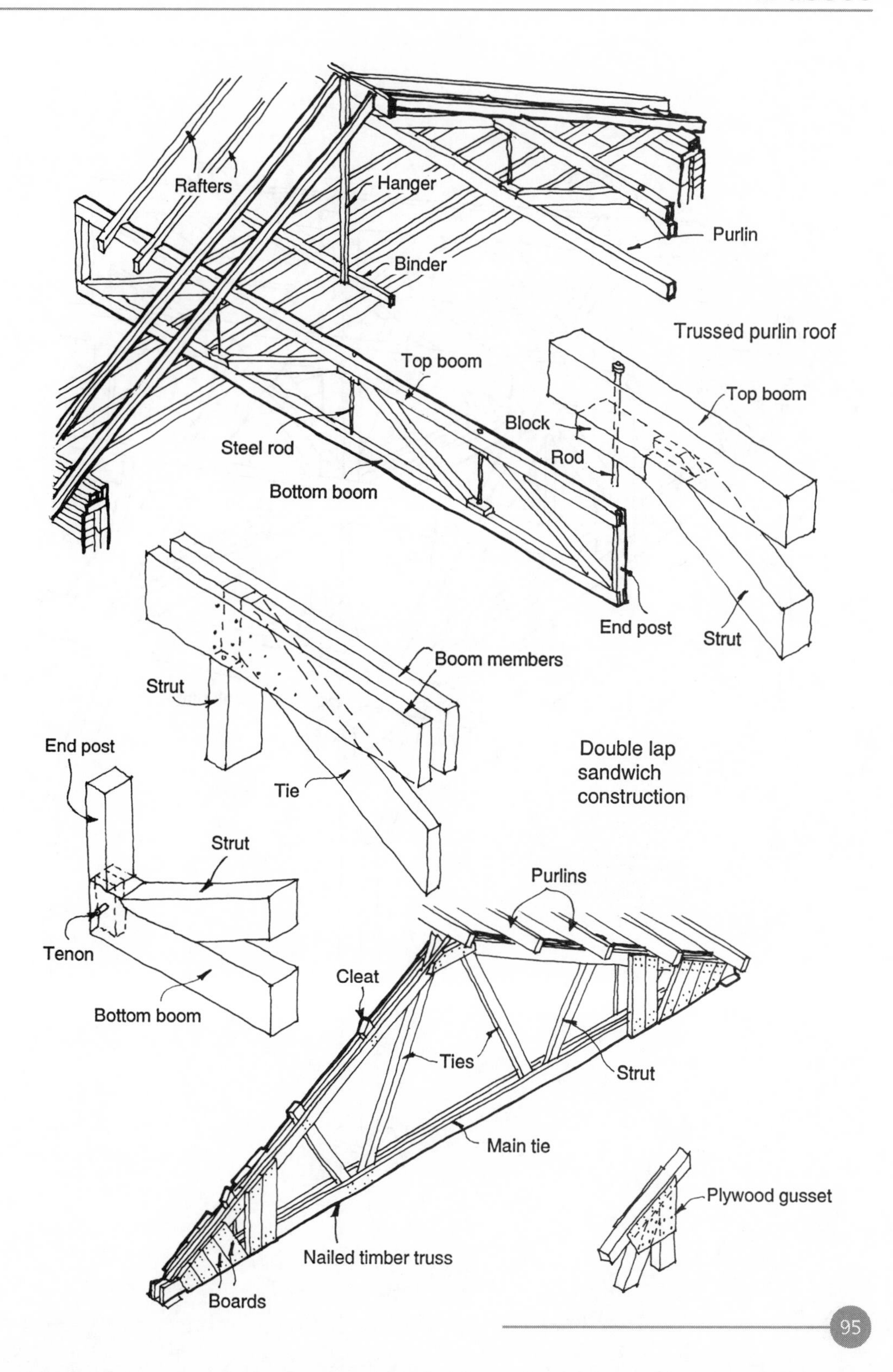
Rafters
Hanger
Purlin
Binder
Trussed purlin roof
Top boom
Top boom
Block
Rod
Steel rod
Bottom boom
End post
Strut
Boom members
Strut
End post
Double lap
sandwich
construction
Tie
Strut
Purlins
Tenon
Cleat
Bottom boom
Ties
Strut
Main tie
Plywood gusset
Nailed timber truss
Boards

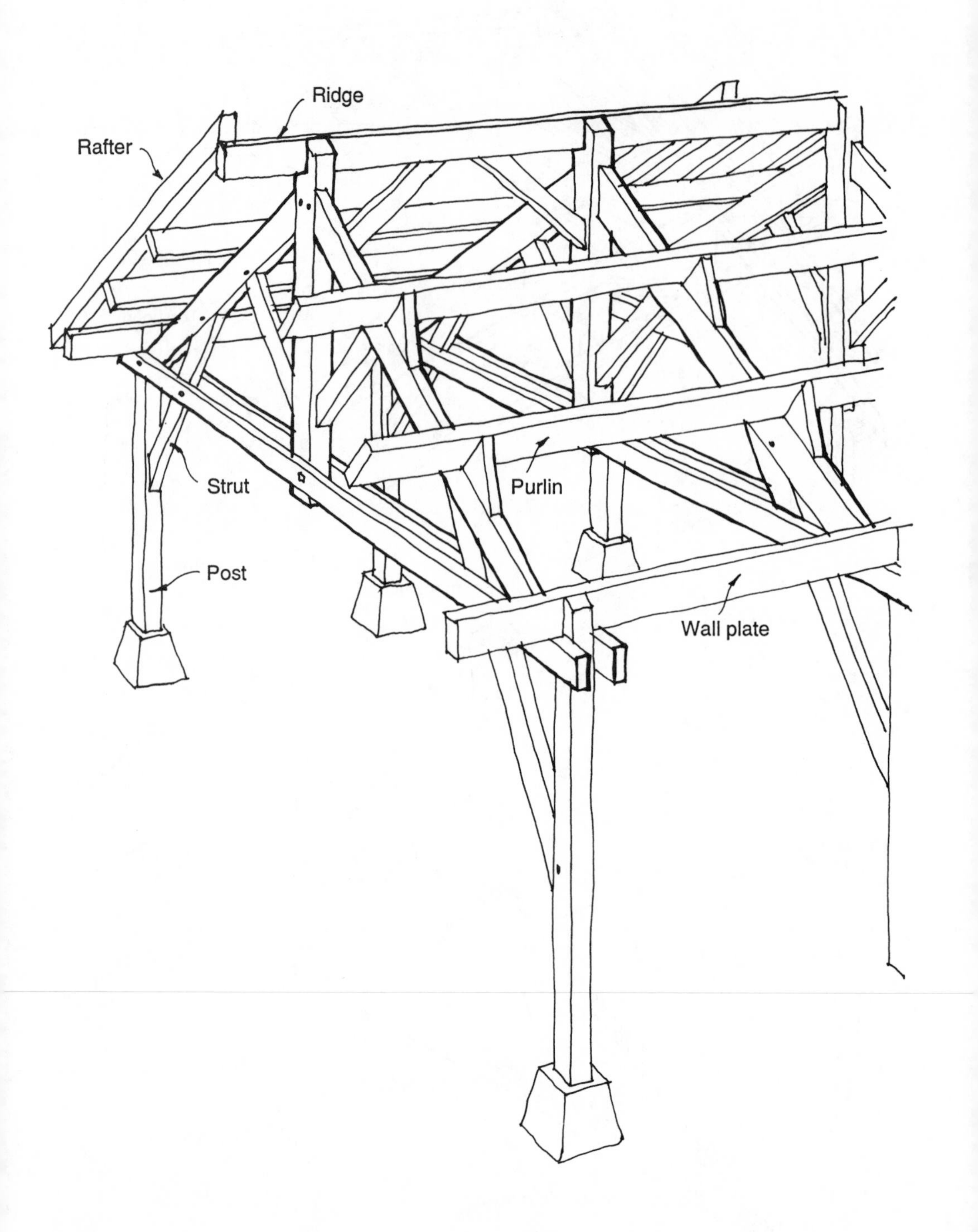
Ridge
Rafter
Strut
Purlin
Post
Wall plate

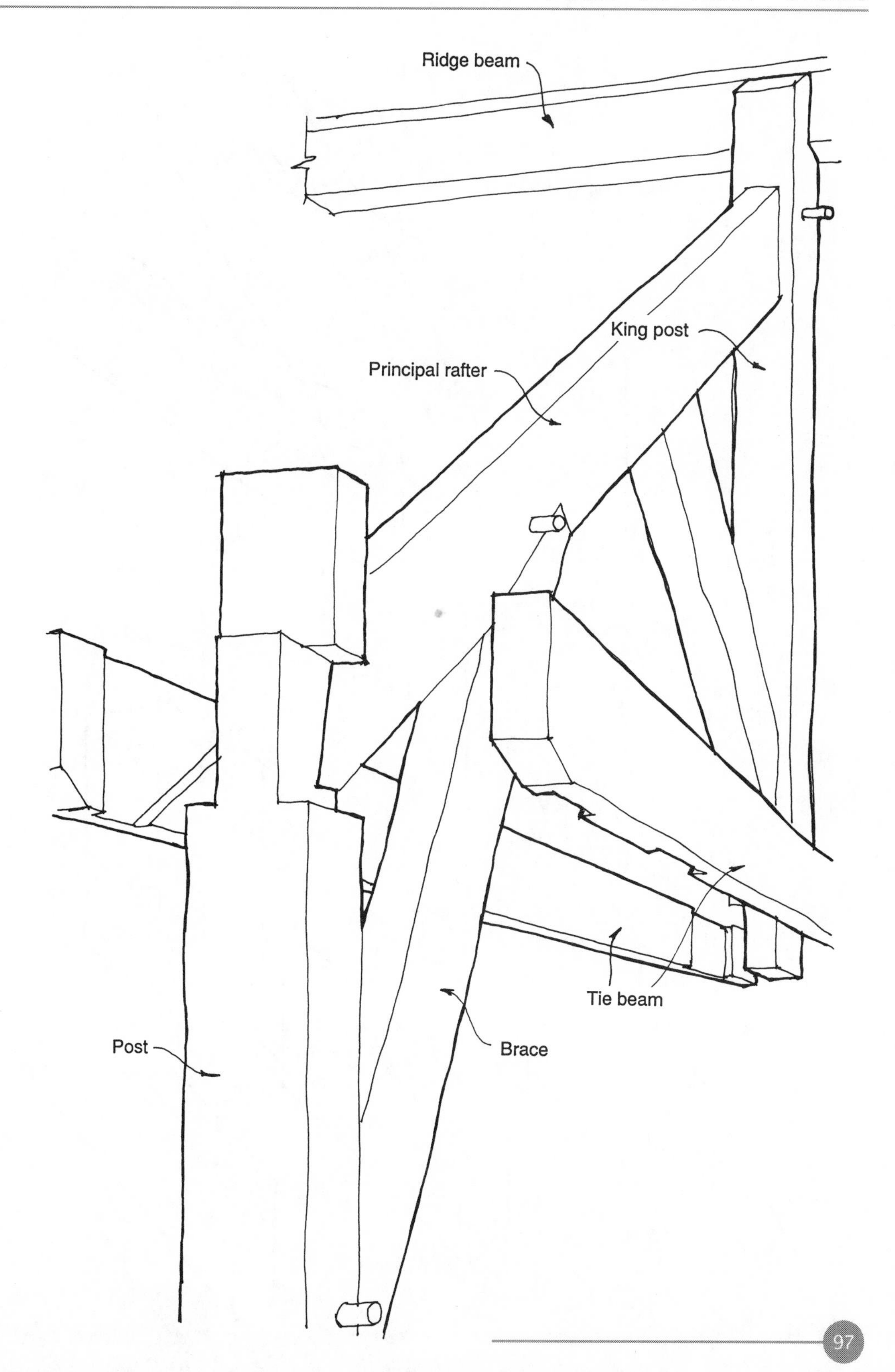
Ridge beam
King post
Principal rafter
Tie beam
Post
Brace

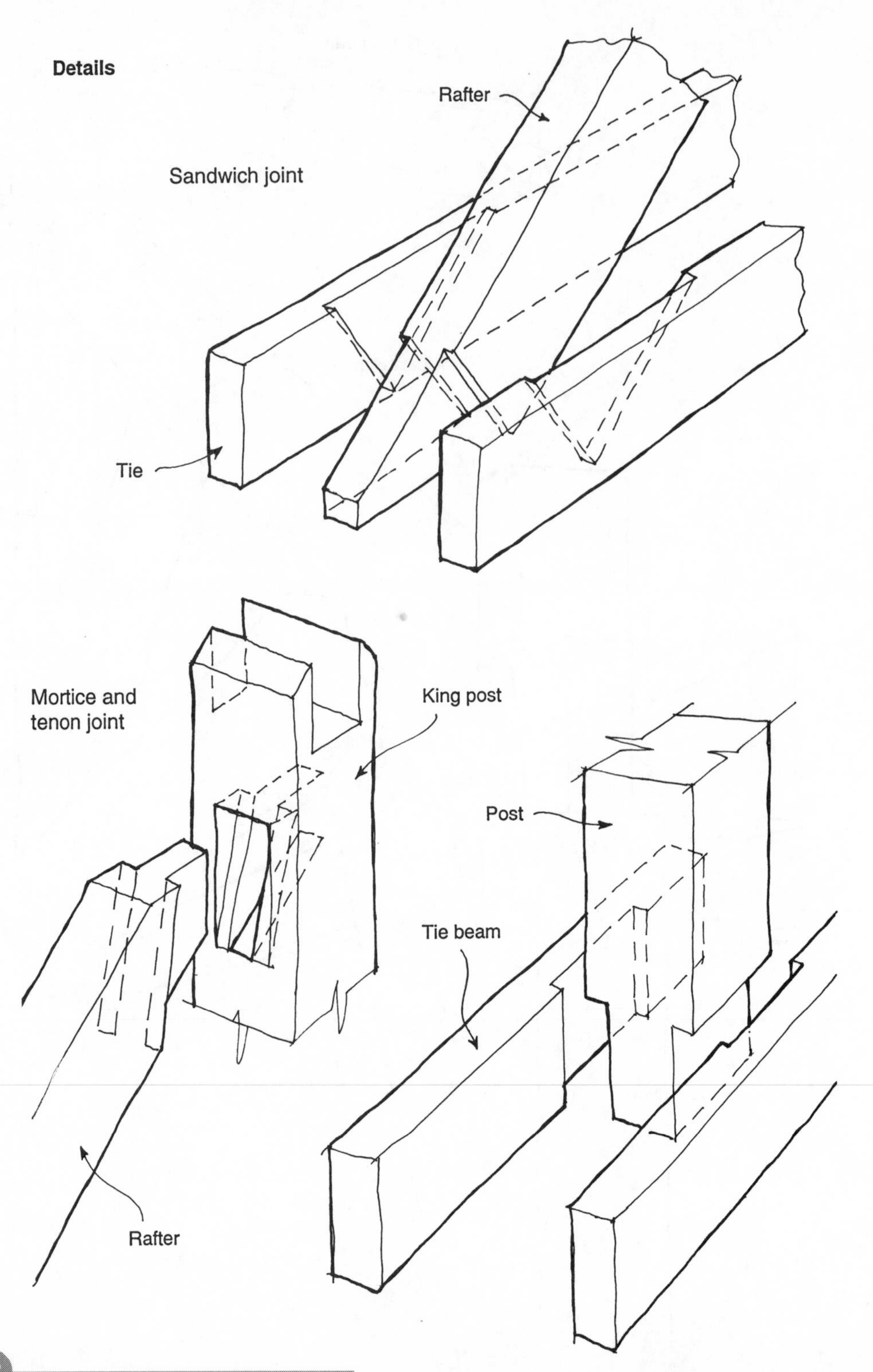
Details
Rafter
Sandwich joint
Tie
Mortice and
tenon joint
King post
Post
Tie beam
Rafter

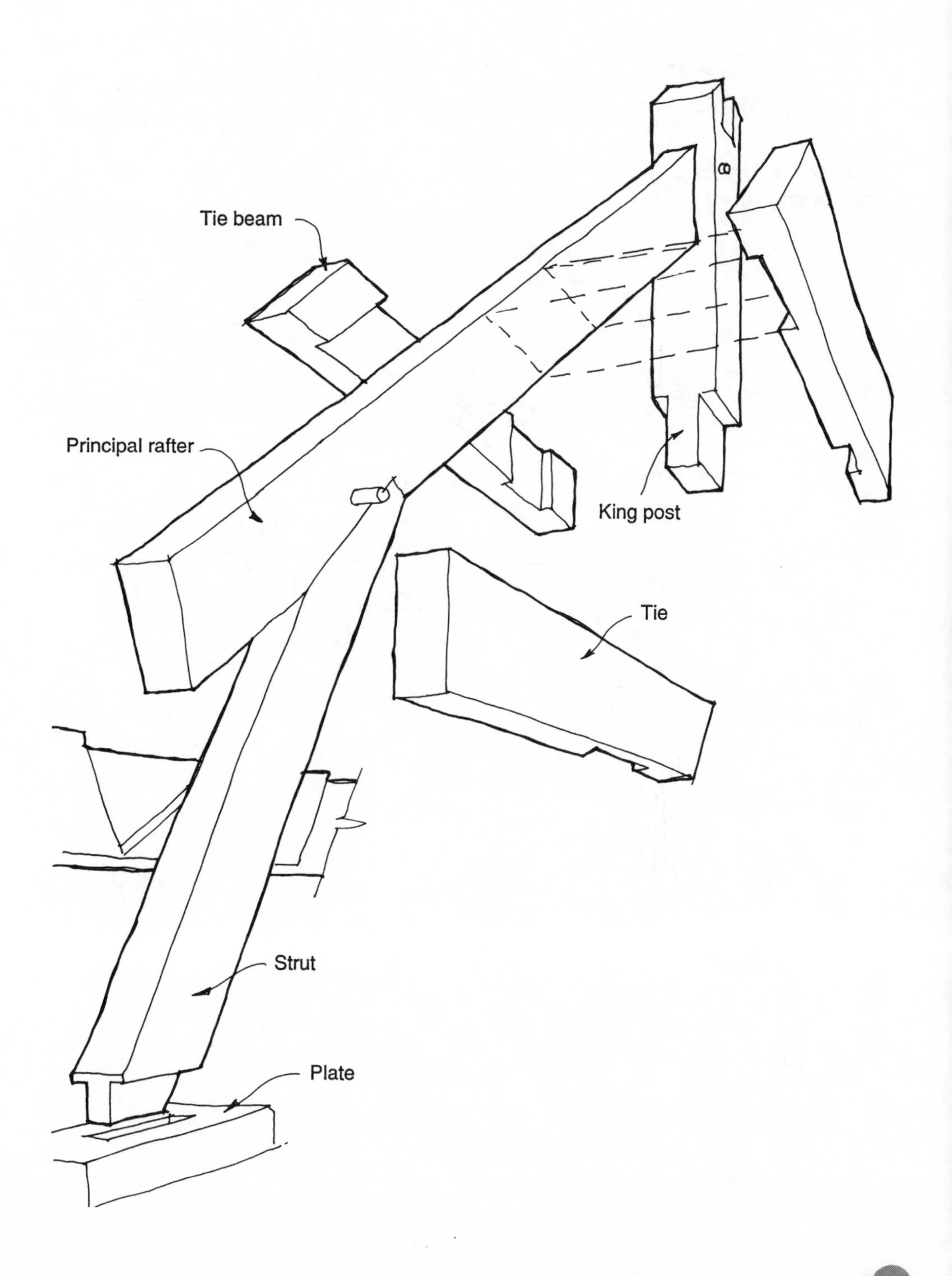
Tie beam
Principal rafter
King post
Tie
Strut
Plate

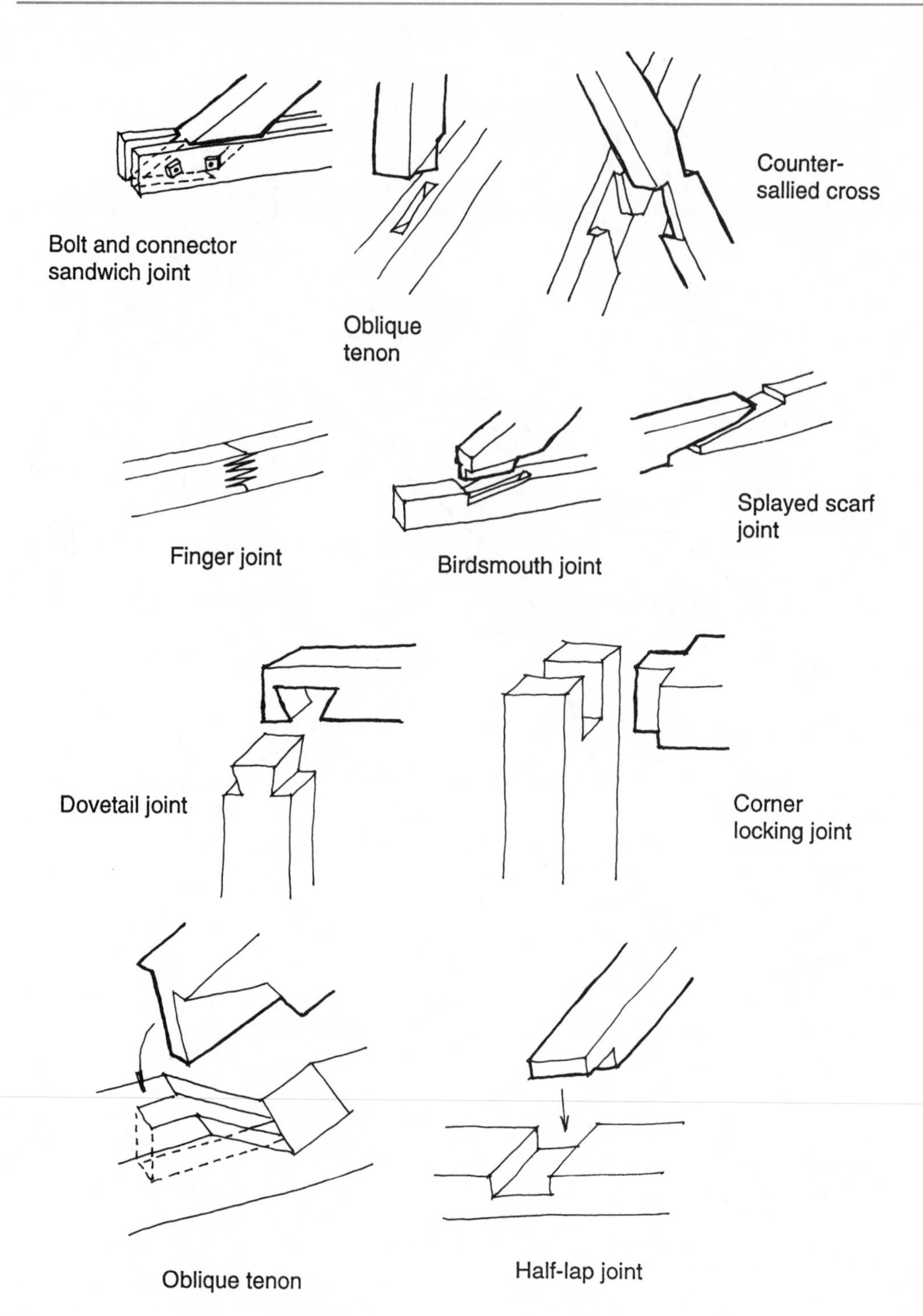
Bolt and connector sandwich joint
Oblique tenon
Counter-sallied cross
Finger joint
Birdsmouth joint
Splayed scarf joint
Dovetail joint
Corner locking joint
Oblique tenon
Half-lap joint

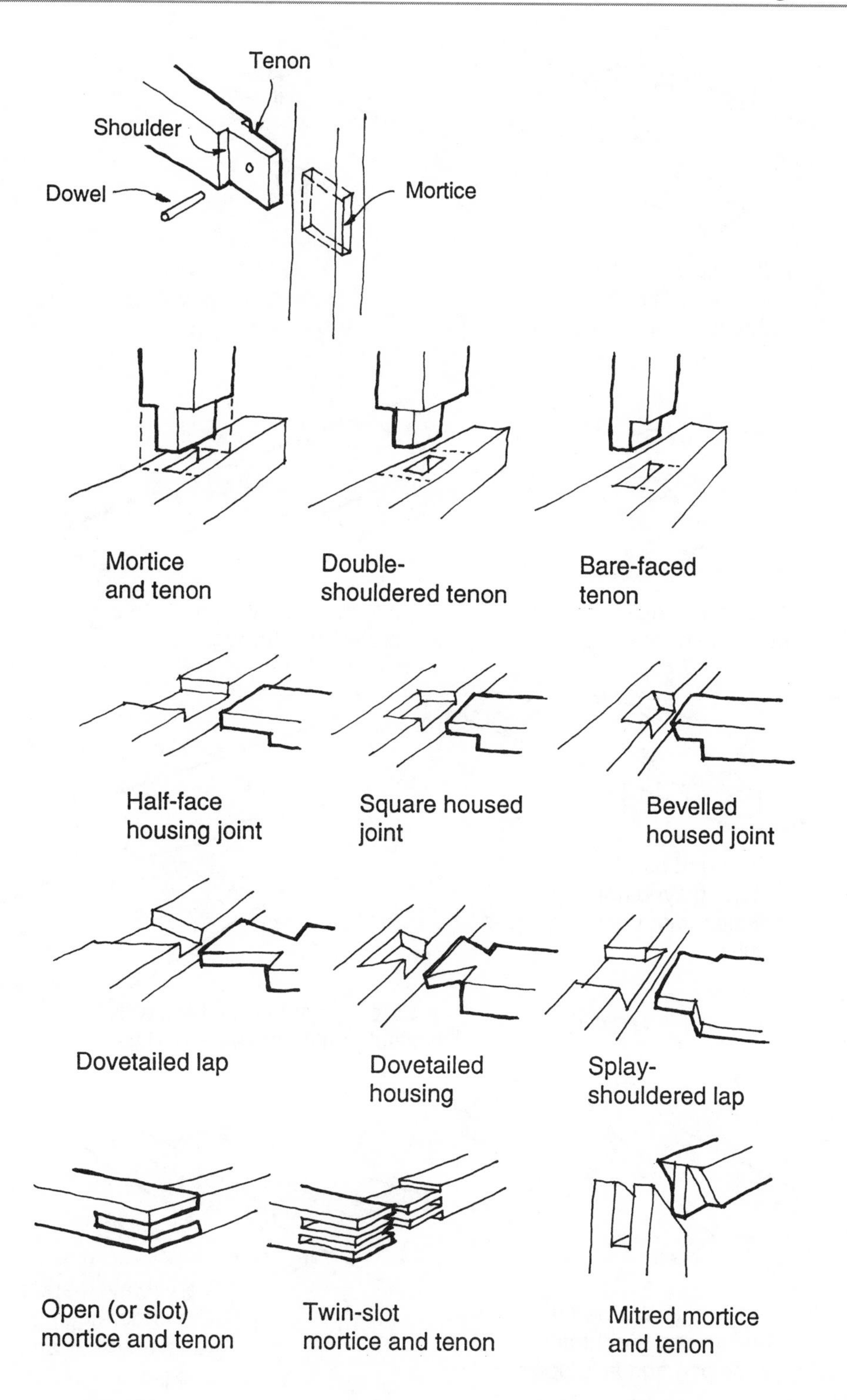

Mortice and tenon

Double-shouldered tenon

Bare-faced tenon

Half-face housing joint

Square housed joint

Bevelled housed joint

Dovetailed lap

Dovetailed housing

Splay-shouldered lap

Open (or slot) mortice and tenon

Twin-slot mortice and tenon

Mitred mortice and tenon

Traditional scarf joints

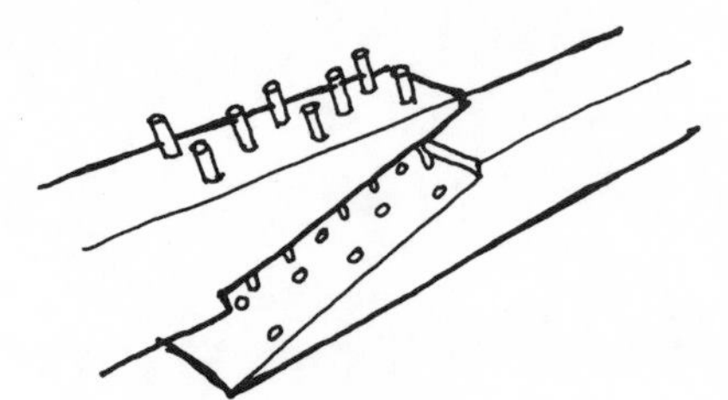

Stop-splayed with square under-squinted abutments and face pegs

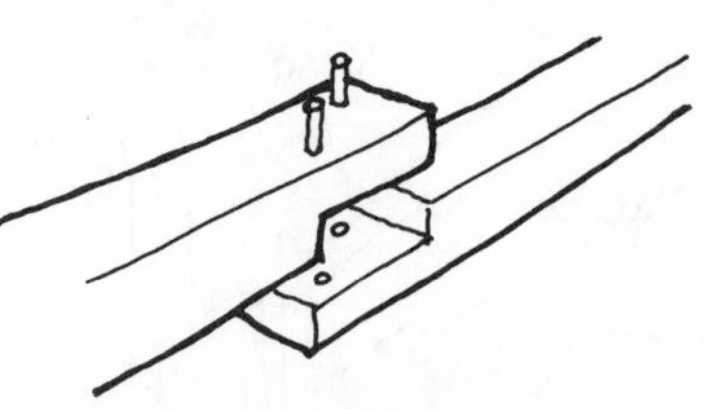

Edge-halved with square abutments and face pegs

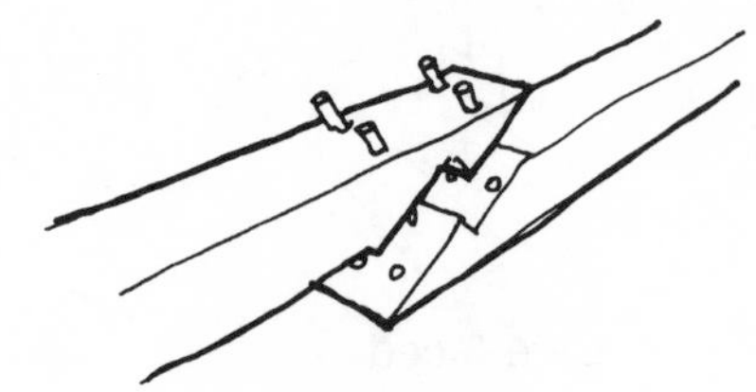

Through-splayed and tabled with face pegs

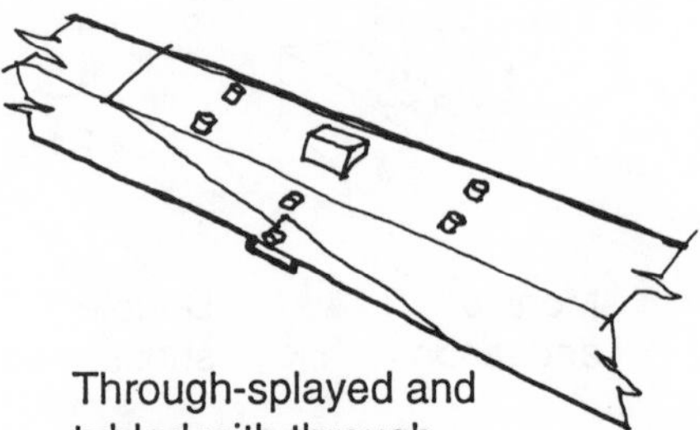

Through-splayed and tabled with through-tenon tabling

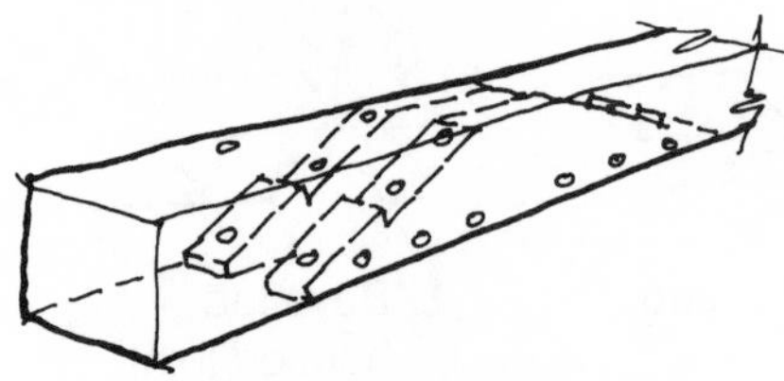

Four-part scarf with two stop-splayed and tabled 'fishes' with under-squinted butts

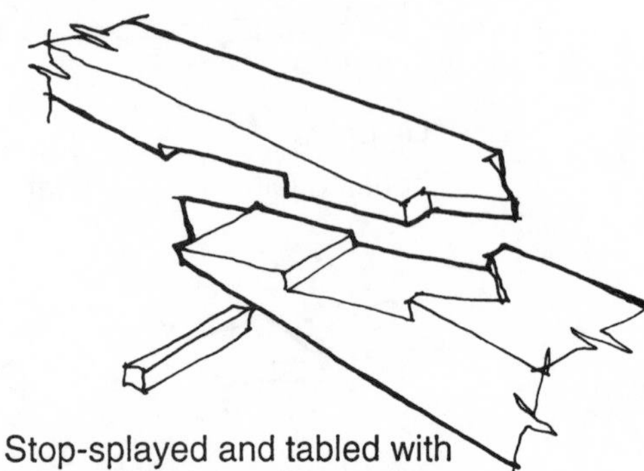

Stop-splayed and tabled with inset abutment salies under-squinted with transverse key

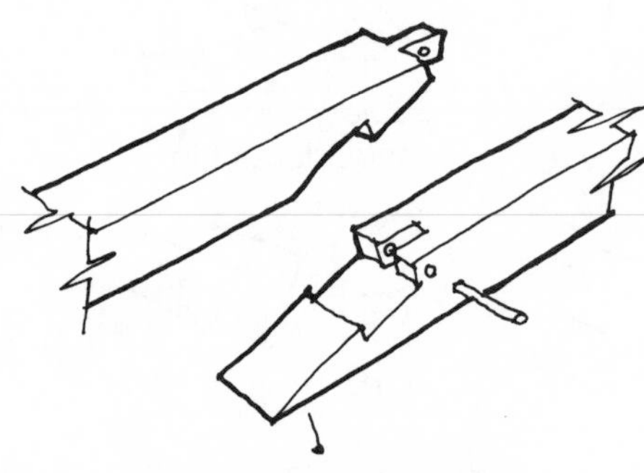

Splayed and tabled with bridled upper abutment edge peg and face spike

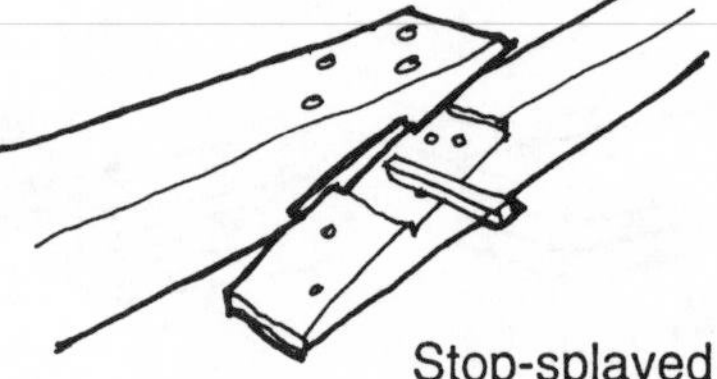

Stop-splayed and tabled with under-squinted transverse key and face pegs

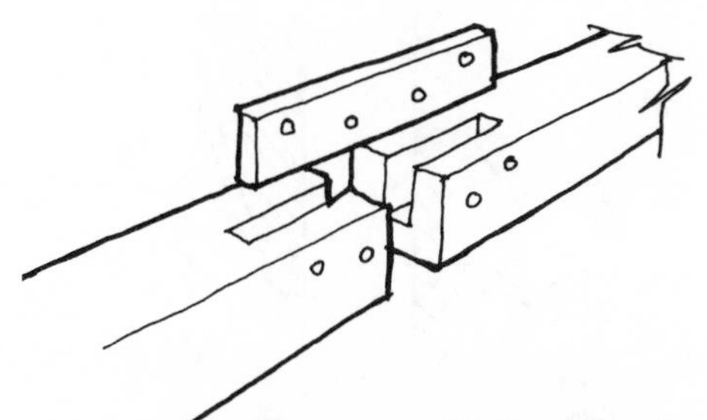

Three-part 'fished' scarf with square and vertical abutments

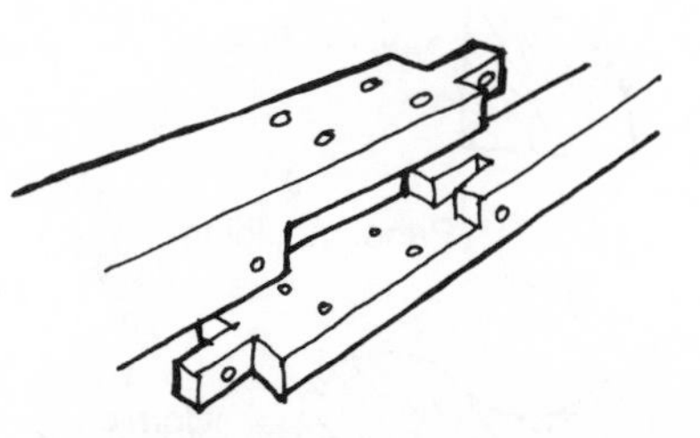

Edge-halved and stop-splayed with bridled abutments

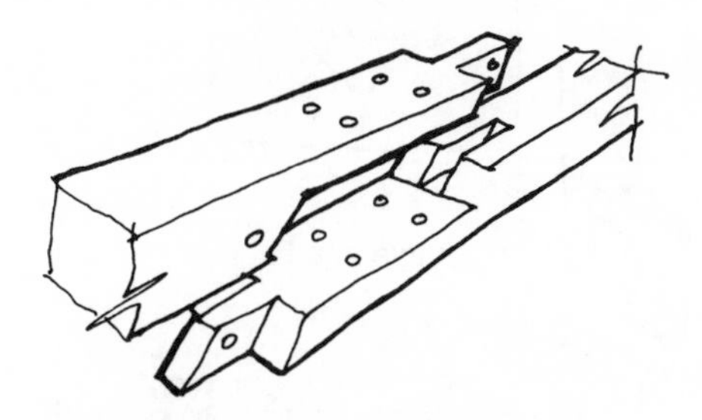

Edge-halved and bridled with over-squinted abutments

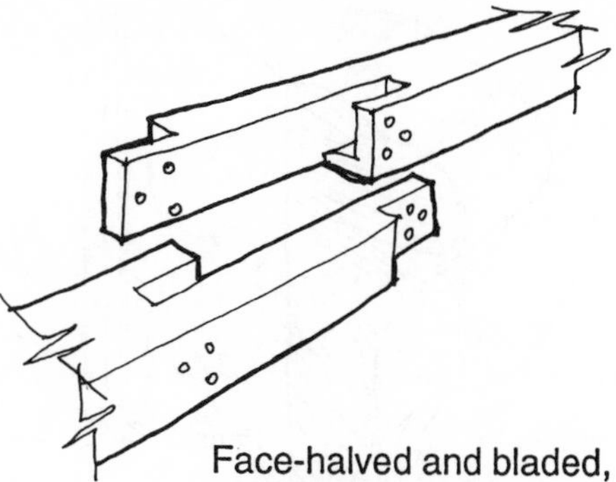

Face-halved and bladed, one blade housed, edge pegs

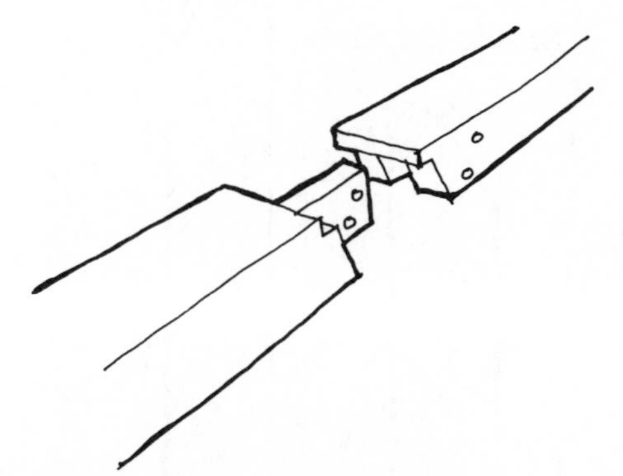

Straight bridling of three-quarter depth, with squinted abutments, overlipped face, edge pegs

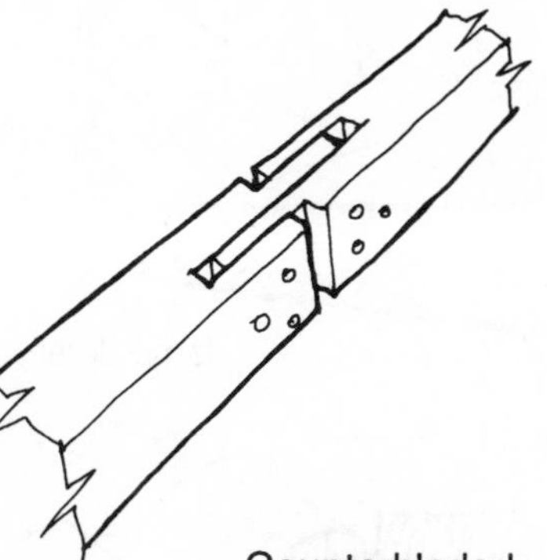

Counterbladed, face-halved with edge pegs

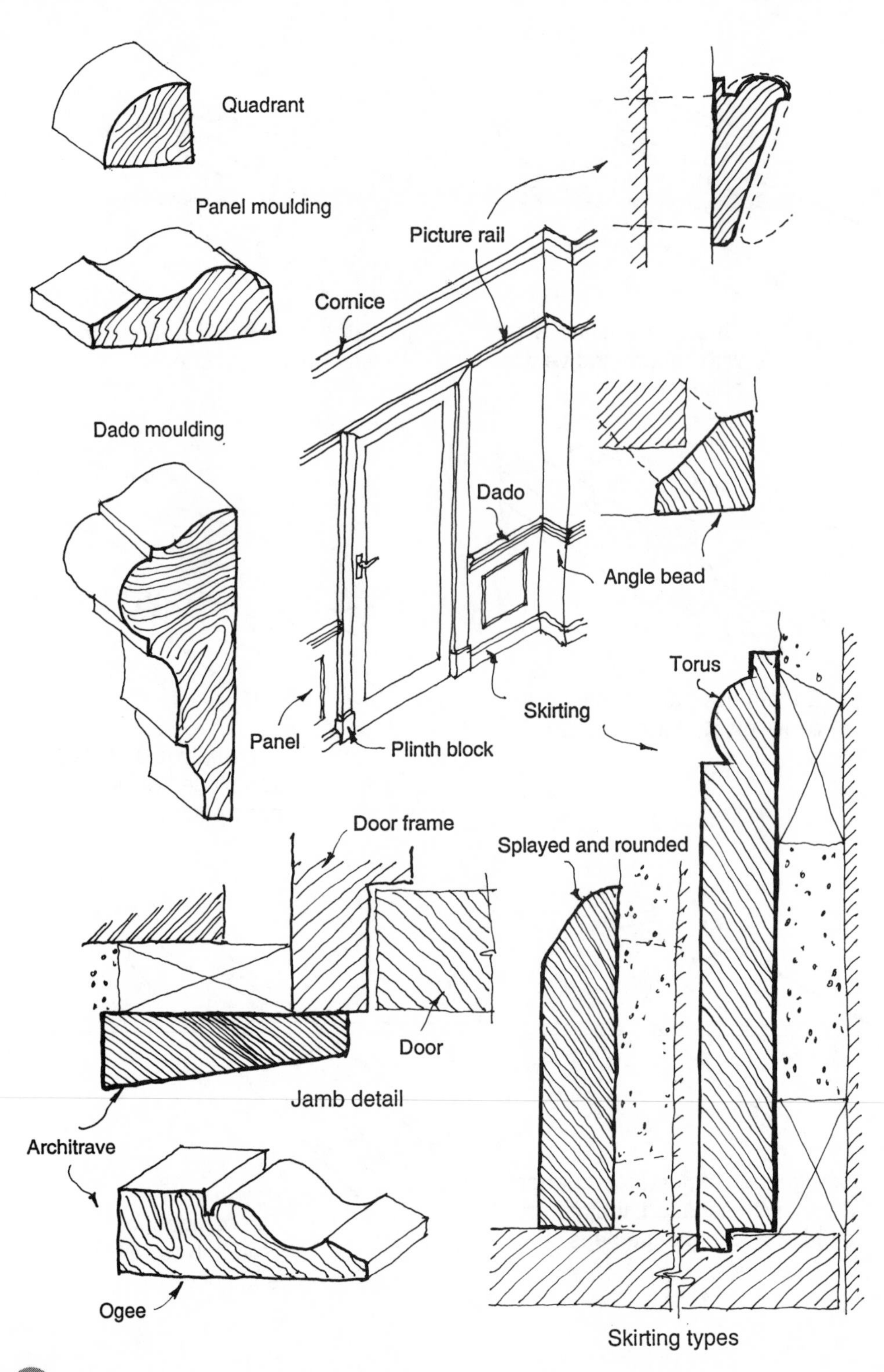
Quadrant
Panel moulding
Picture rail
Cornice
Dado moulding
Dado
Angle bead
Torus
Skirting
Panel
Plinth block
Door frame
Splayed and rounded
Door
Jamb detail
Architrave
Ogee
Skirting types

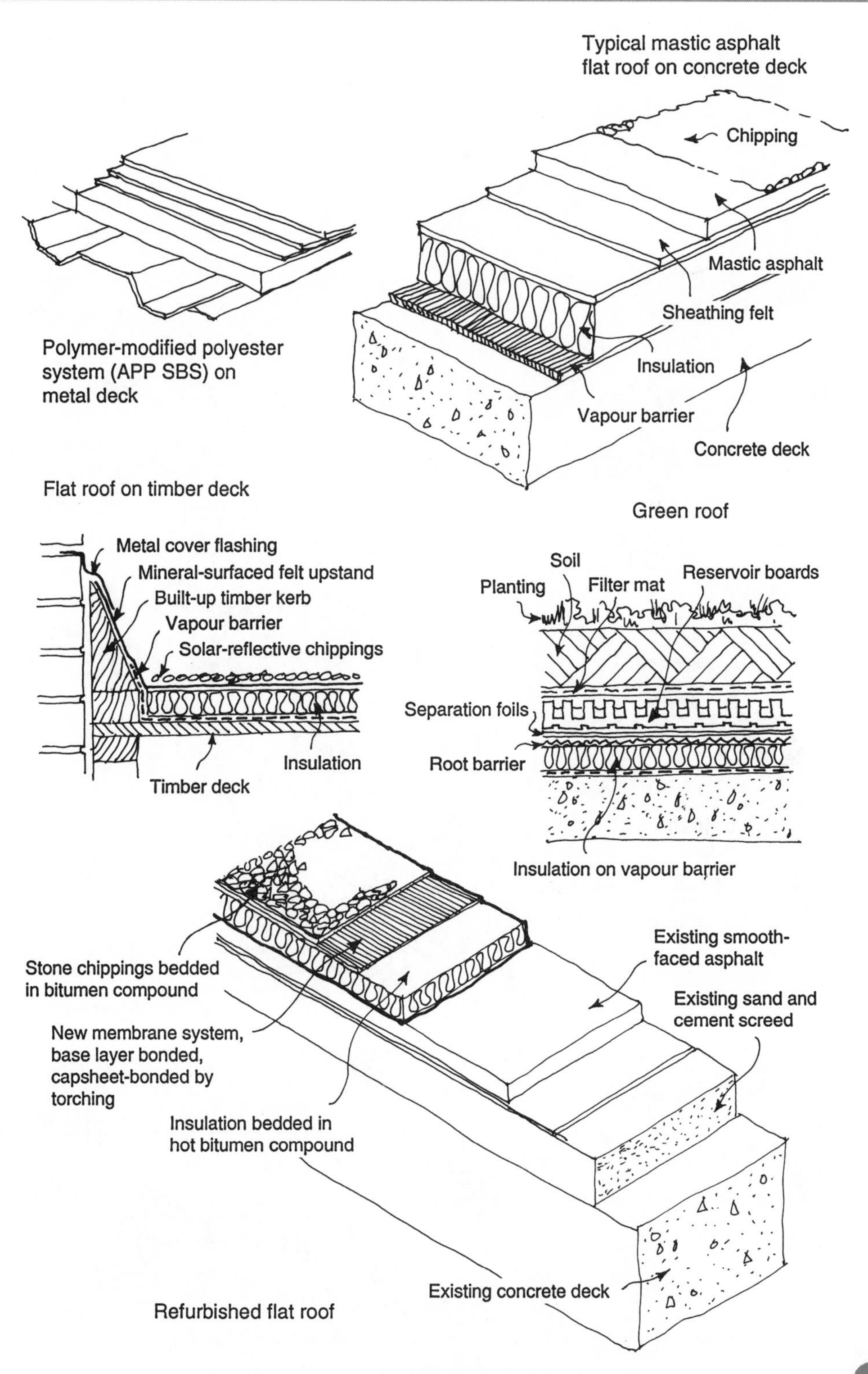
Typical mastic asphalt flat roof on concrete deck
Chipping
Mastic asphalt
Sheathing felt
Insulation
Vapour barrier
Concrete deck
Polymer-modified polyester system (APP SBS) on metal deck
Flat roof on timber deck
Metal cover flashing
Mineral-surfaced felt upstand
Built-up timber kerb
Vapour barrier
Solar-reflective chippings
Insulation
Timber deck
Green roof
Soil
Planting
Filter mat
Reservoir boards
Separation foils
Root barrier
Insulation on vapour barrier
Stone chippings bedded in bitumen compound
New membrane system, base layer bonded, capsheet-bonded by torching
Insulation bedded in hot bitumen compound
Existing smooth-faced asphalt
Existing sand and cement screed
Existing concrete deck
Refurbished flat roof

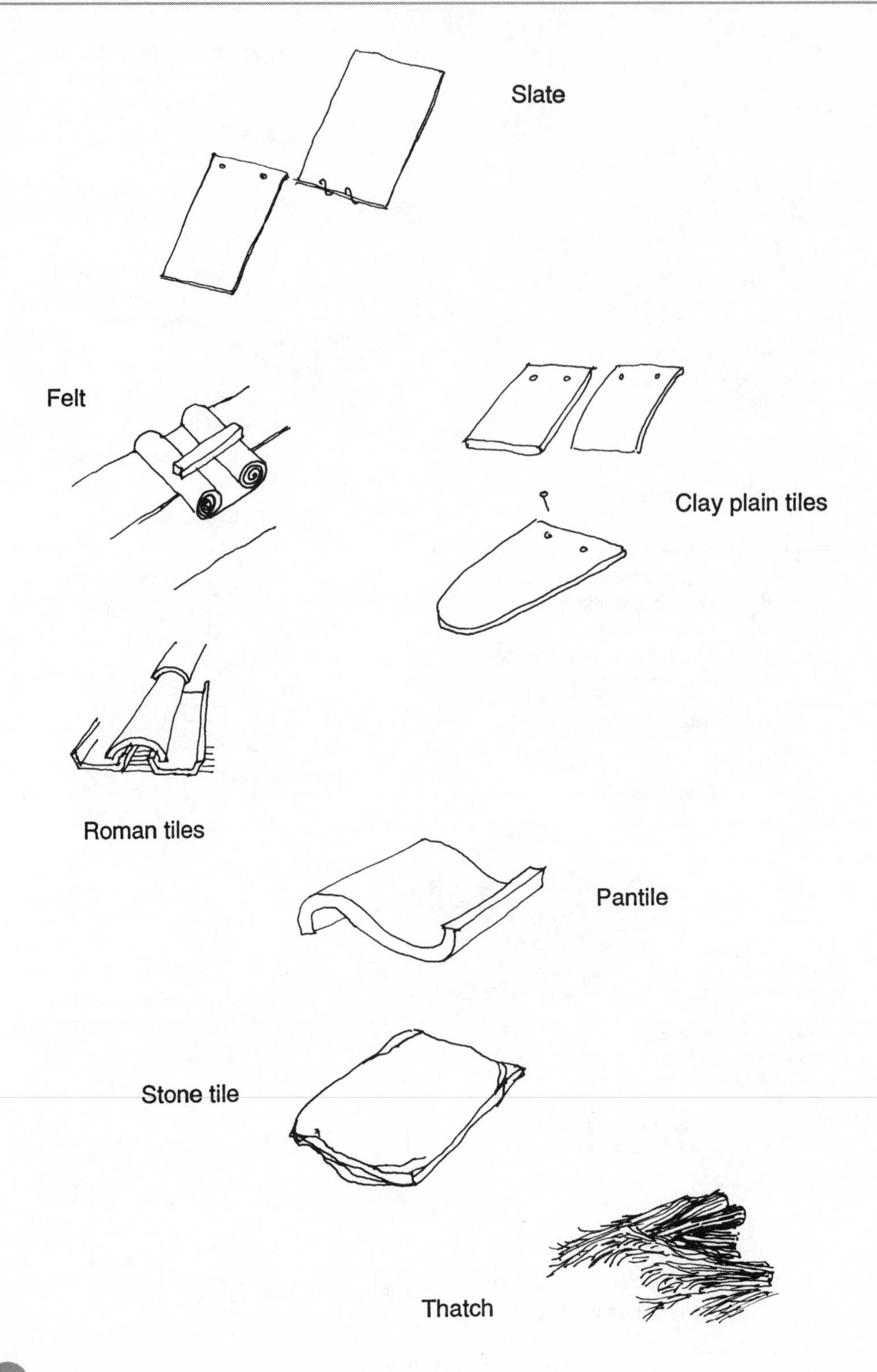
Slate
Felt
Clay plain tiles
Roman tiles
Pantile
Stone tile
Thatch

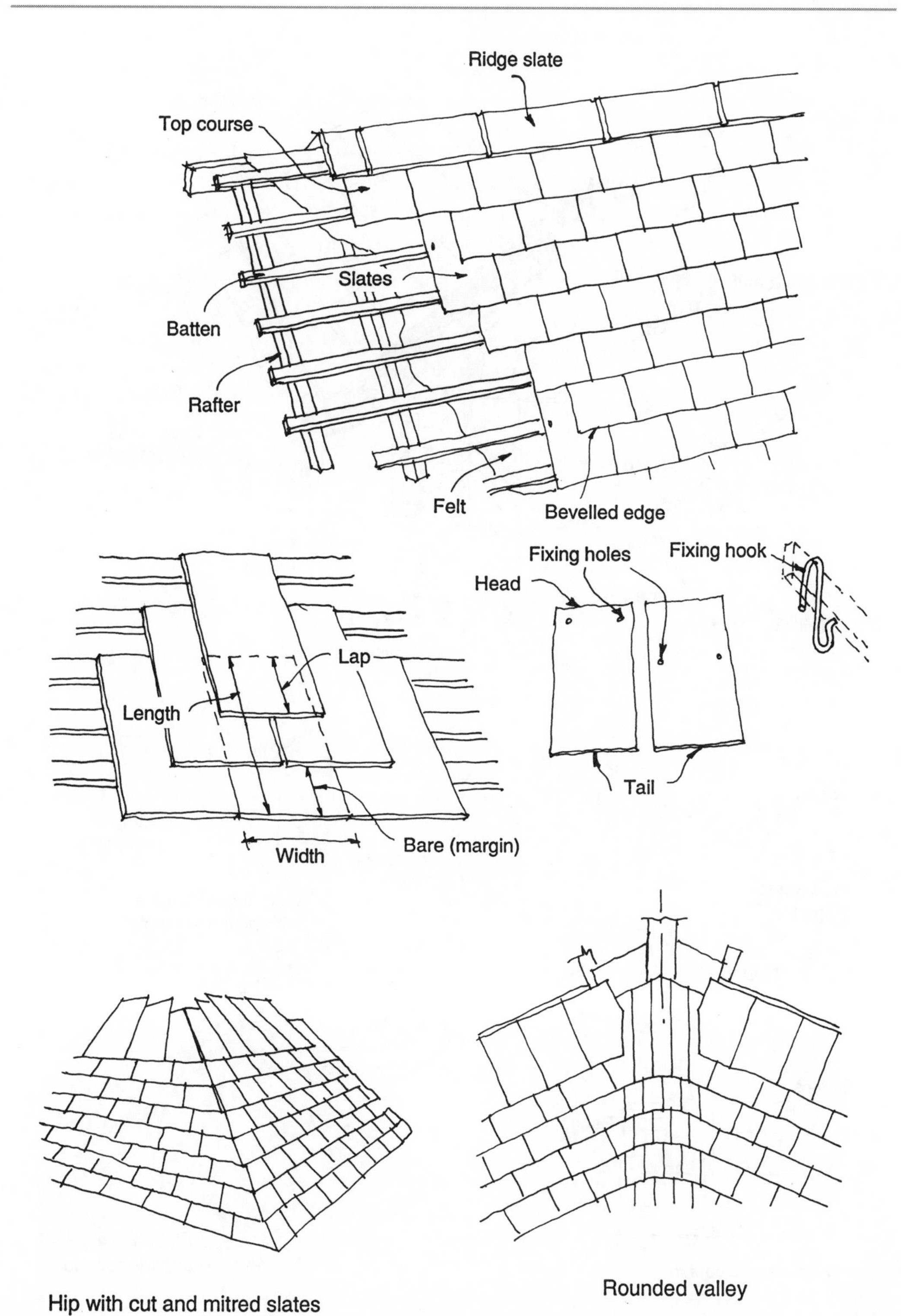

Hip with cut and mitred slates

Rounded valley

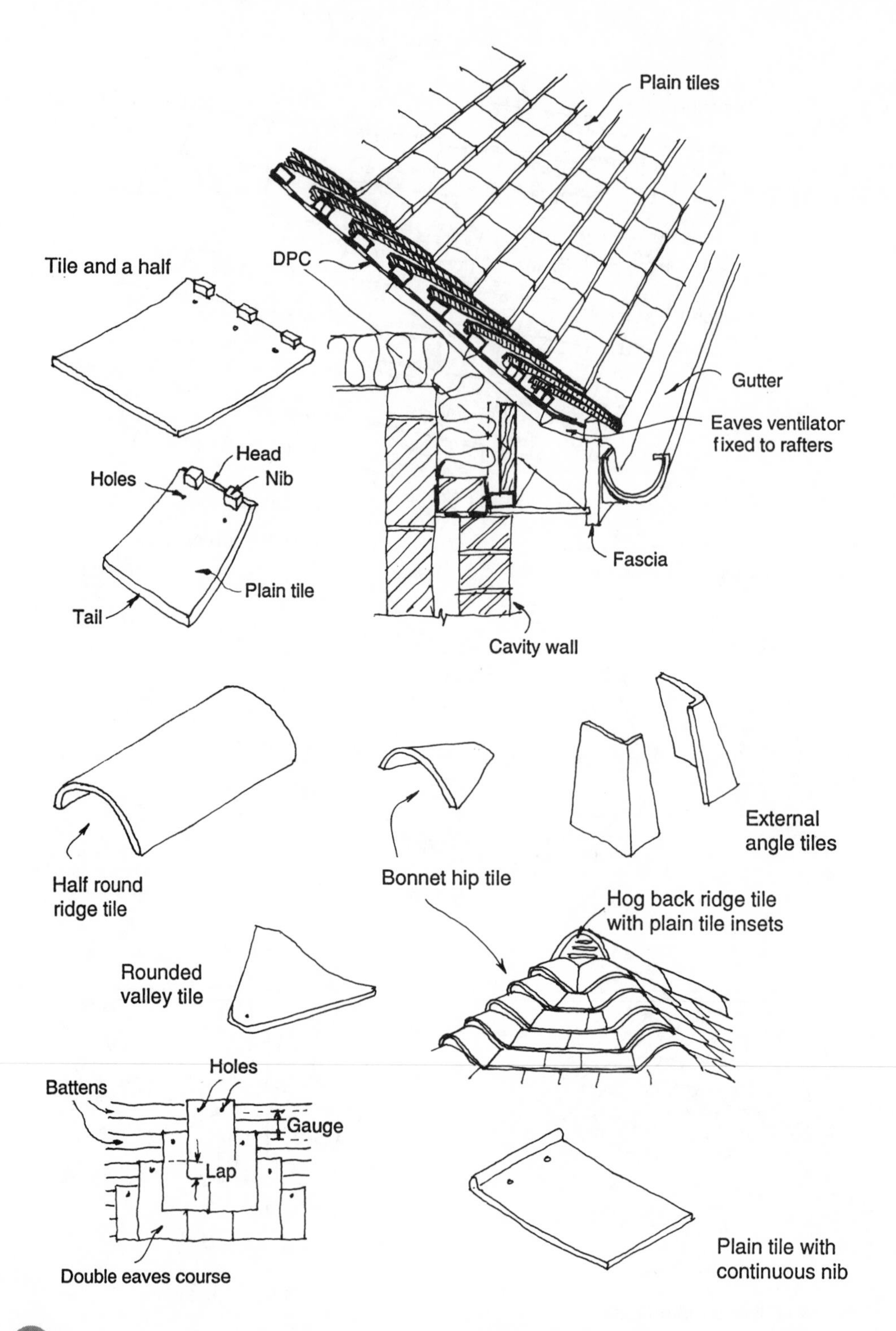
Plain tiles
Tile and a half
DPC
Gutter
Eaves ventilator fixed to rafters
Head
Holes
Nib
Fascia
Plain tile
Tail
Cavity wall
External angle tiles
Half round ridge tile
Bonnet hip tile
Hog back ridge tile with plain tile insets
Rounded valley tile
Holes
Battens
Gauge
Lap
Double eaves course
Plain tile with continuous nib

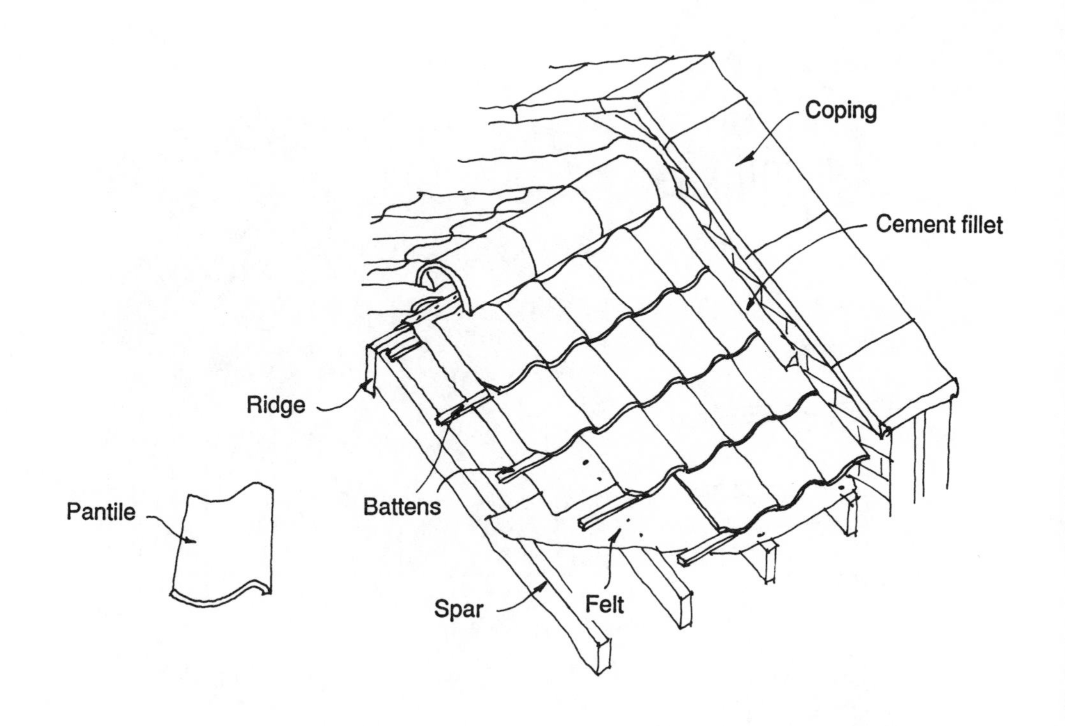

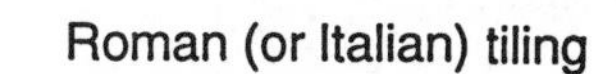
Roman (or Italian) tiling

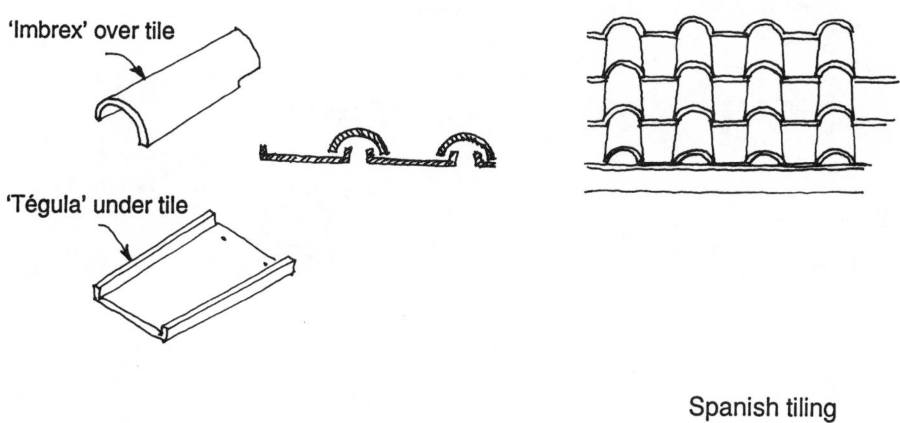

Spanish tiling

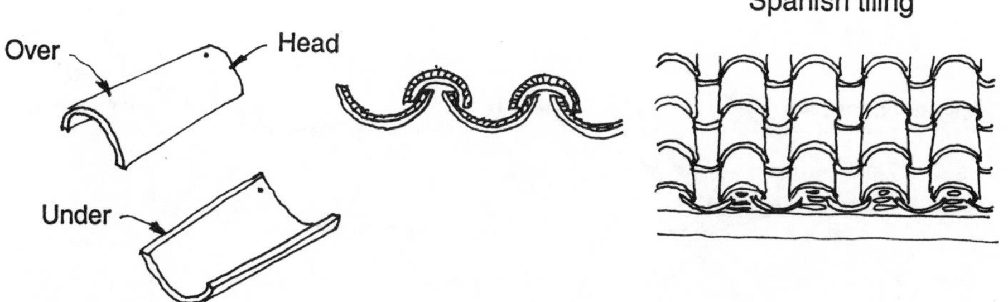

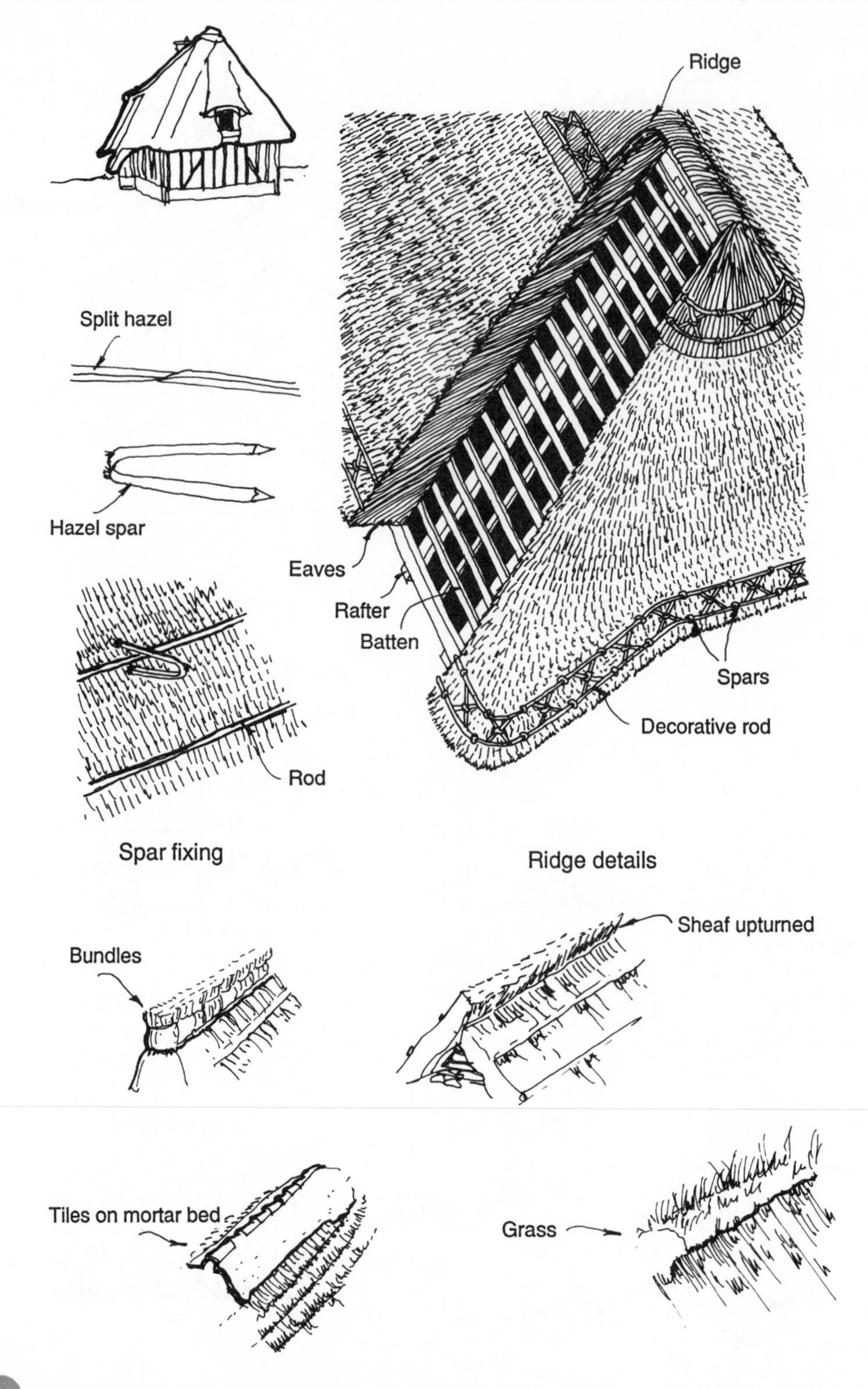
Ridge
Split hazel
Hazel spar
Eaves
Rafter
Batten
Spars
Decorative rod
Rod
Spar fixing
Ridge details
Sheaf upturned
Bundles
Tiles on mortar bed
Grass

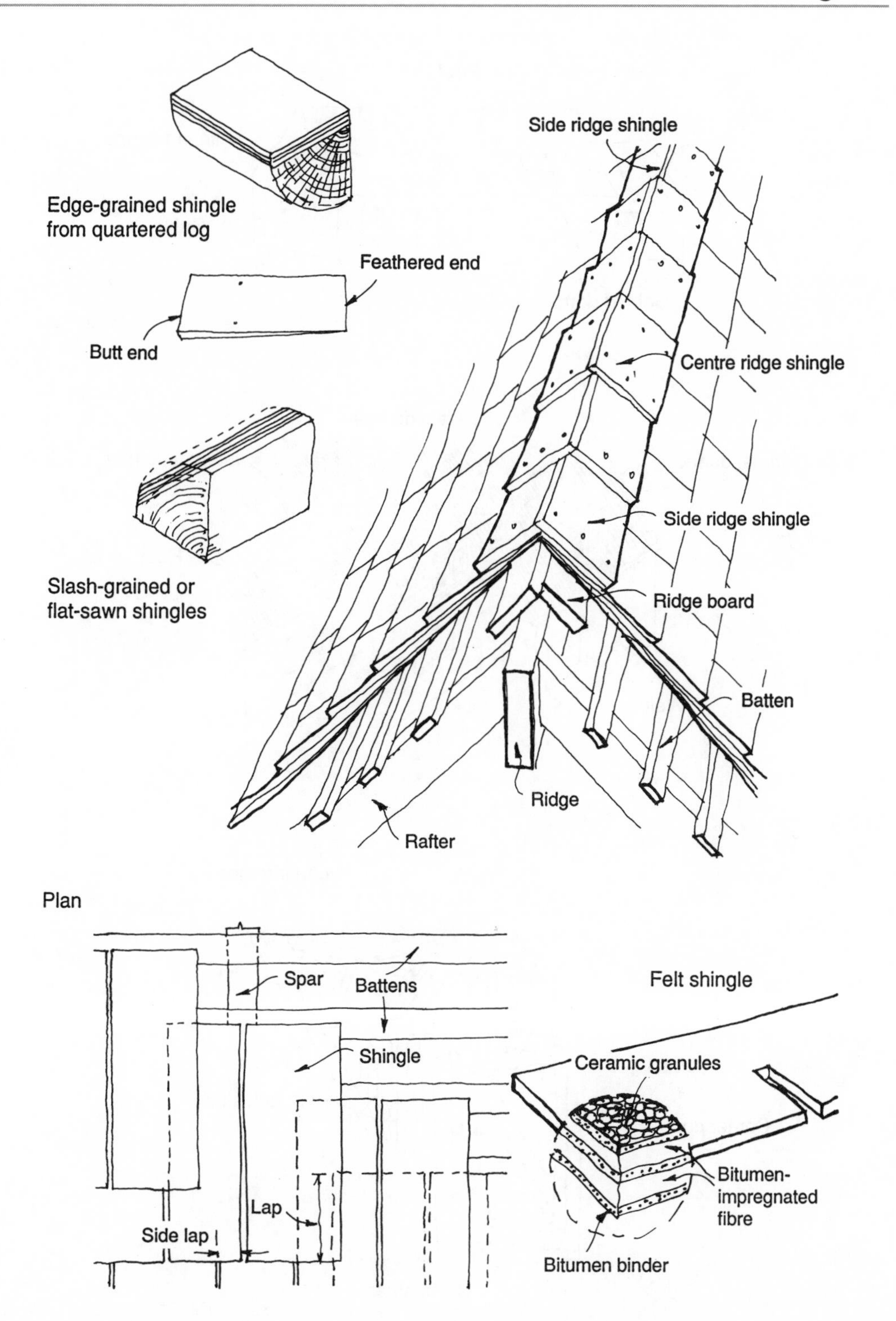
Edge-grained shingle
from quartered log
Feathered end
Butt end
Slash-grained or
flat-sawn shingles
Side ridge shingle
Centre ridge shingle
Side ridge shingle
Ridge board
Batten
Ridge
Rafter
Plan
Spar
Battens
Shingle
Lap
Side lap
Felt shingle
Ceramic granules
Bitumen-
impregnated
fibre
Bitumen binder

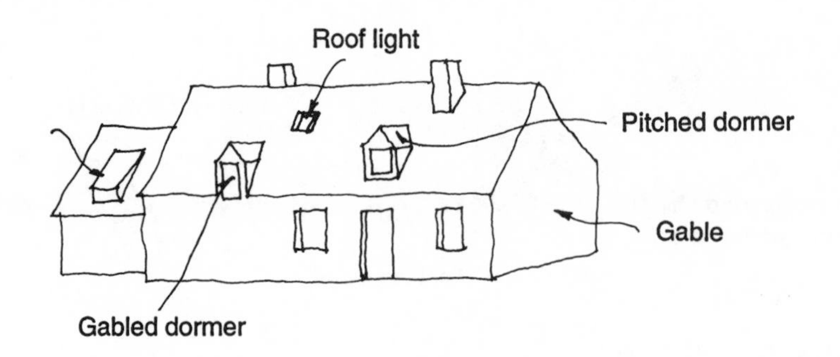
Roof light
Pitched dormer
Gable
Gabled dormer

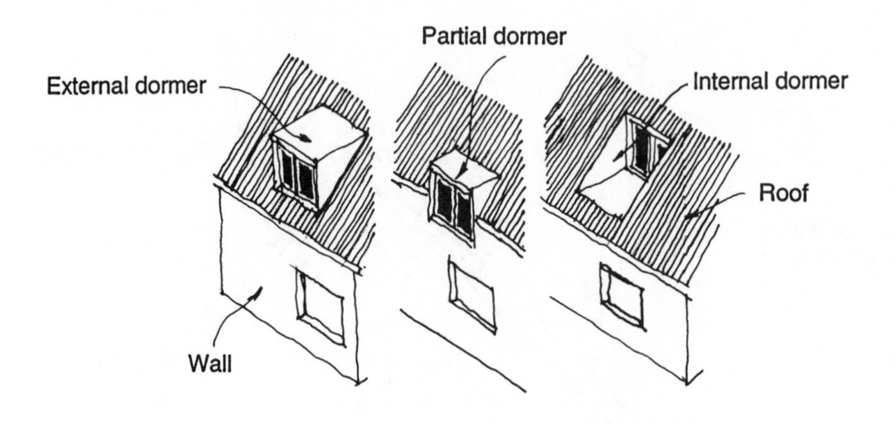
Partial dormer
External dormer
Internal dormer
Roof
Wall

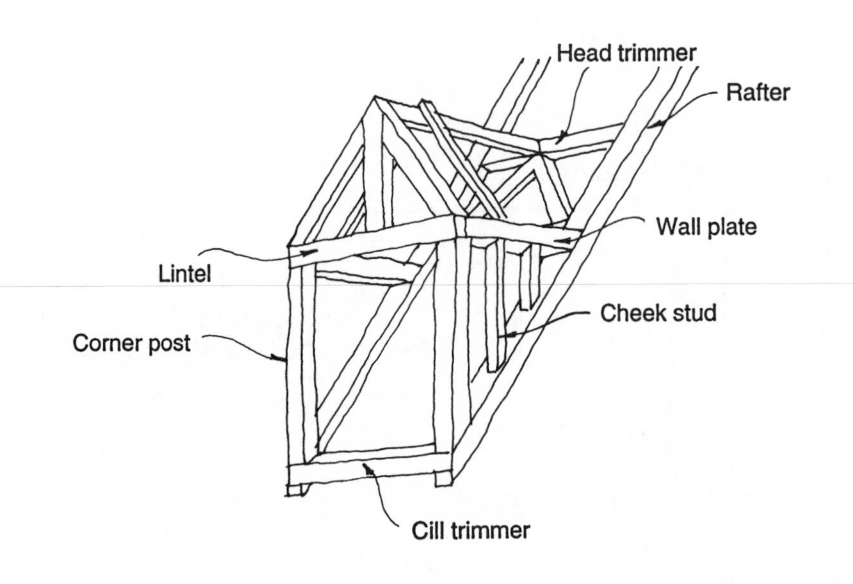
Head trimmer
Rafter
Wall plate
Lintel
Cheek stud
Corner post
Cill trimmer

Polygonal piended

Lead roofed rectangular

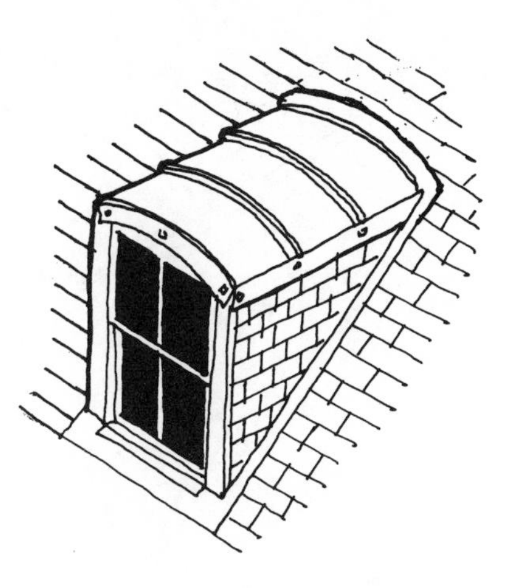

Rectangular

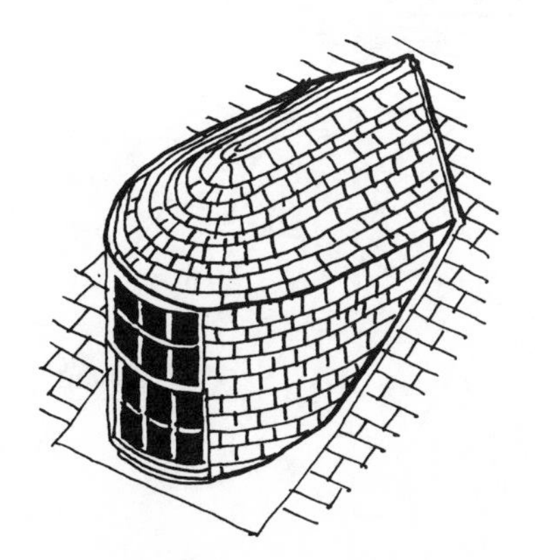

Bowed front

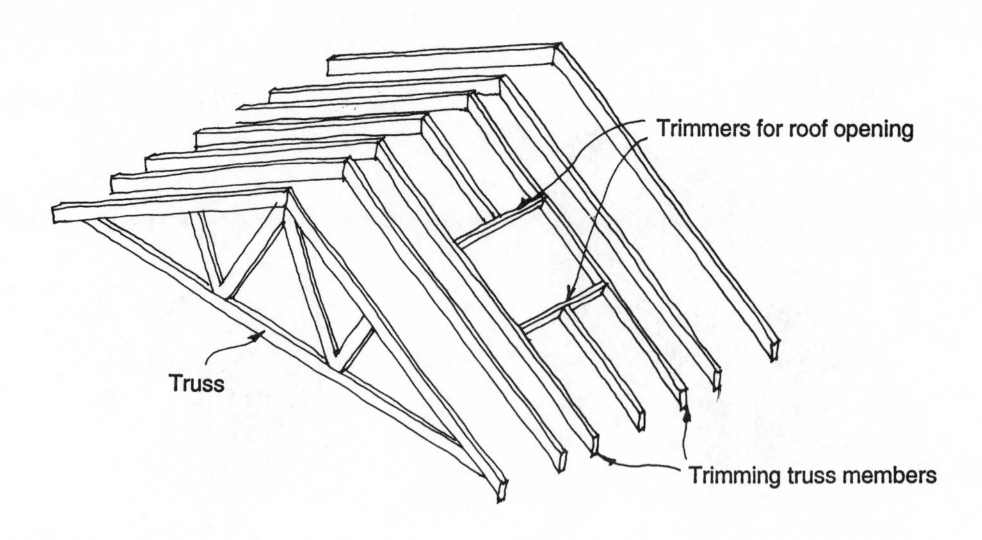

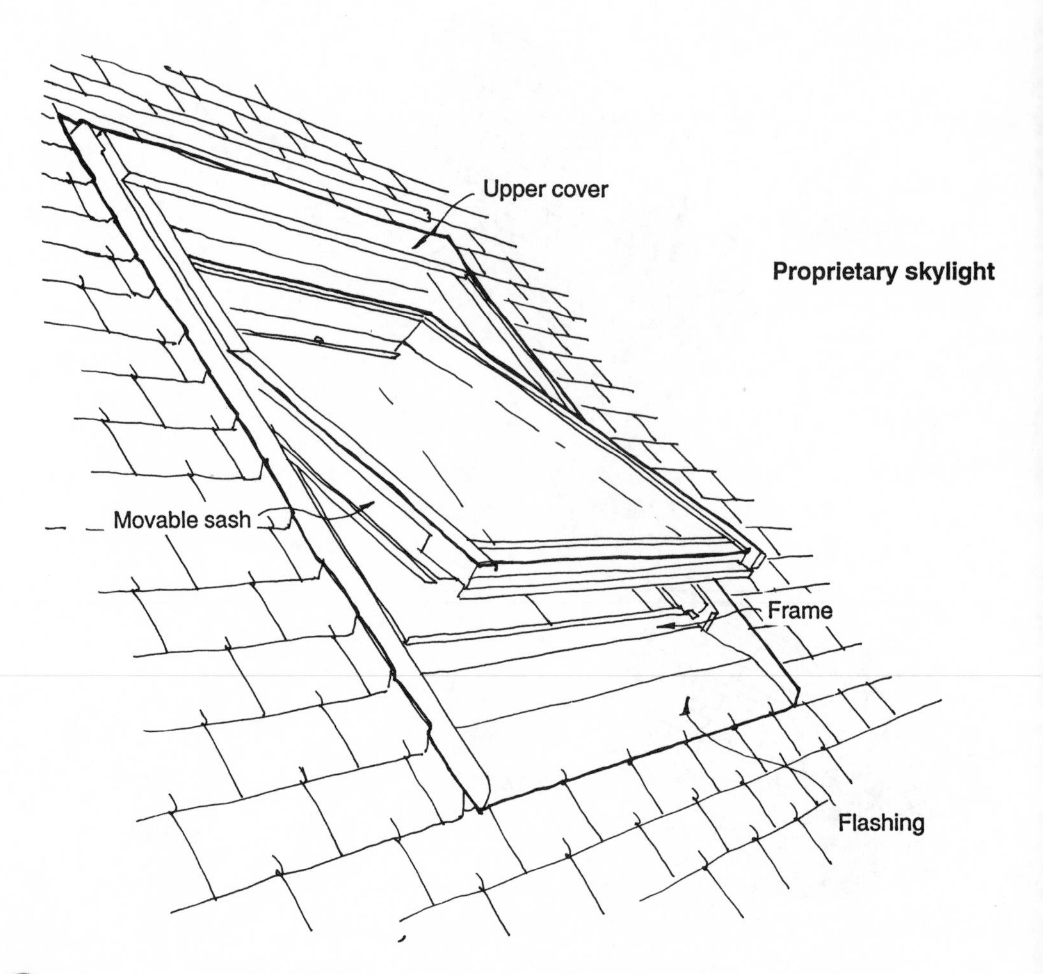

Proprietary skylight

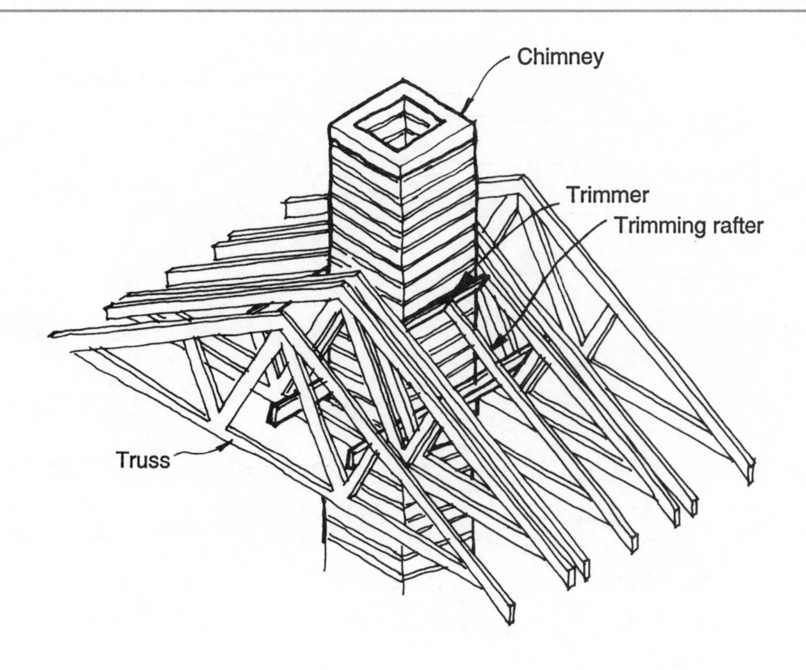

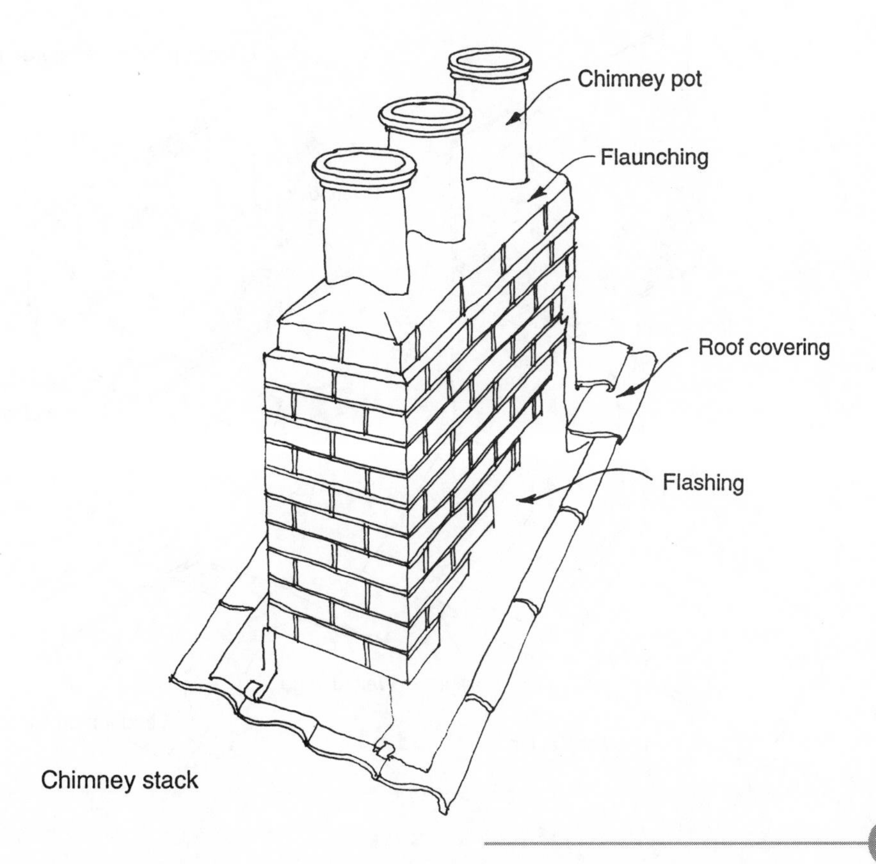

Chimney stack

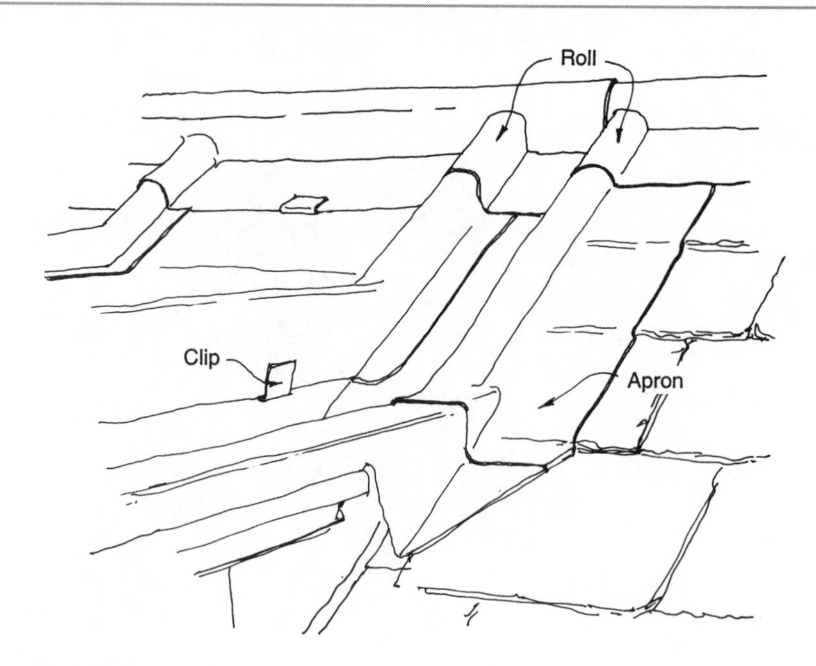

Bossed wood-cored roll

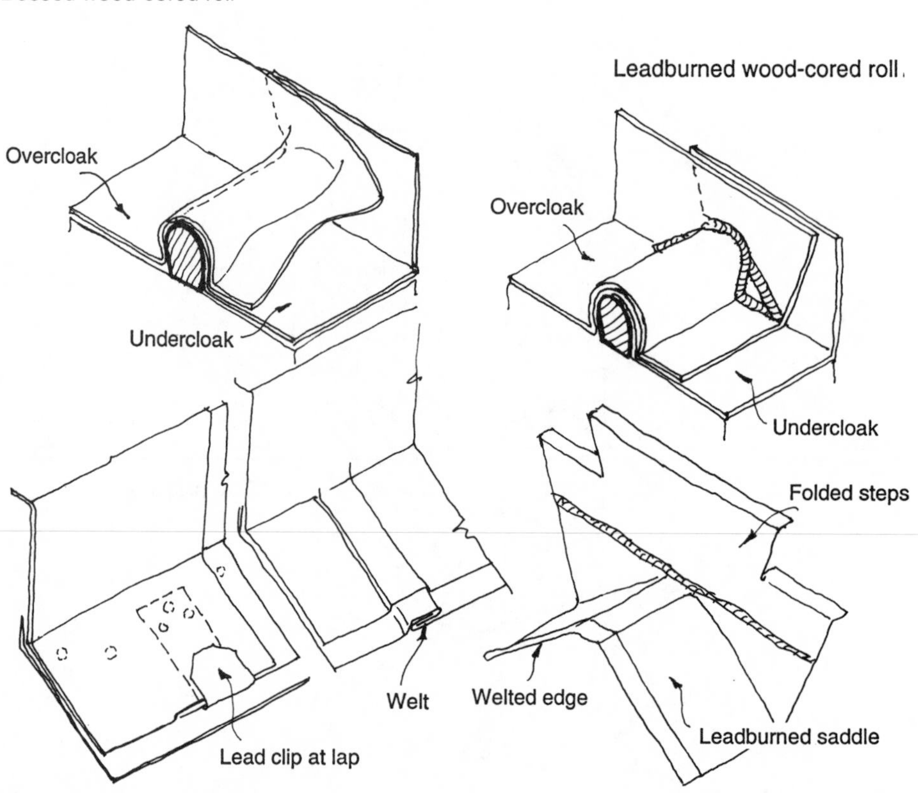

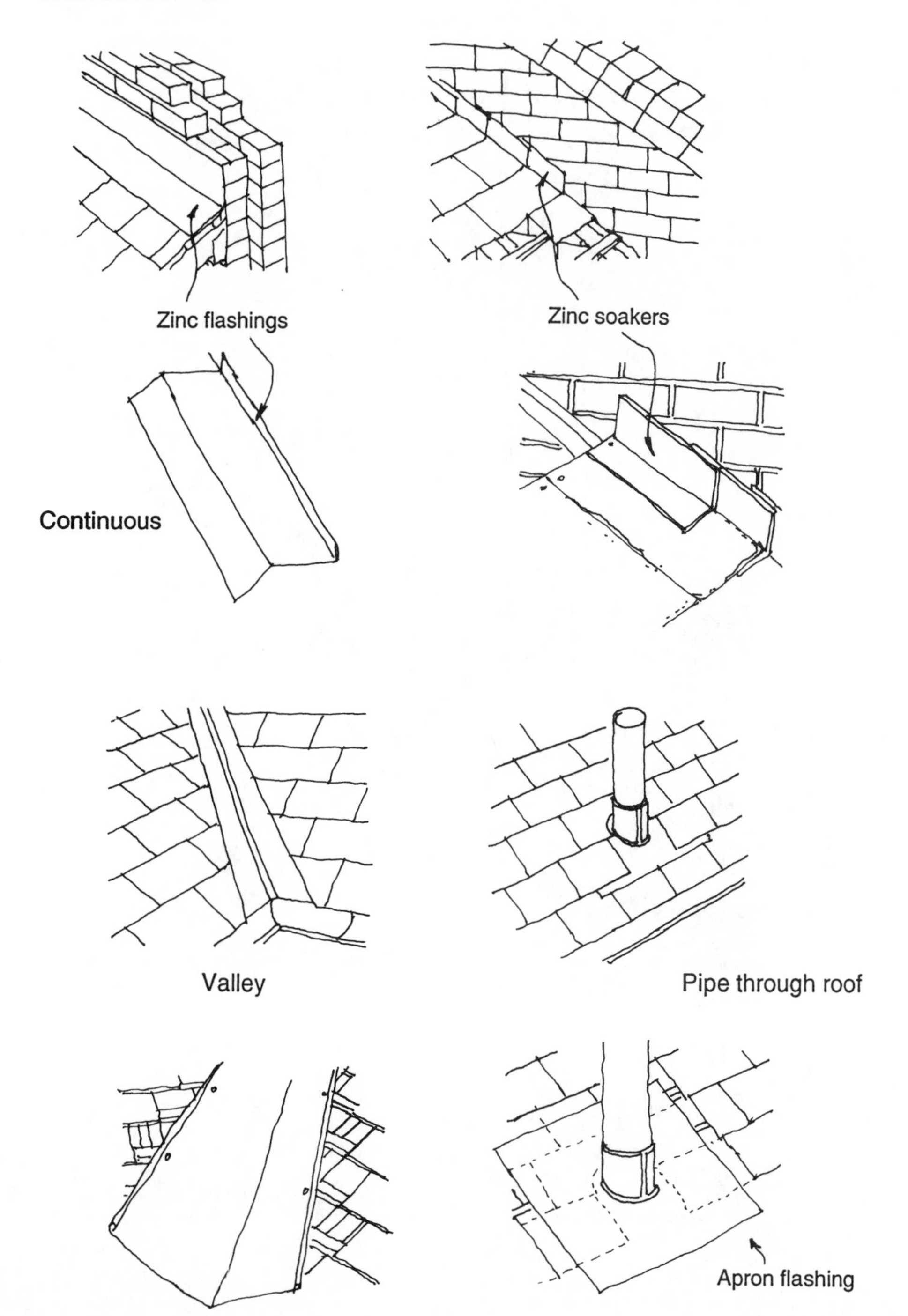
Junction roof/wall
Zinc flashings
Zinc soakers
Continuous
Valley
Pipe through roof
Apron flashing
Valley liner zinc sheet

Lightening conductor

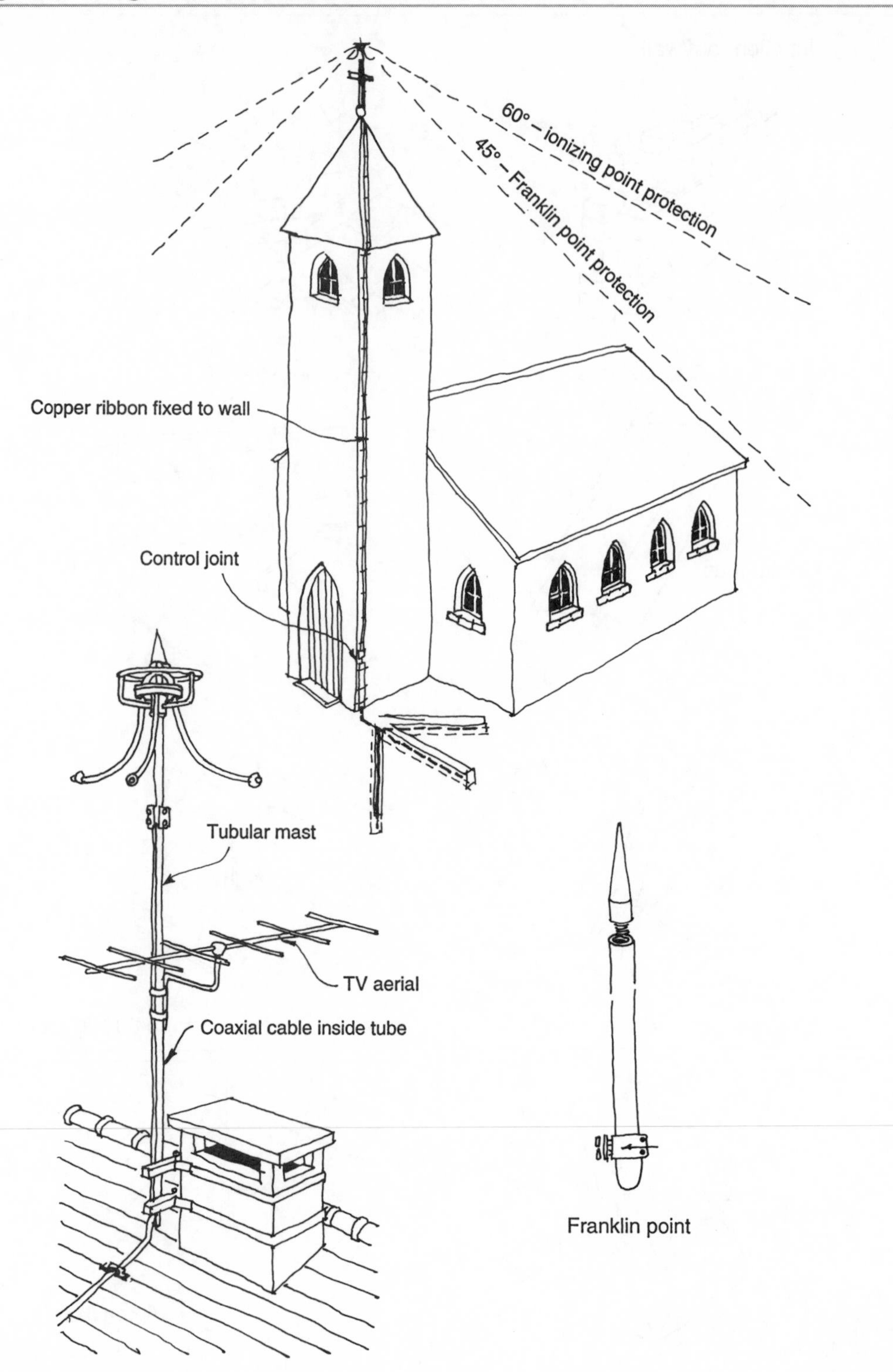

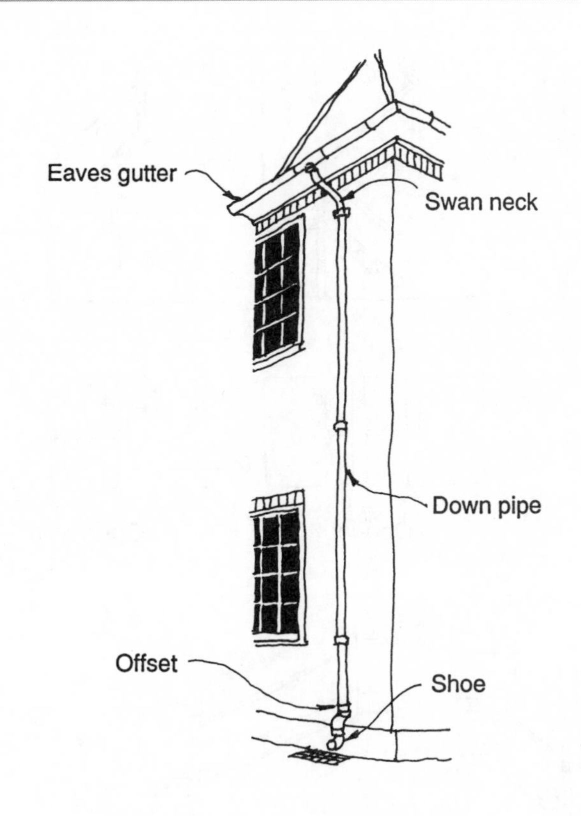
Eaves gutter
Swan neck
Down pipe
Offset
Shoe

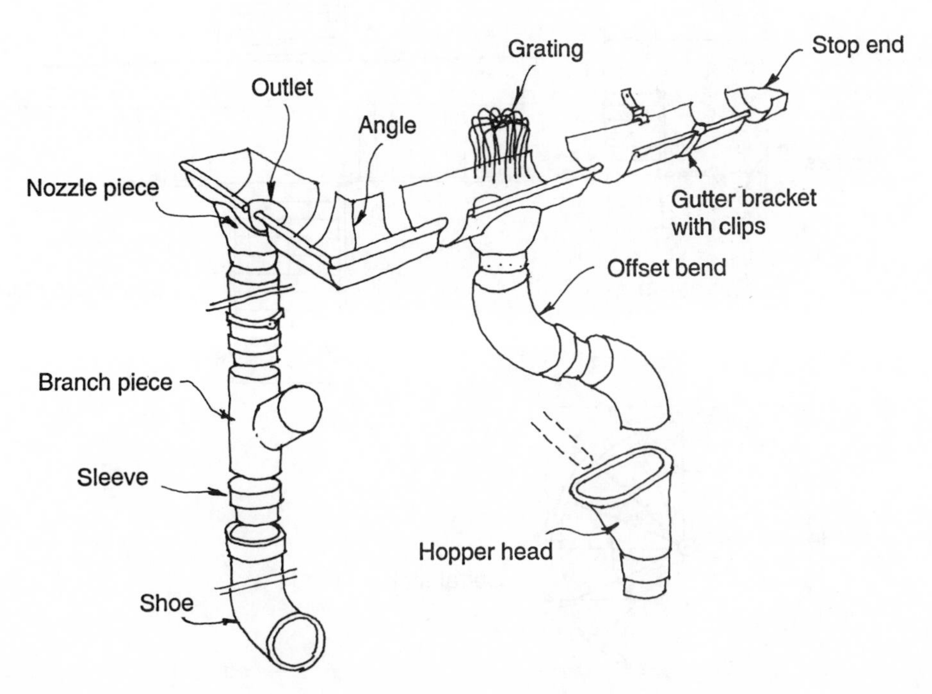
Grating
Stop end
Outlet
Angle
Nozzle piece
Gutter bracket with clips
Offset bend
Branch piece
Sleeve
Hopper head
Shoe

Types of staircases

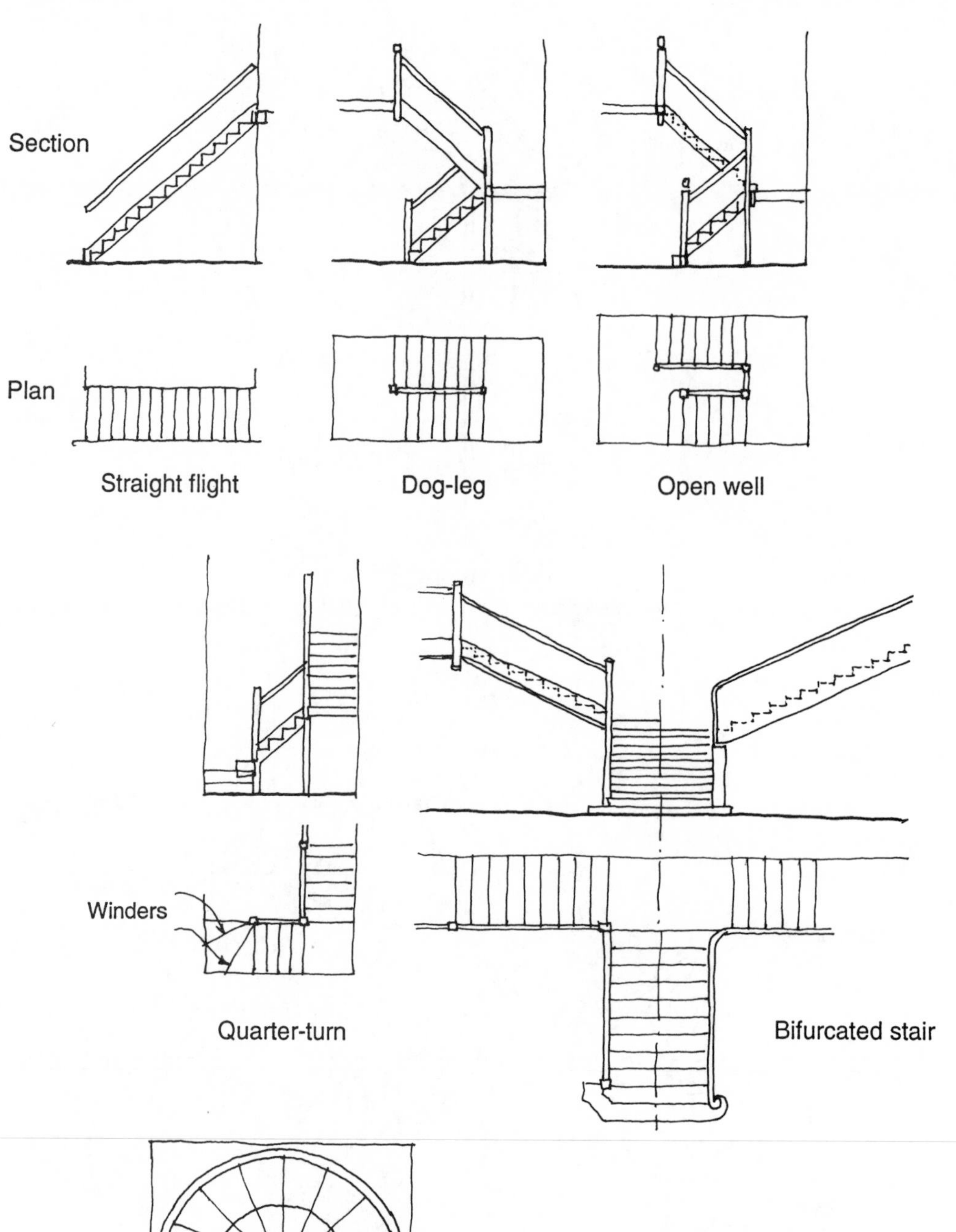

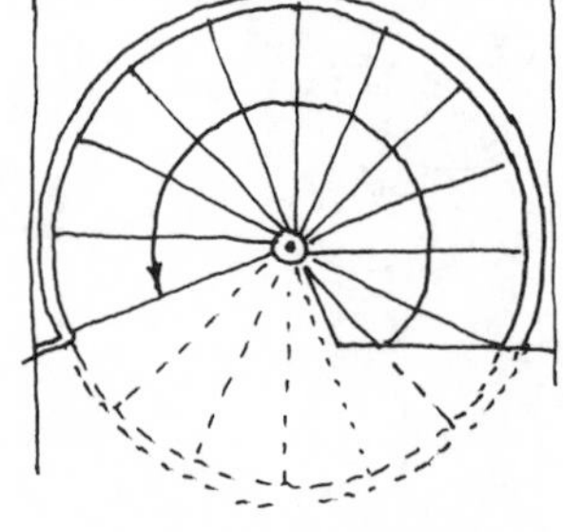

Spiral stair

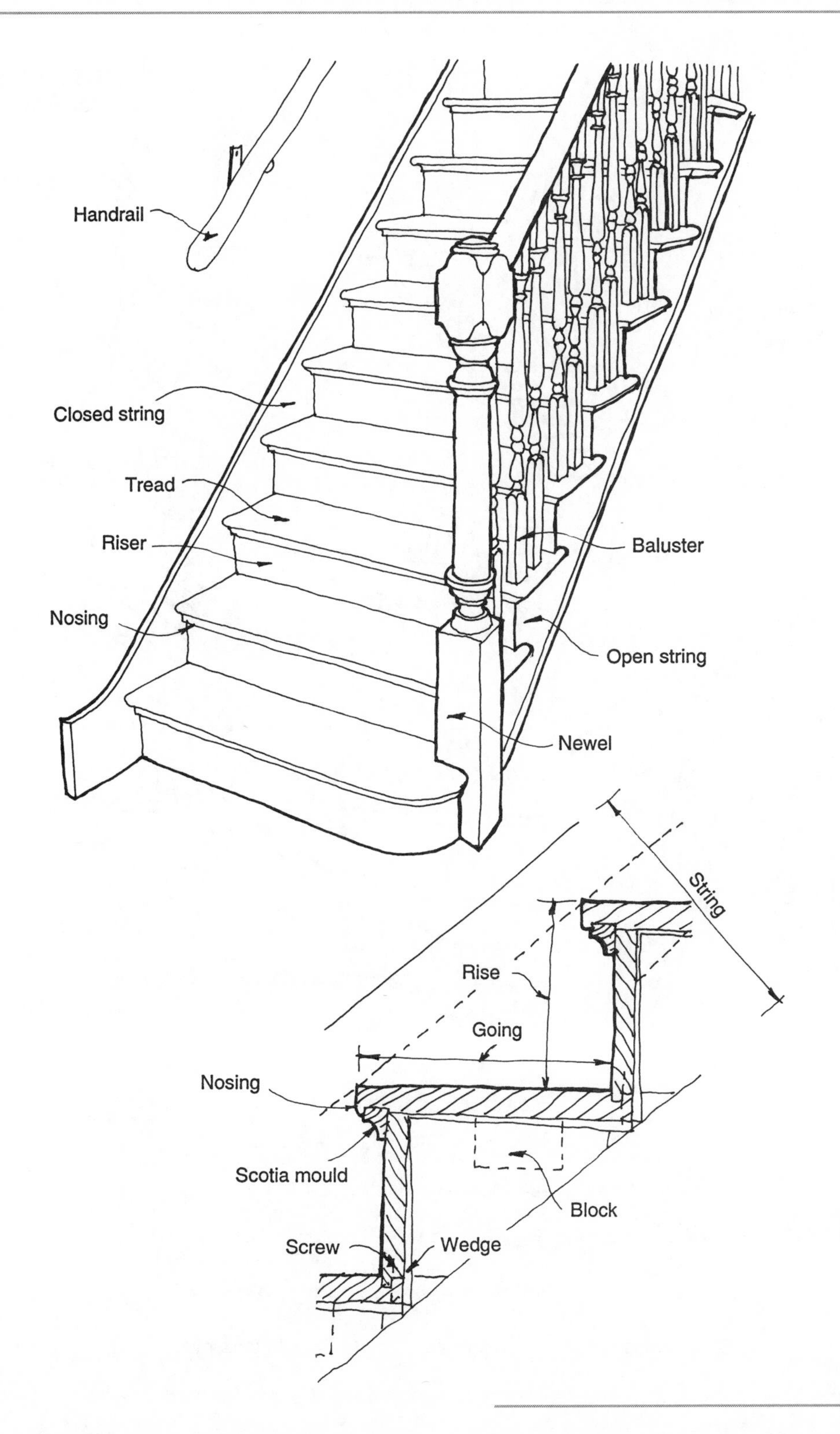
Handrail
Closed string
Tread
Riser
Nosing
Baluster
Open string
Newel
String
Rise
Going
Nosing
Scotia mould
Block
Screw
Wedge

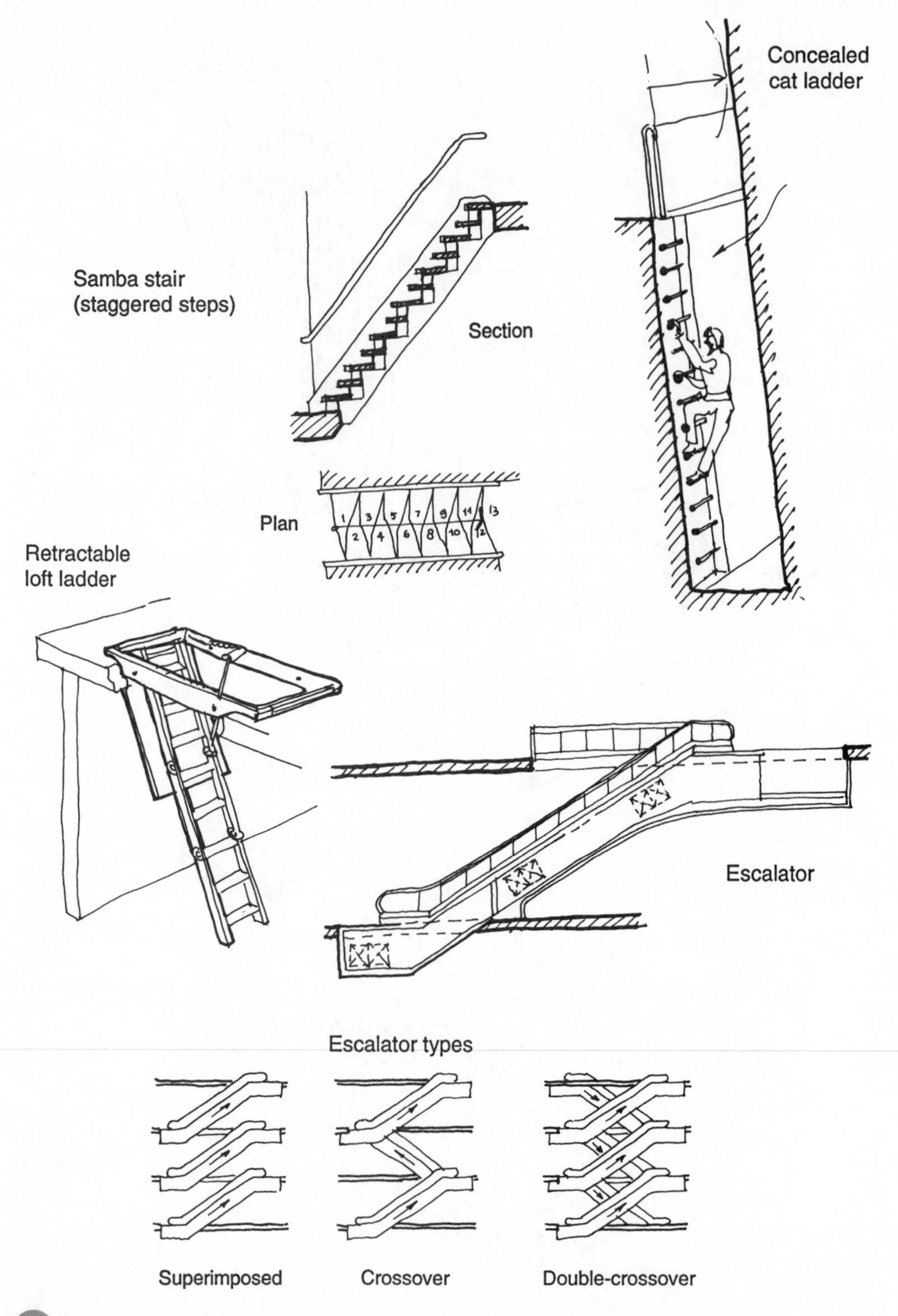
Concealed cat ladder
Samba stair (staggered steps)
Section
Plan
1
2
3
4
5
6
7
8
9
10
11
12
13
Retractable loft ladder
Escalator
Escalator types
Superimposed
Crossover
Double-crossover

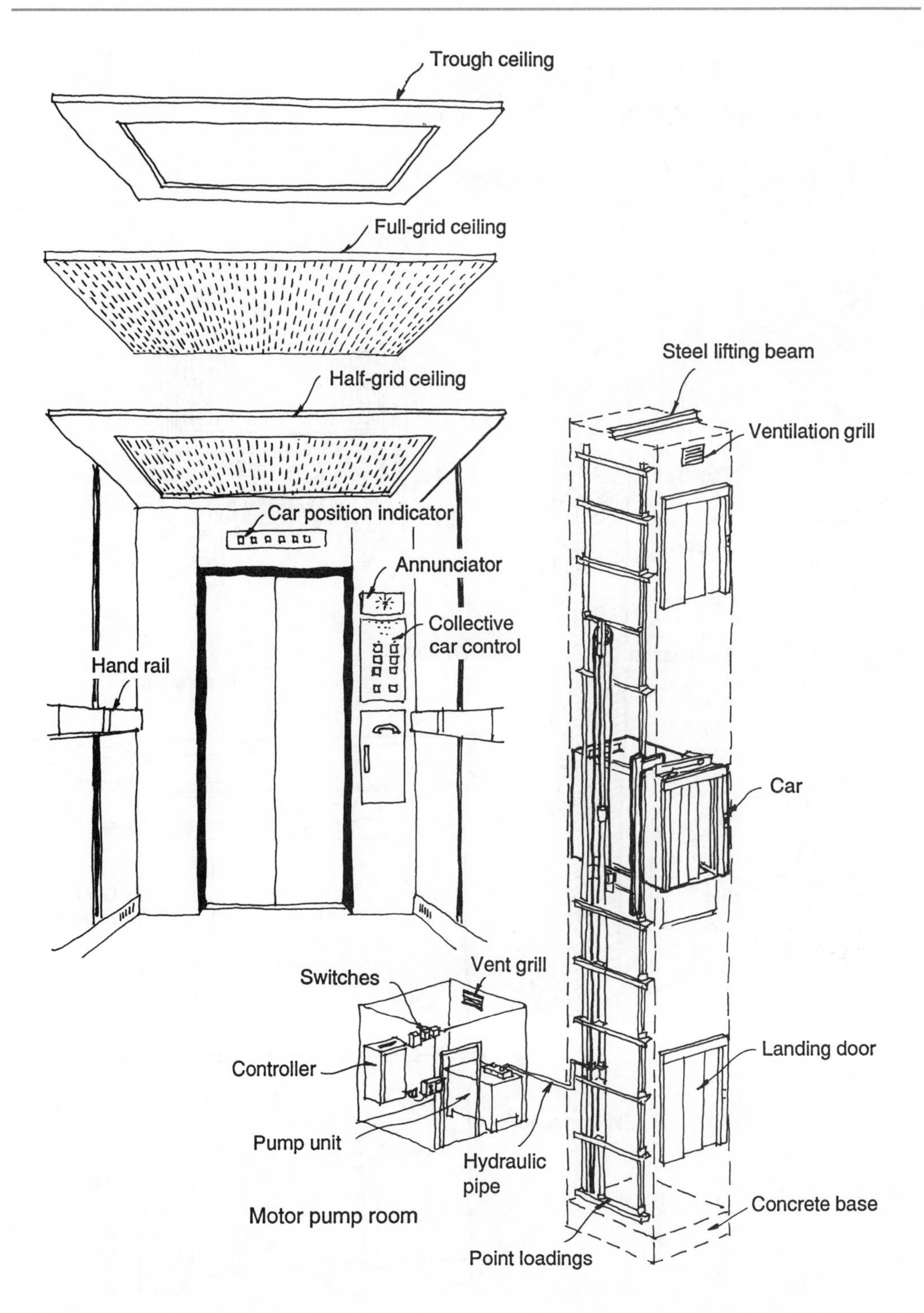

Hydraulic lift well

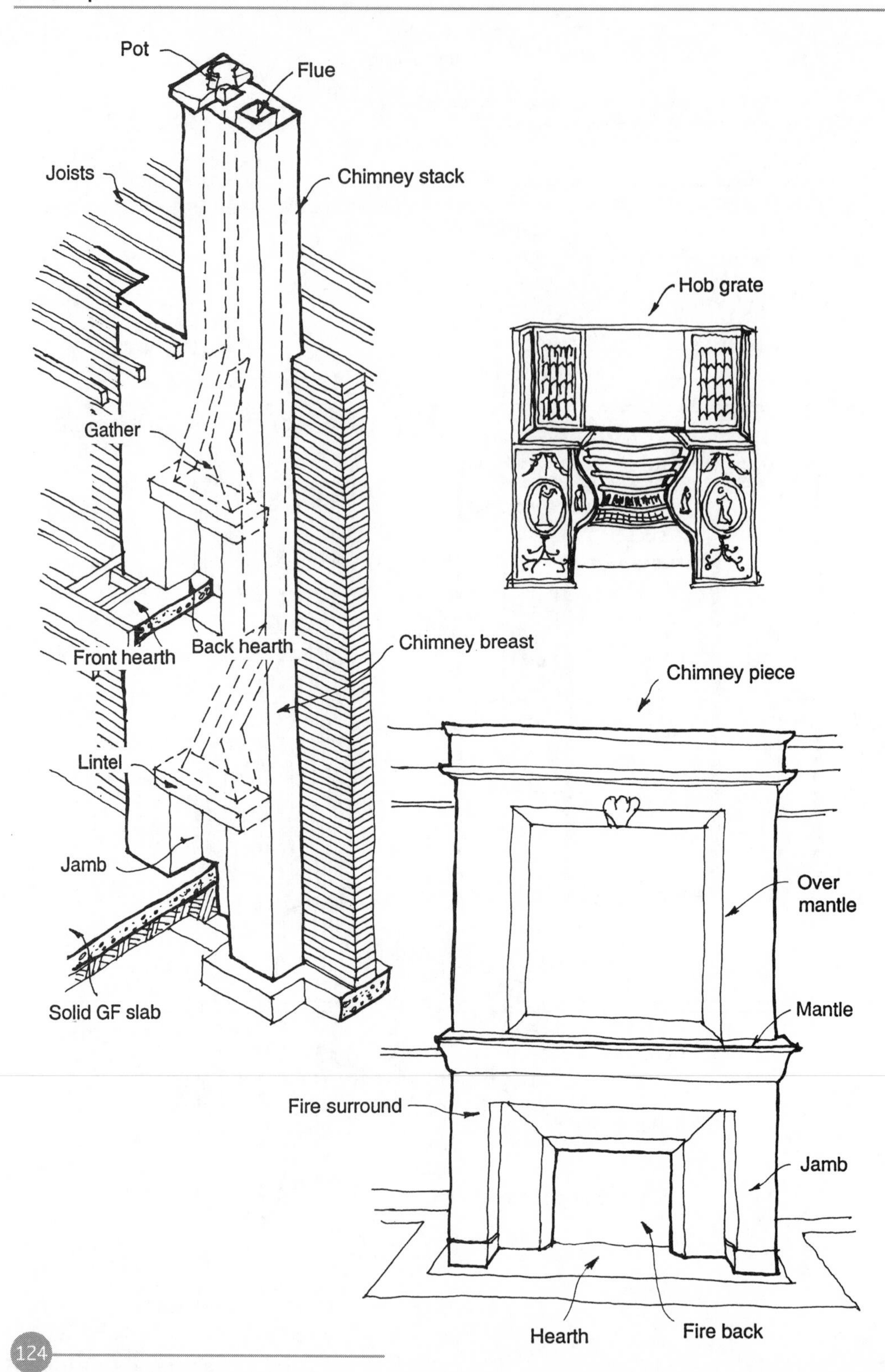
Pot
Flue
Joists
Chimney stack
Hob grate
Gather
Front hearth
Back hearth
Chimney breast
Chimney piece
Lintel
Jamb
Over mantle
Solid GF slab
Mantle
Fire surround
Jamb
Hearth
Fire back

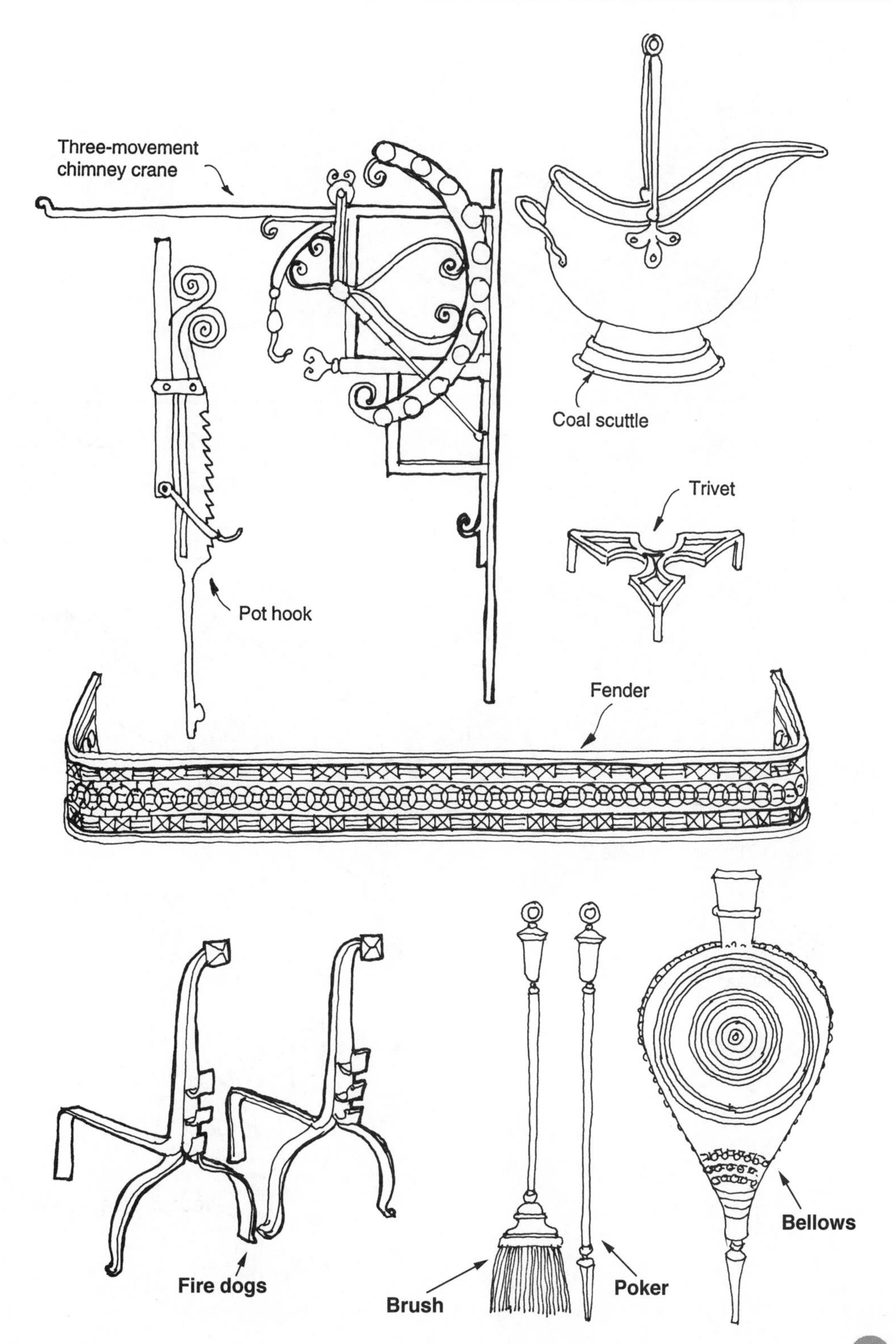
Three-movement
chimney crane
Coal scuttle
Trivet
Pot hook
Fender
Fire dogs
Brush
Poker
Bellows

Timber floor construction

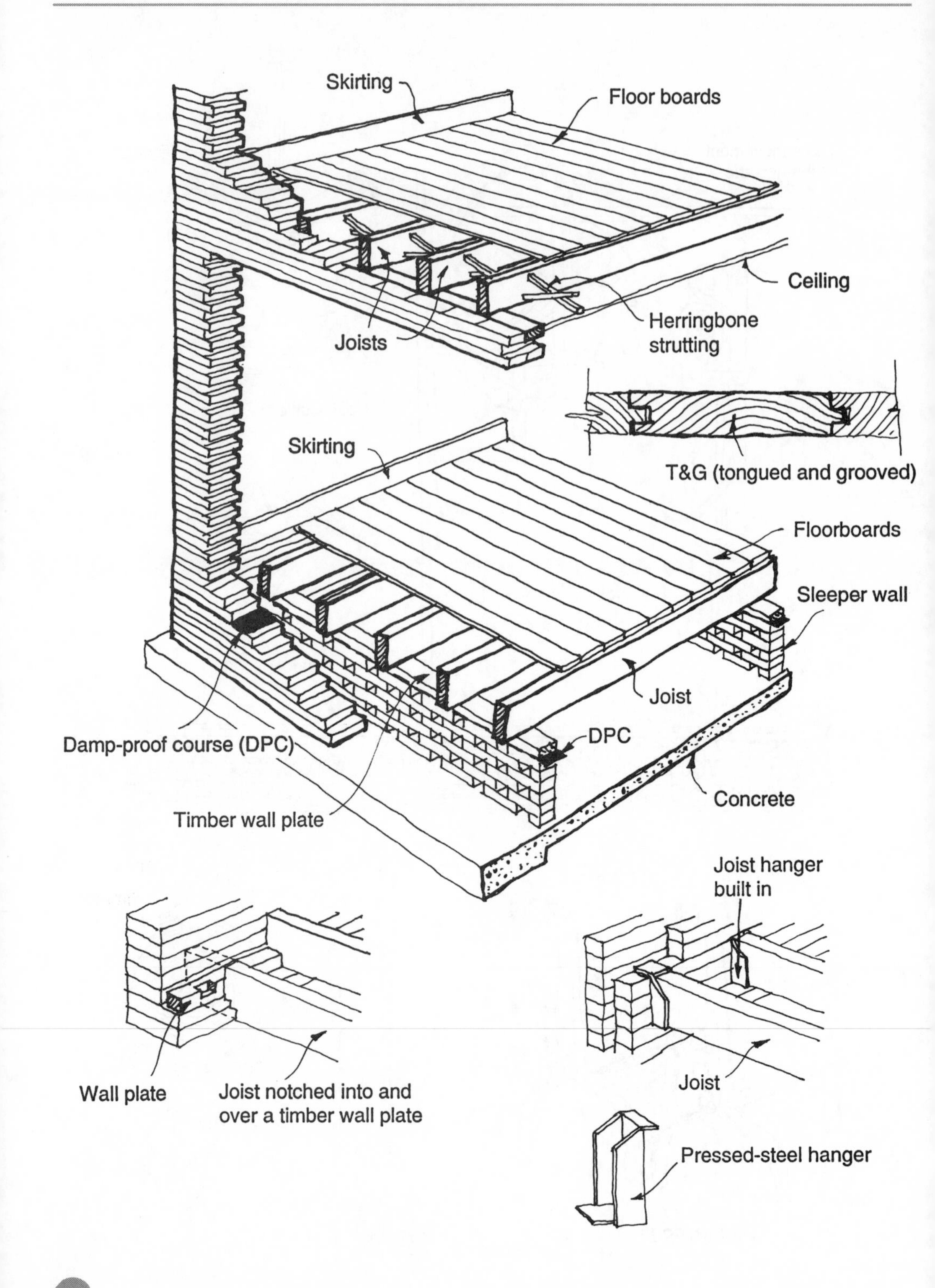

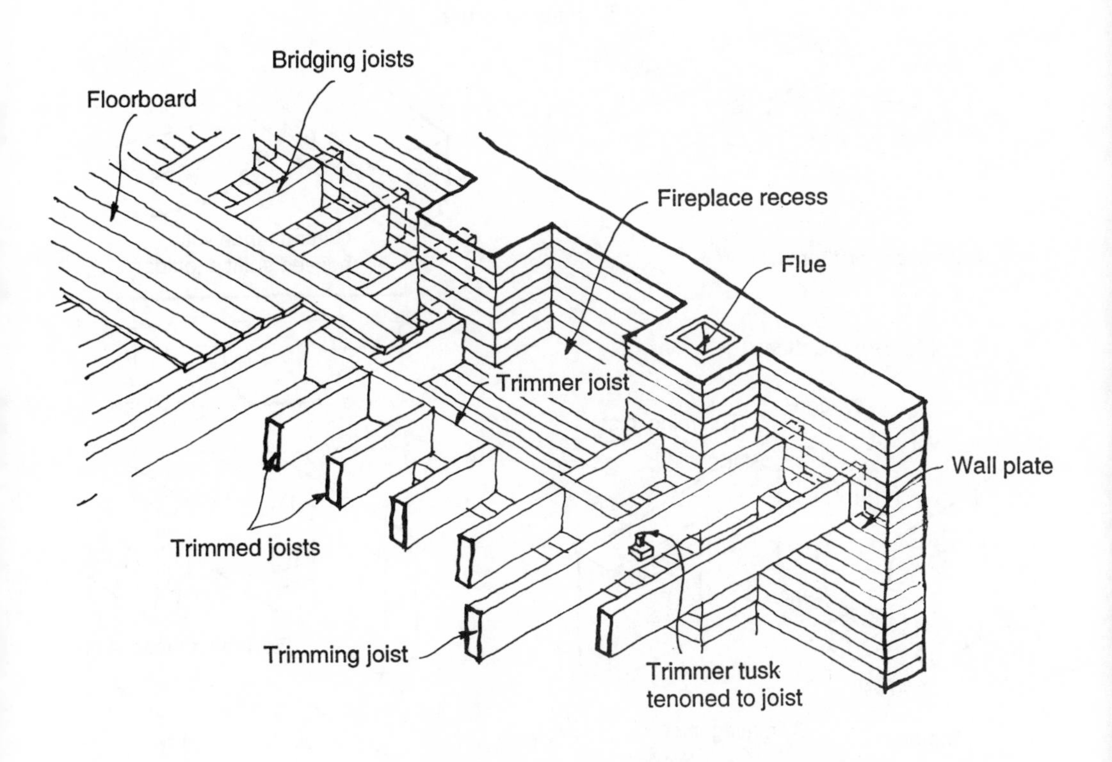
Bridging joists
Floorboard
Fireplace recess
Flue
Trimmer joist
Wall plate
Trimmed joists
Trimming joist
Trimmer tusk
tenoned to joist

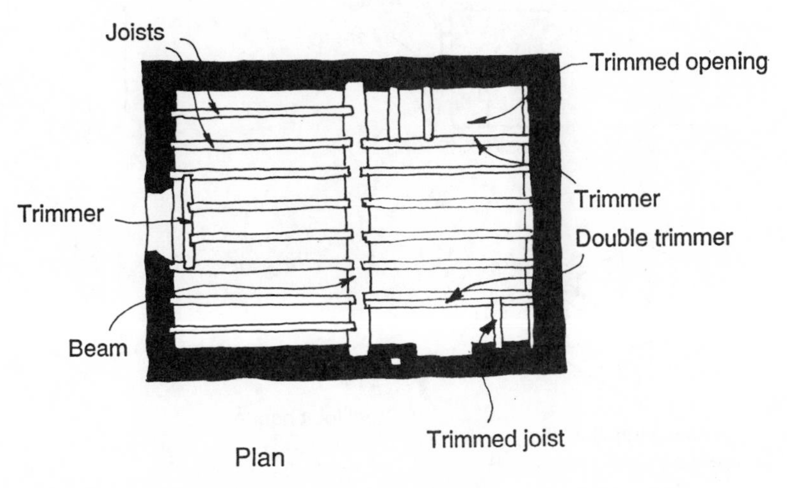
Joists
Trimmed opening
Trimmer
Trimmer
Double trimmer
Beam
Trimmed joist

Plan

Traditional joints

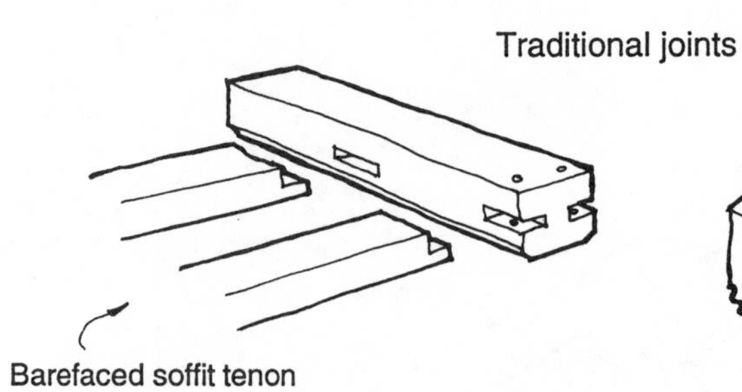

Barefaced soffit tenon

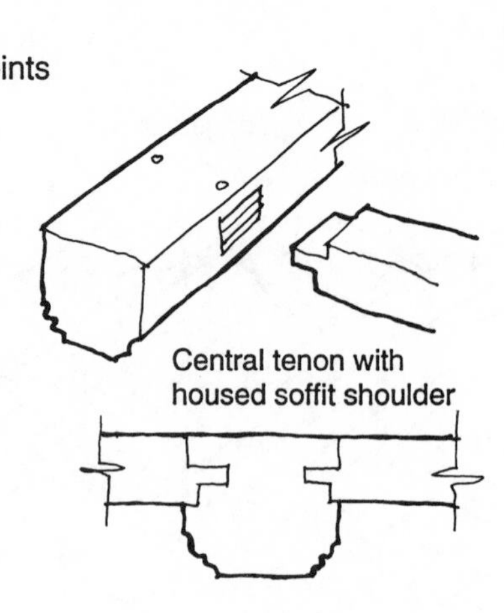

Central tenon with housed soffit shoulder

Modern practice

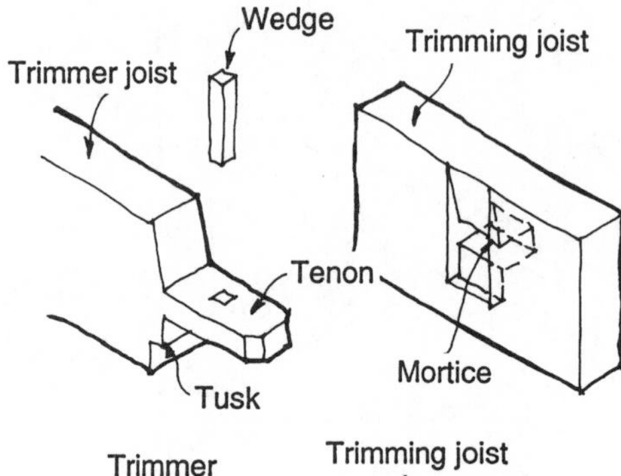

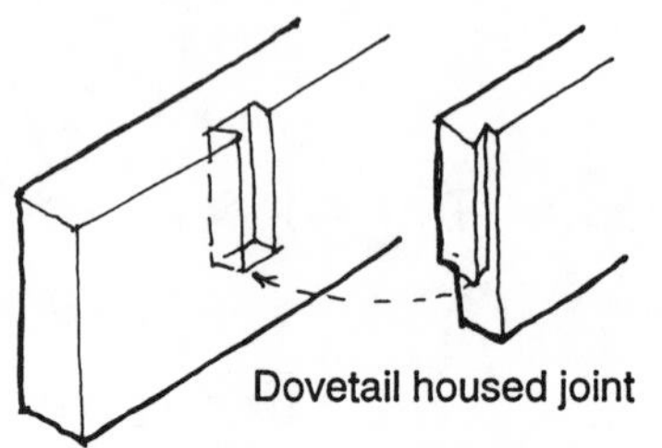

Dovetail housed joint

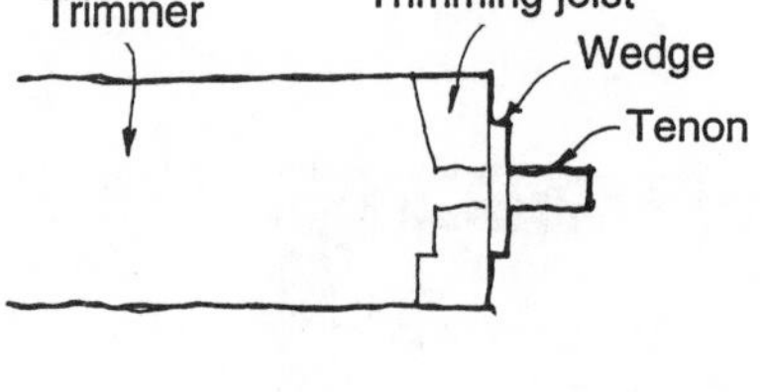

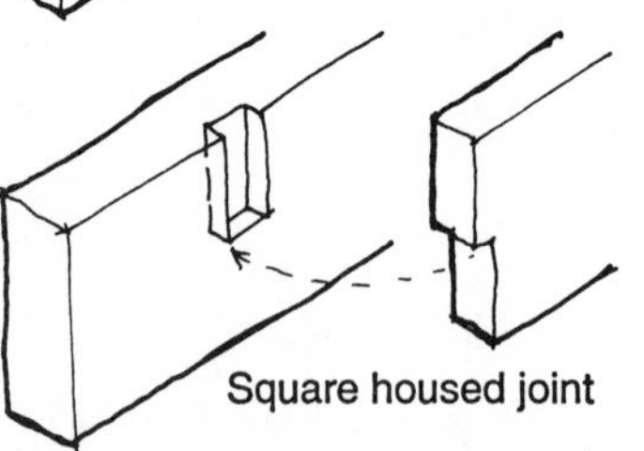

Square housed joint

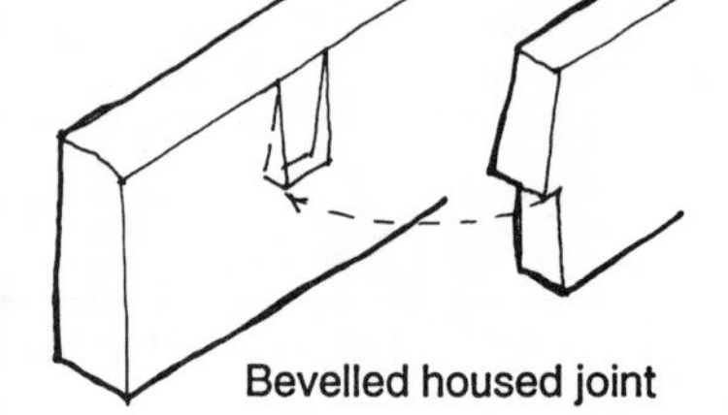

Bevelled housed joint

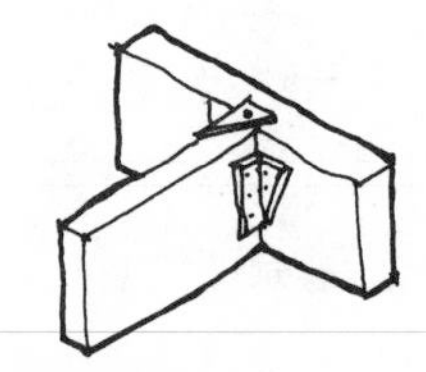

Trimming junction with pressed steel fixing plates

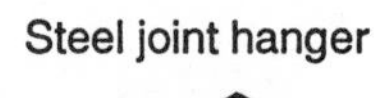

Steel joint hanger

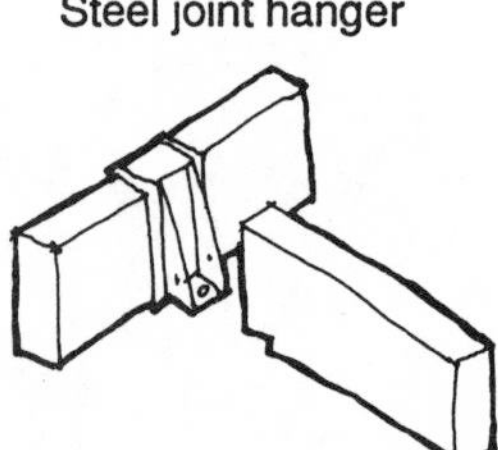

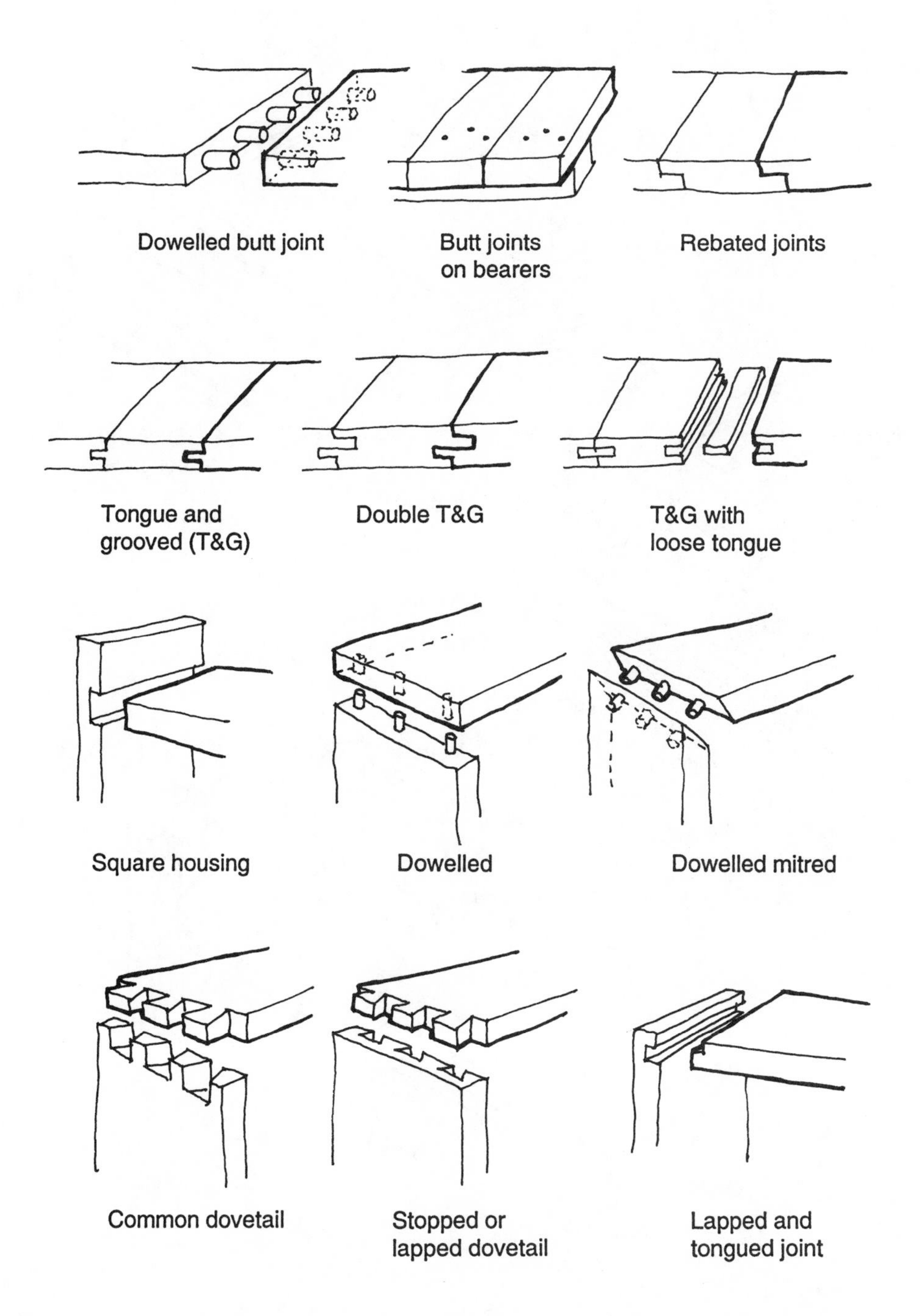
Dowelled butt joint
Butt joints on bearers
Rebated joints
Tongue and grooved (T&G)
Double T&G
T&G with loose tongue
Square housing
Dowelled
Dowelled mitred
Common dovetail
Stopped or lapped dovetail
Lapped and tongued joint

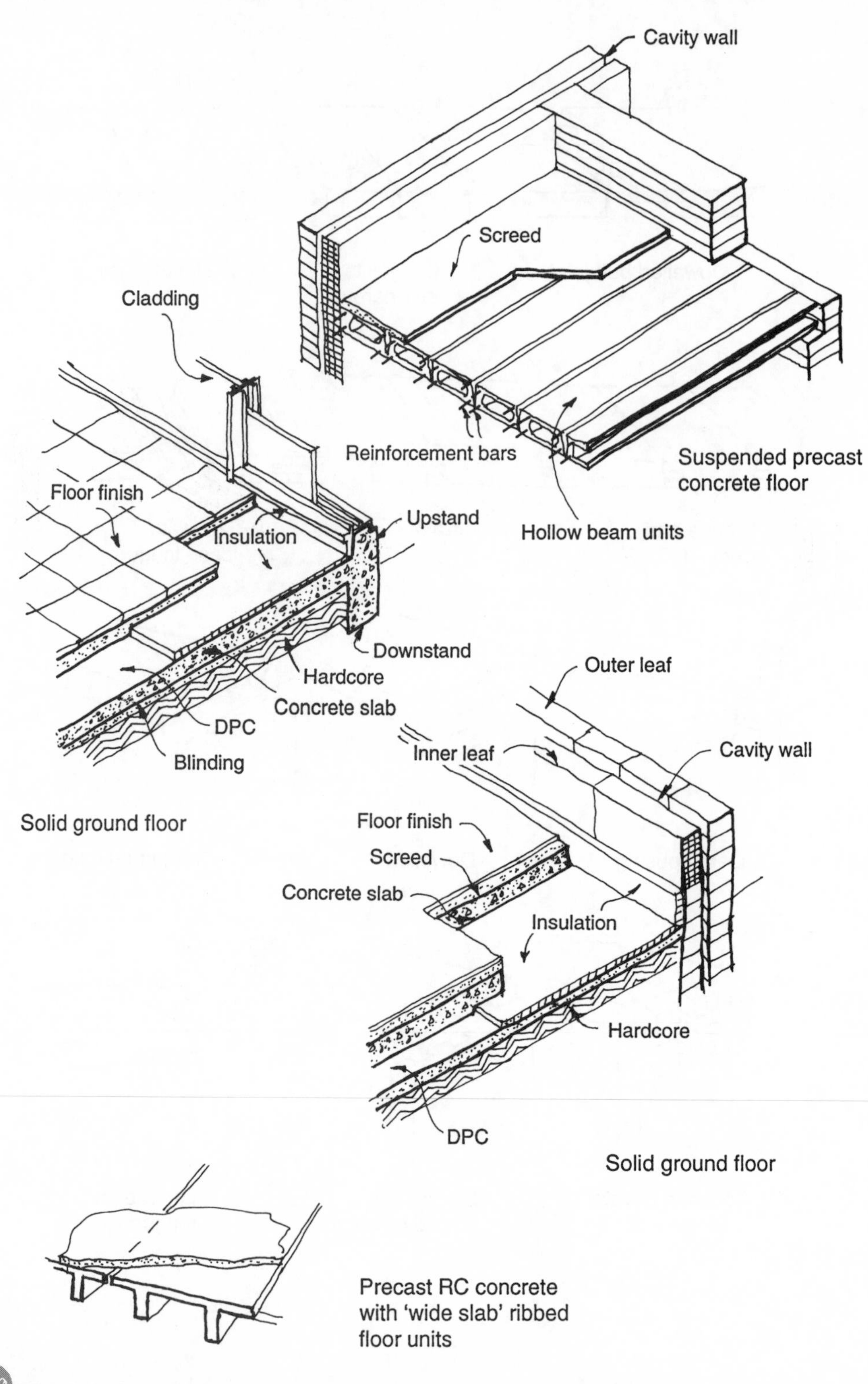

Solid ground floor

Solid ground floor

Precast RC concrete with 'wide slab' ribbed floor units

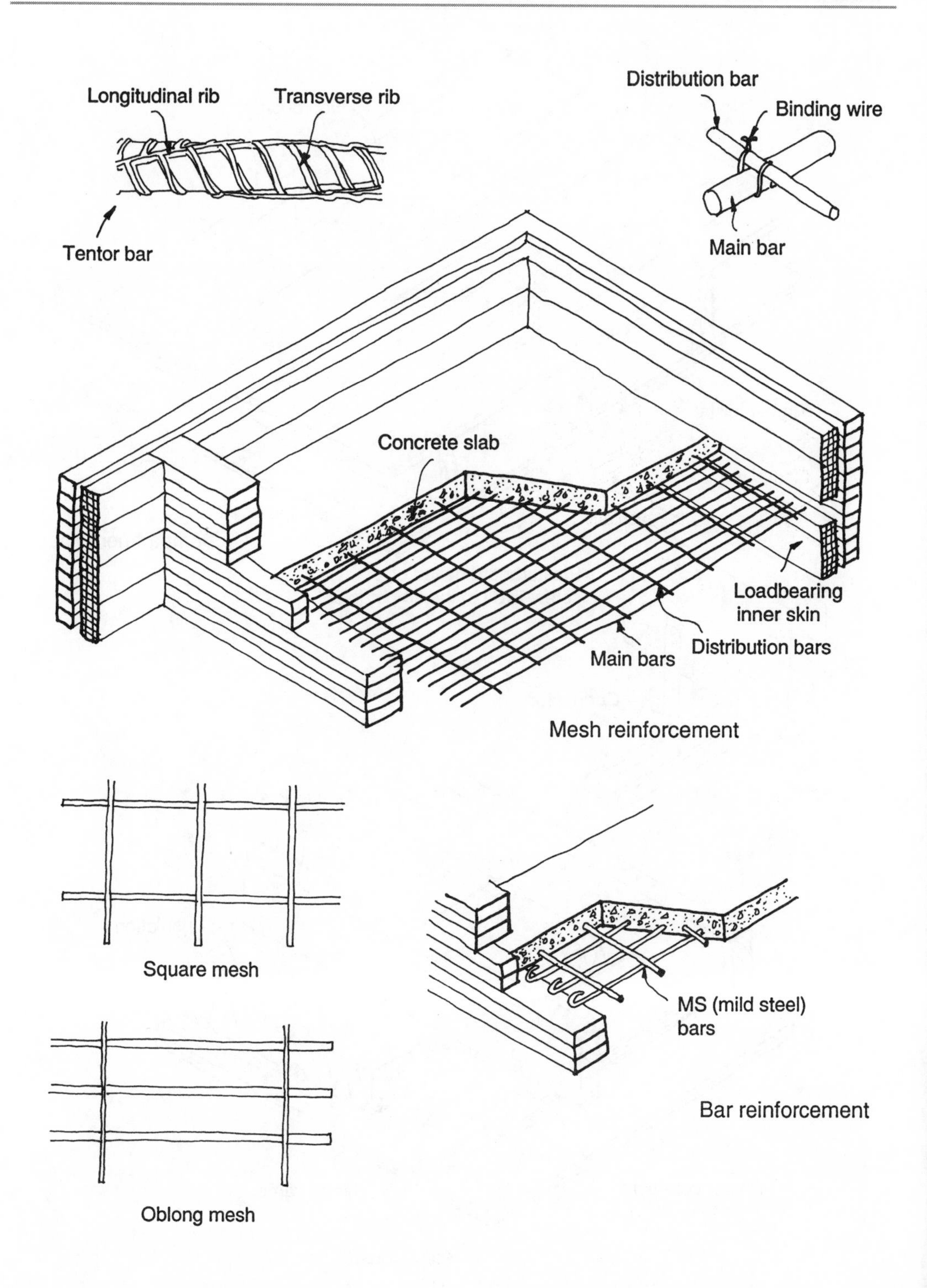
Longitudinal rib
Transverse rib
Tentor bar
Distribution bar
Binding wire
Main bar
Concrete slab
Loadbearing inner skin
Distribution bars
Main bars
Mesh reinforcement
Square mesh
Oblong mesh
MS (mild steel) bars
Bar reinforcement

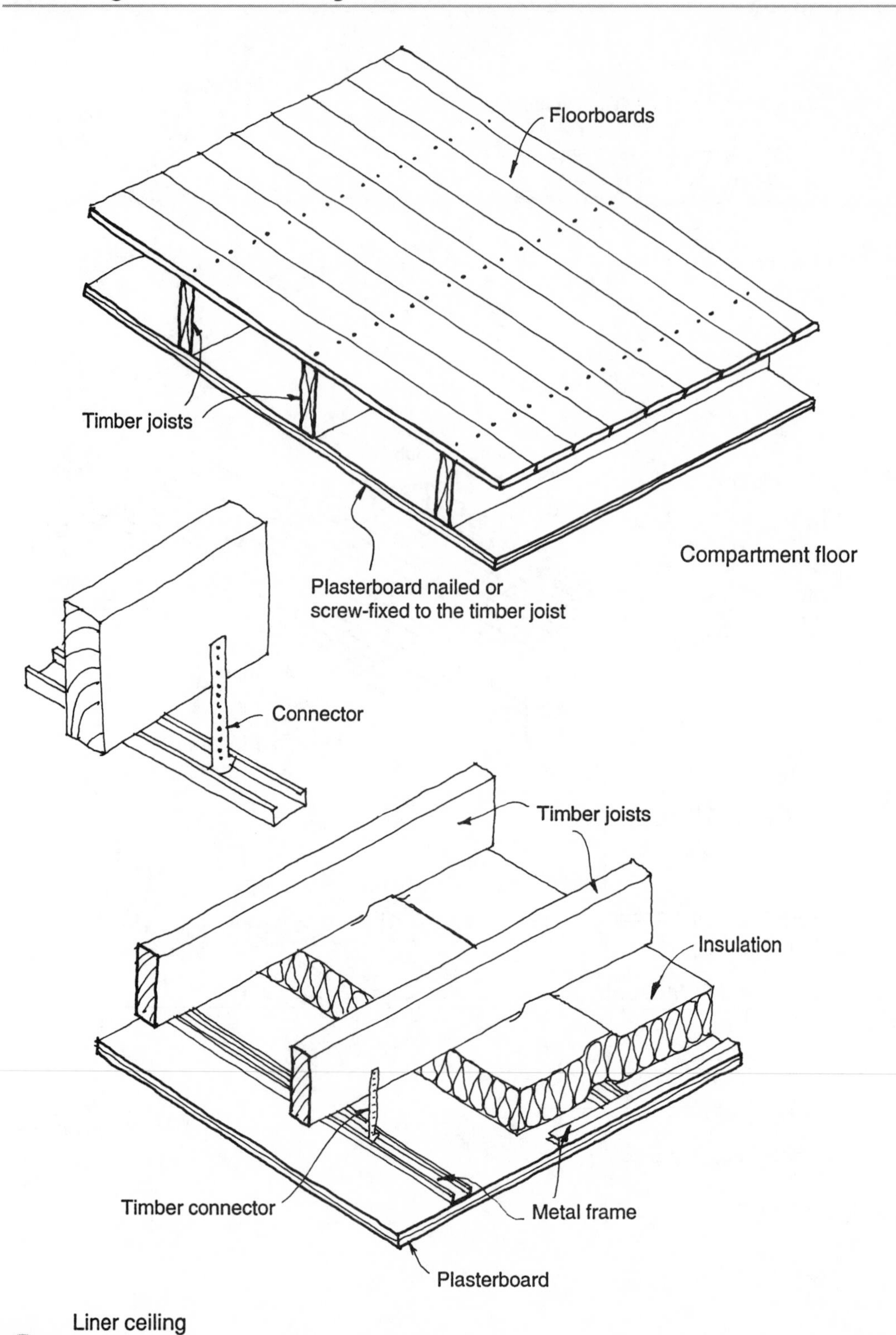
Floorboards
Timber joists
Plasterboard nailed or screw-fixed to the timber joist
Compartment floor
Connector
Timber joists
Insulation
Timber connector
Metal frame
Plasterboard
Liner ceiling

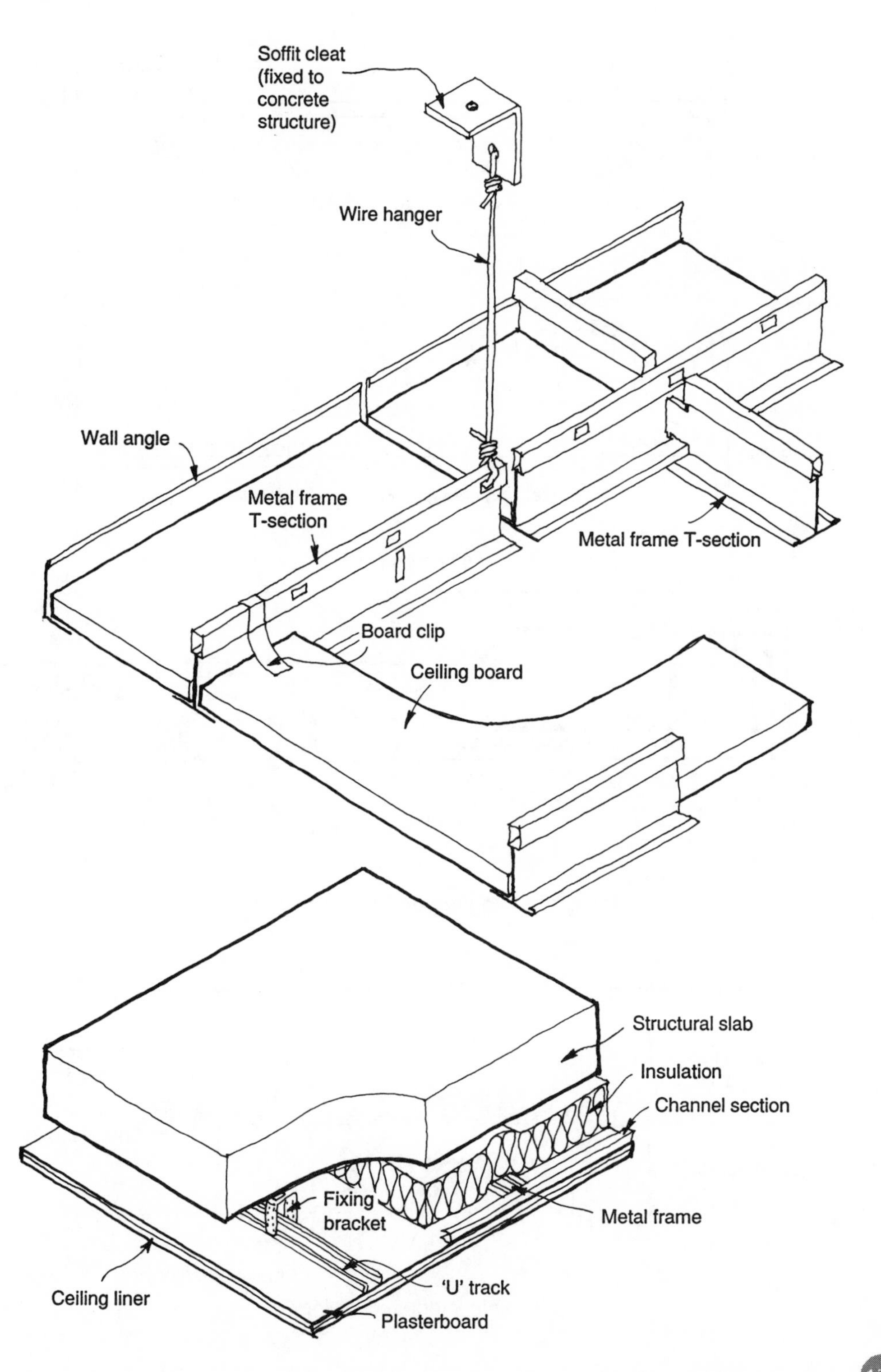
Soffit cleat (fixed to concrete structure)
Wire hanger
Wall angle
Metal frame T-section
Metal frame T-section
Board clip
Ceiling board
Structural slab
Insulation
Channel section
Fixing bracket
Metal frame
Ceiling liner
'U' track
Plasterboard

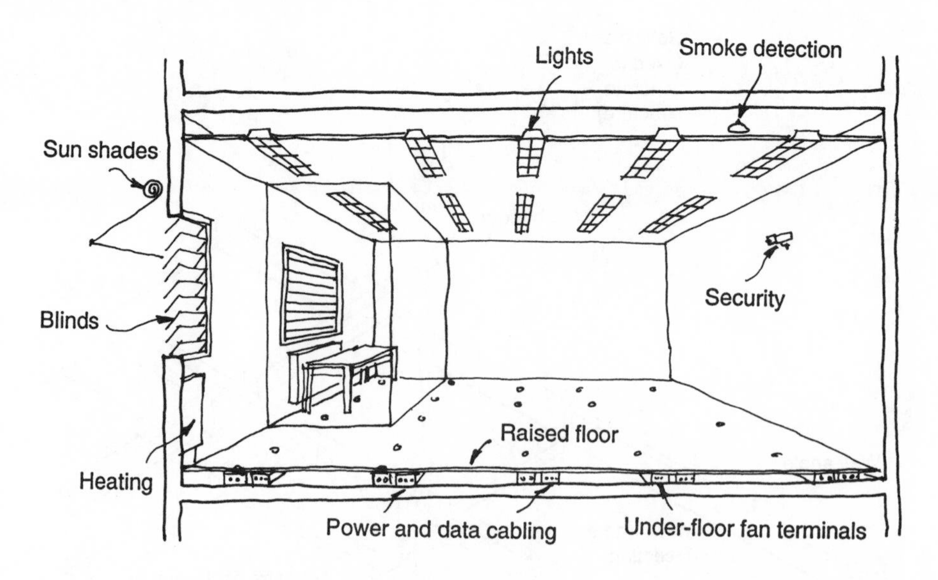

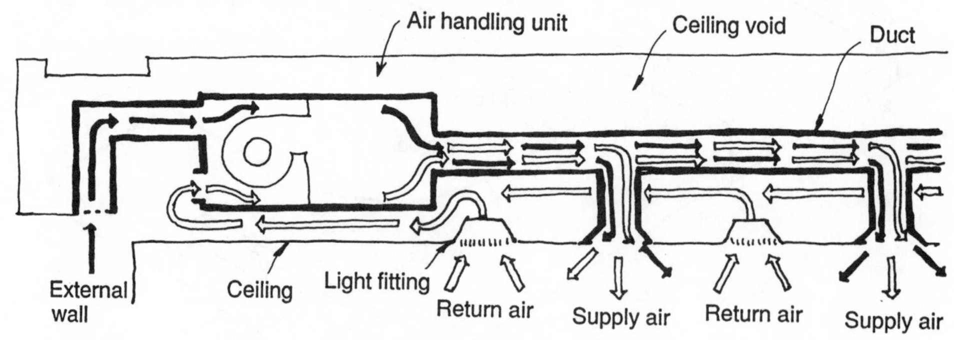

Integrated lighting and air conditioning

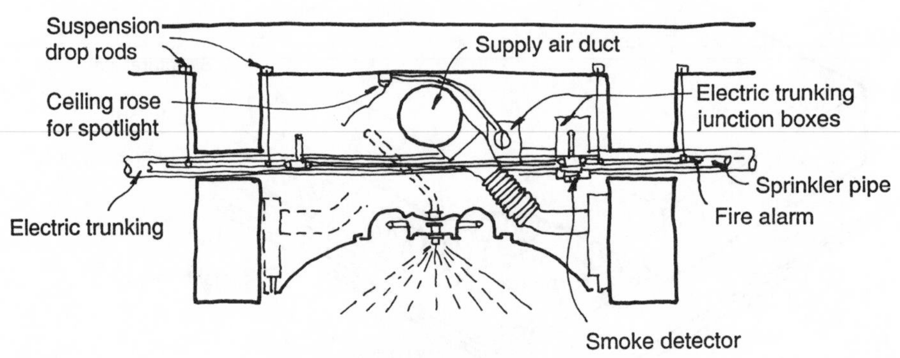

Sprinkler in integrated ceiling

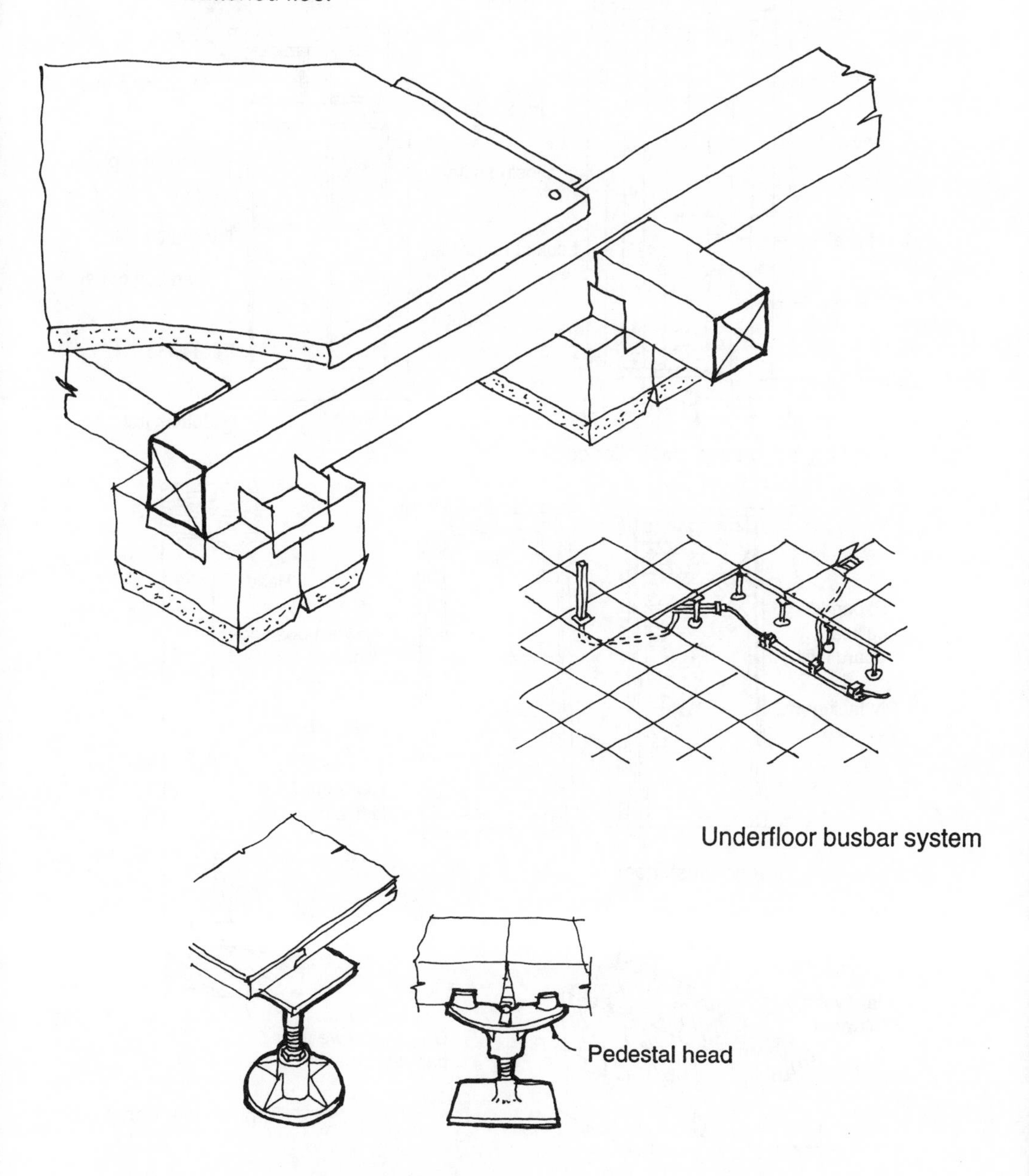
Shallow void
battened floor
Underfloor busbar system
Pedestal head
Deep void platform
floor supports

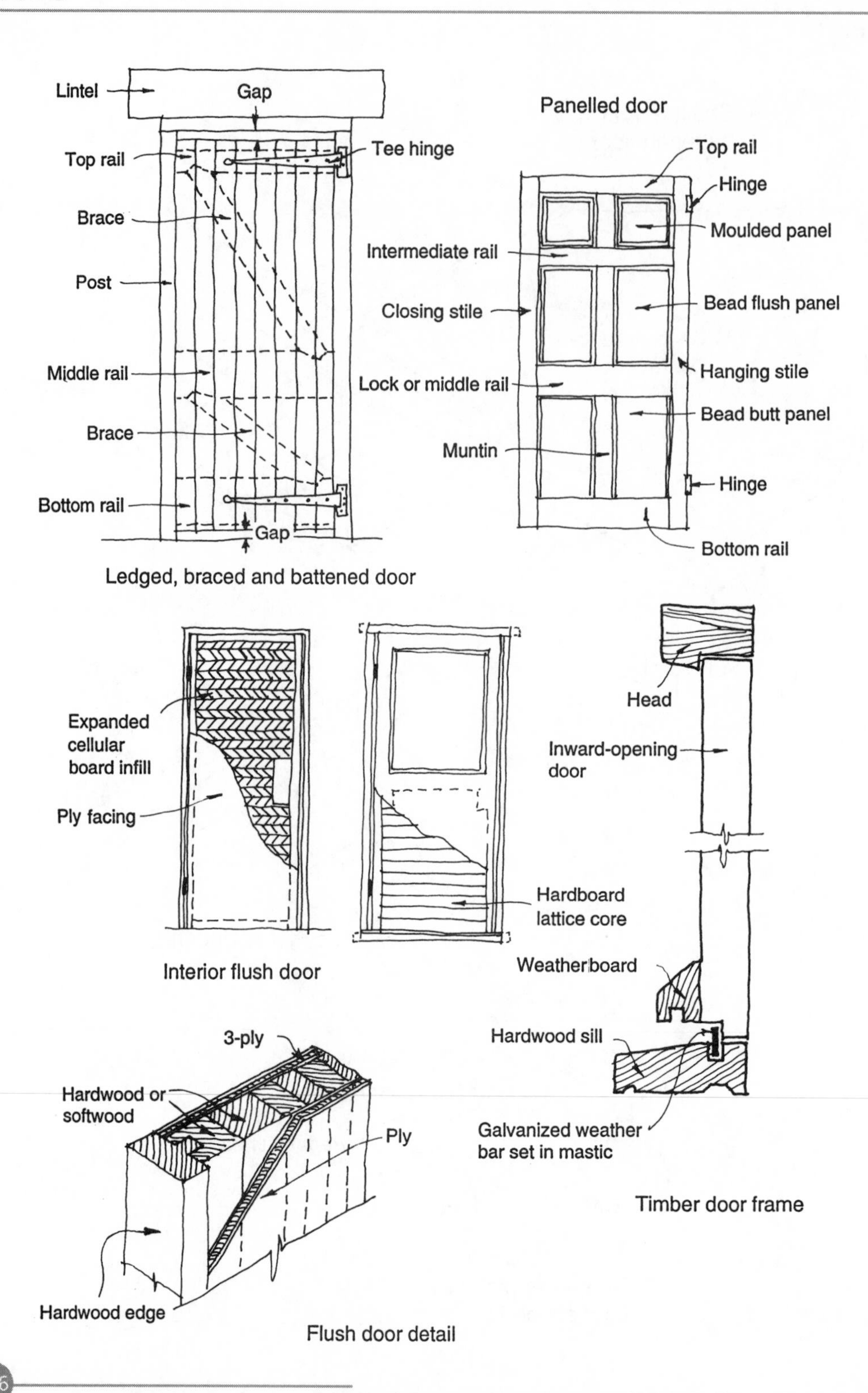

Ledged, braced and battened door

Interior flush door

Timber door frame

Flush door detail

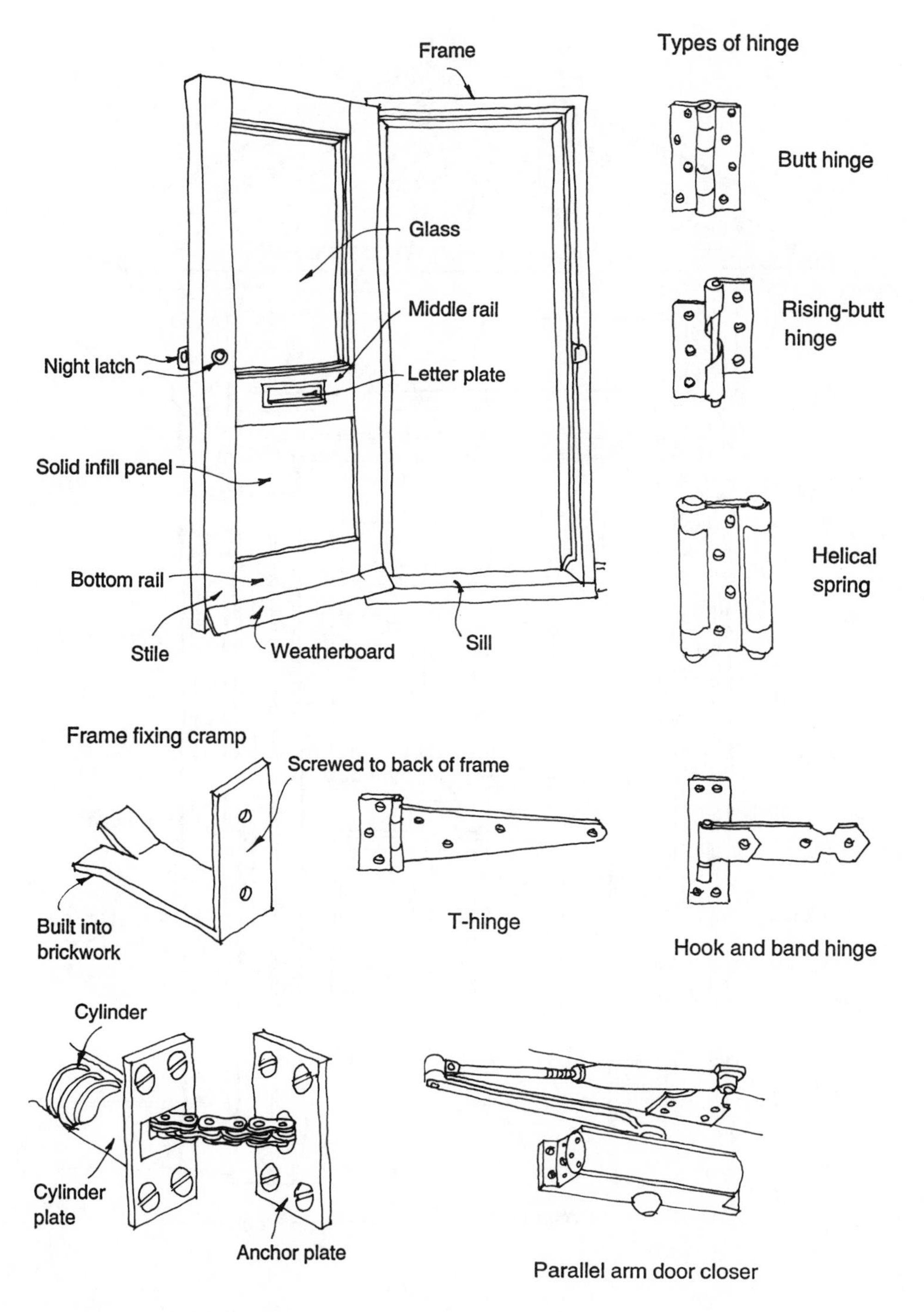
Frame
Types of hinge
Butt hinge
Glass
Rising-butt hinge
Middle rail
Night latch
Letter plate
Solid infill panel
Helical spring
Bottom rail
Stile
Weatherboard
Sill
Frame fixing cramp
Screwed to back of frame
Built into brickwork
T-hinge
Hook and band hinge
Cylinder
Cylinder plate
Anchor plate
Parallel arm door closer
'Perko' door closer

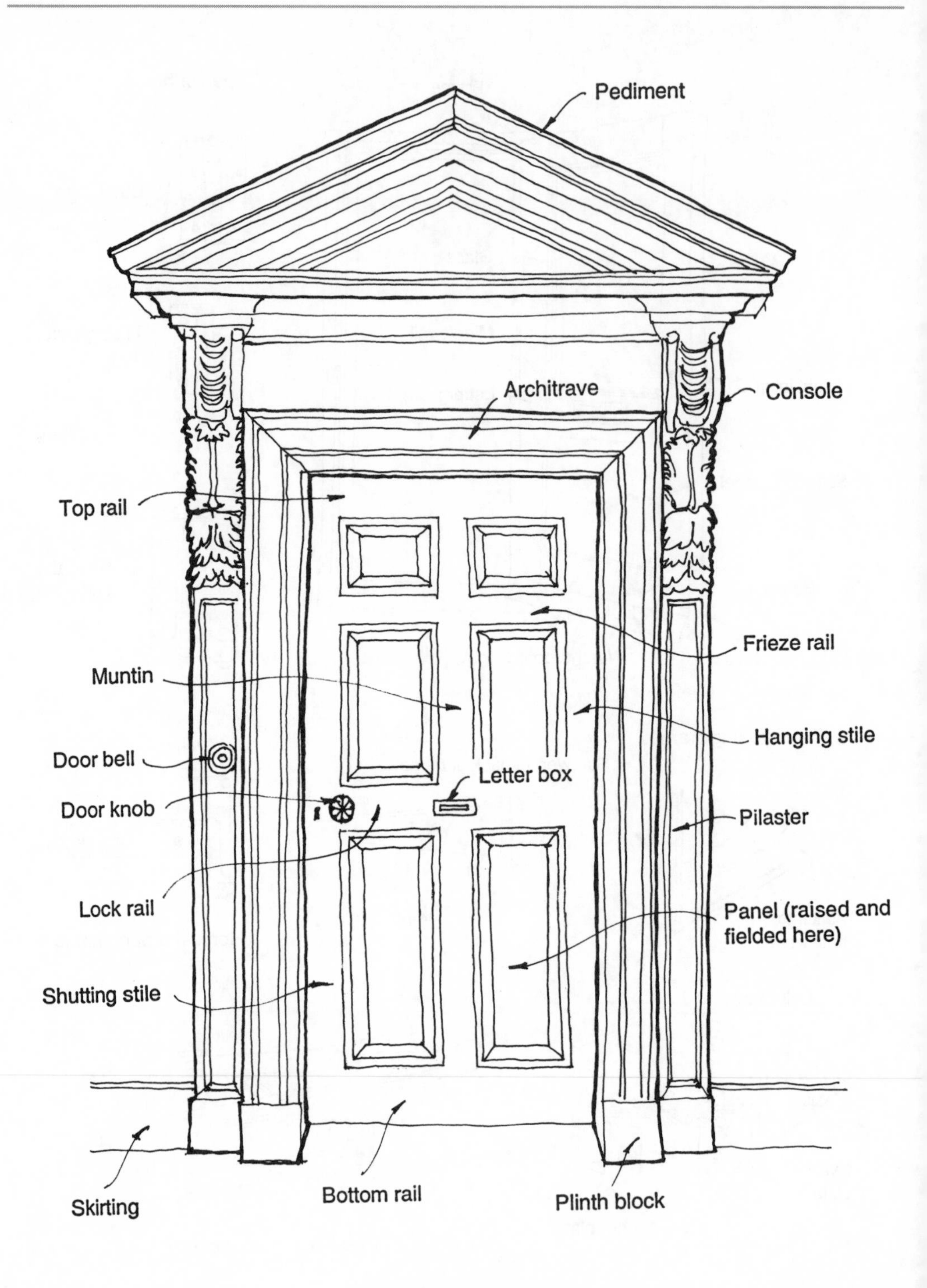

Traditional panel door

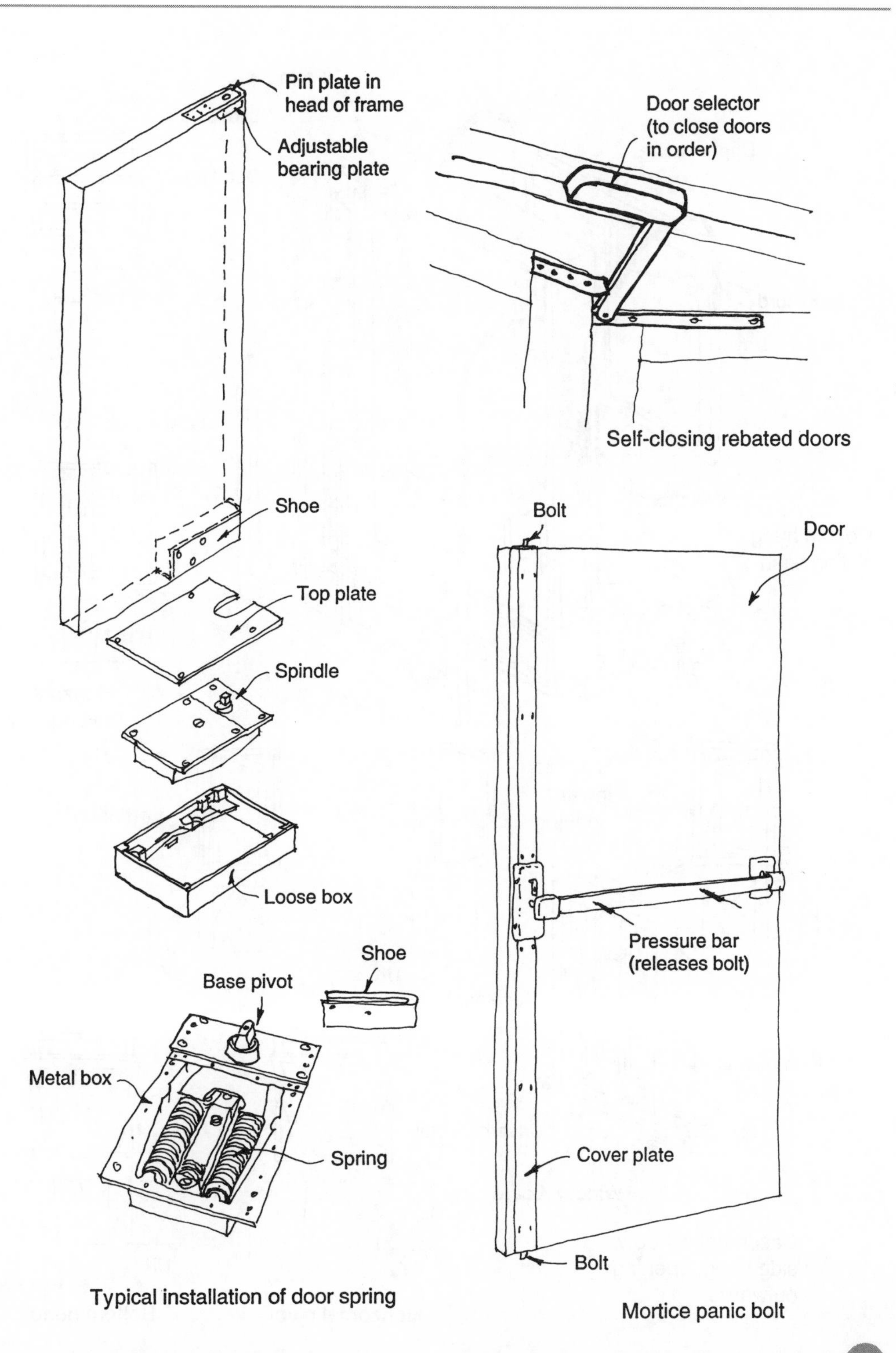

Typical installation of door spring

Self-closing rebated doors

Mortice panic bolt

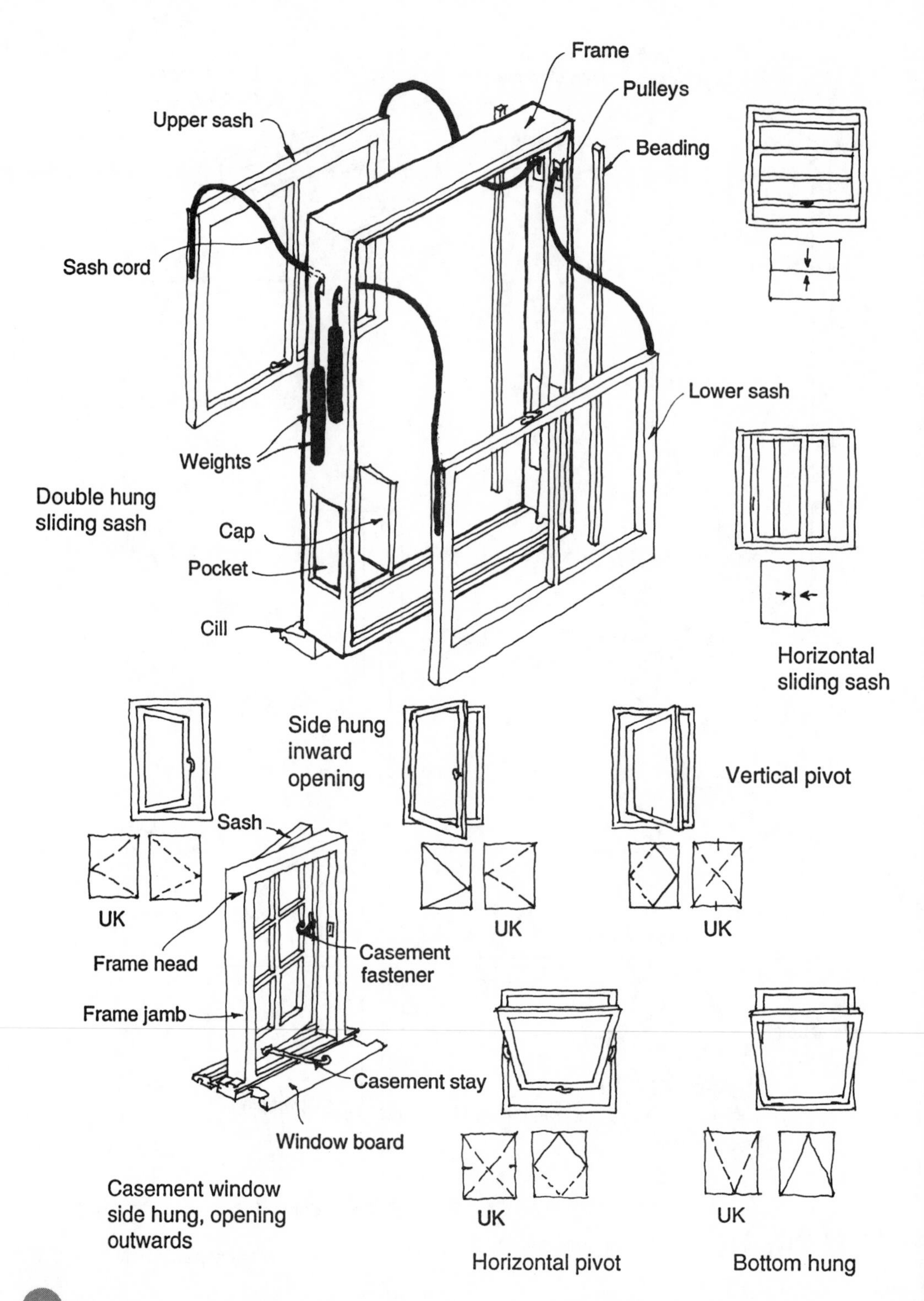
Frame
Pulleys
Upper sash
Beading
Sash cord
Lower sash
Weights
Double hung
sliding sash
Cap
Pocket
Cill
Horizontal
sliding sash
Side hung
inward
opening
Vertical pivot
Sash
UK
UK
UK
Frame head
Casement
fastener
Frame jamb
Casement stay
Window board
Casement window
side hung, opening
outwards
UK
UK
Horizontal pivot
Bottom hung

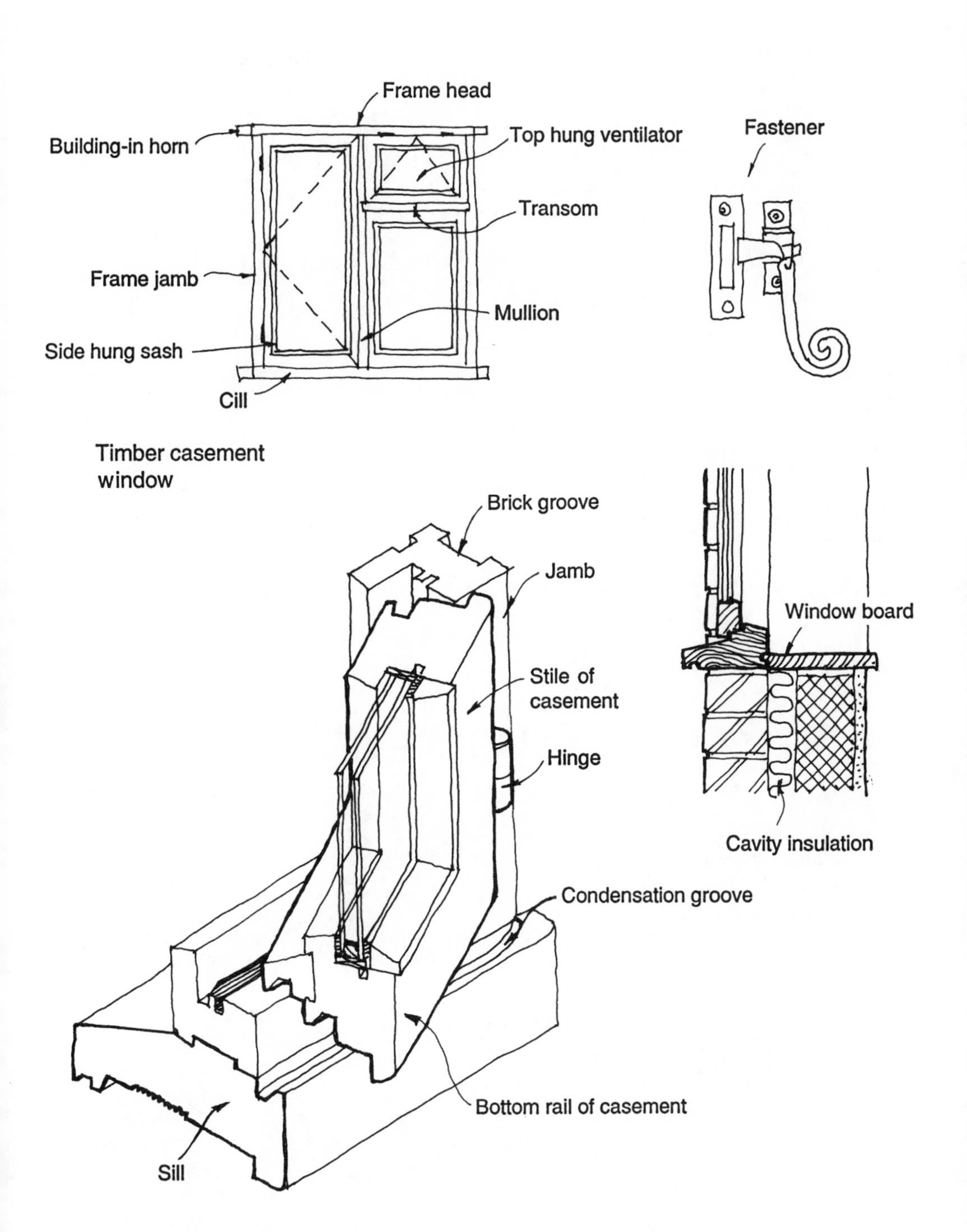
Frame head
Building-in horn
Top hung ventilator
Fastener
Transom
Frame jamb
Mullion
Side hung sash
Cill
Timber casement window
Brick groove
Jamb
Window board
Stile of casement
Hinge
Cavity insulation
Condensation groove
Bottom rail of casement
Sill

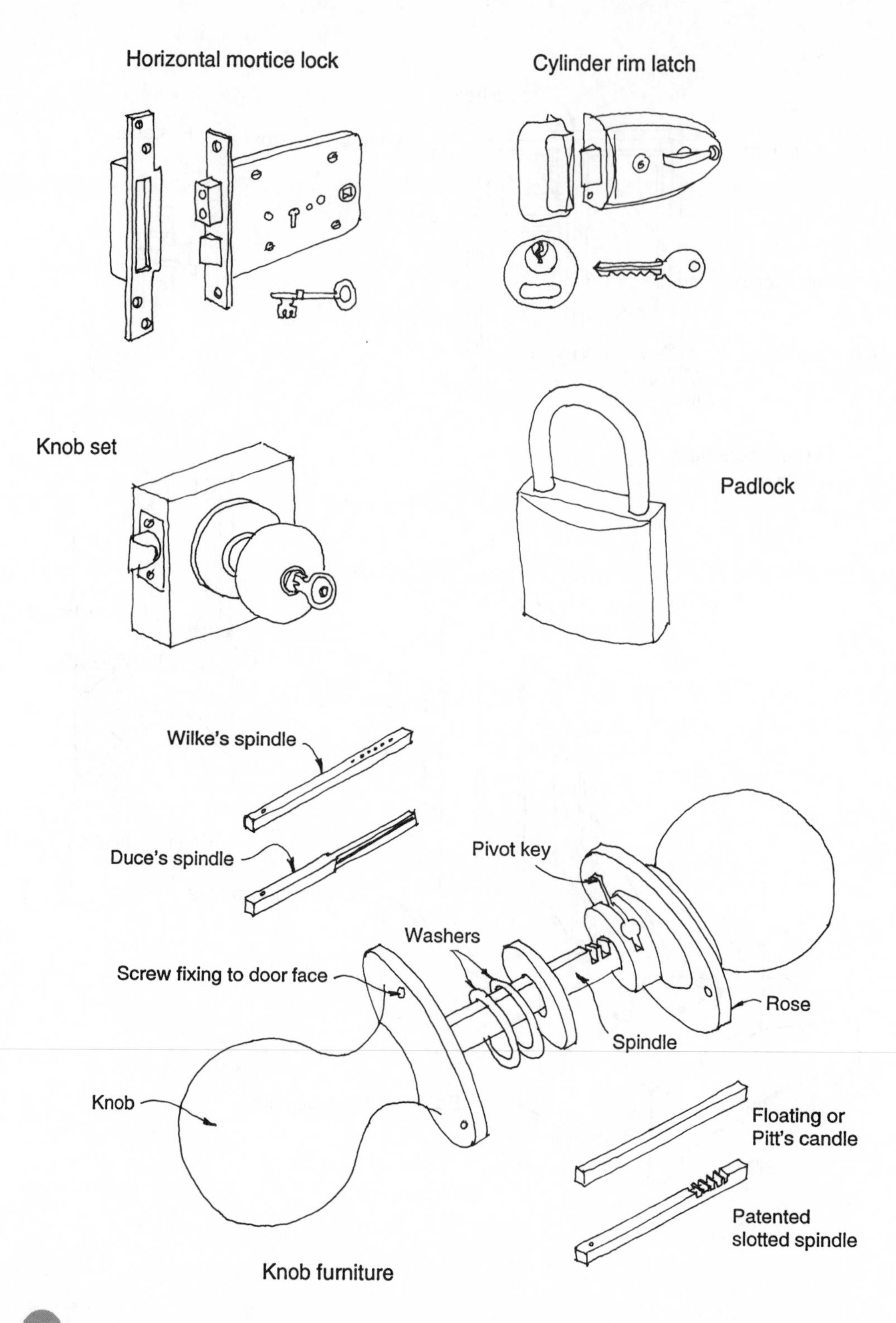
Horizontal mortice lock
Cylinder rim latch
Knob set
Padlock
Wilke's spindle
Duce's spindle
Pivot key
Washers
Screw fixing to door face
Rose
Spindle
Knob
Floating or
Pitt's candle
Patented
slotted spindle
Knob furniture

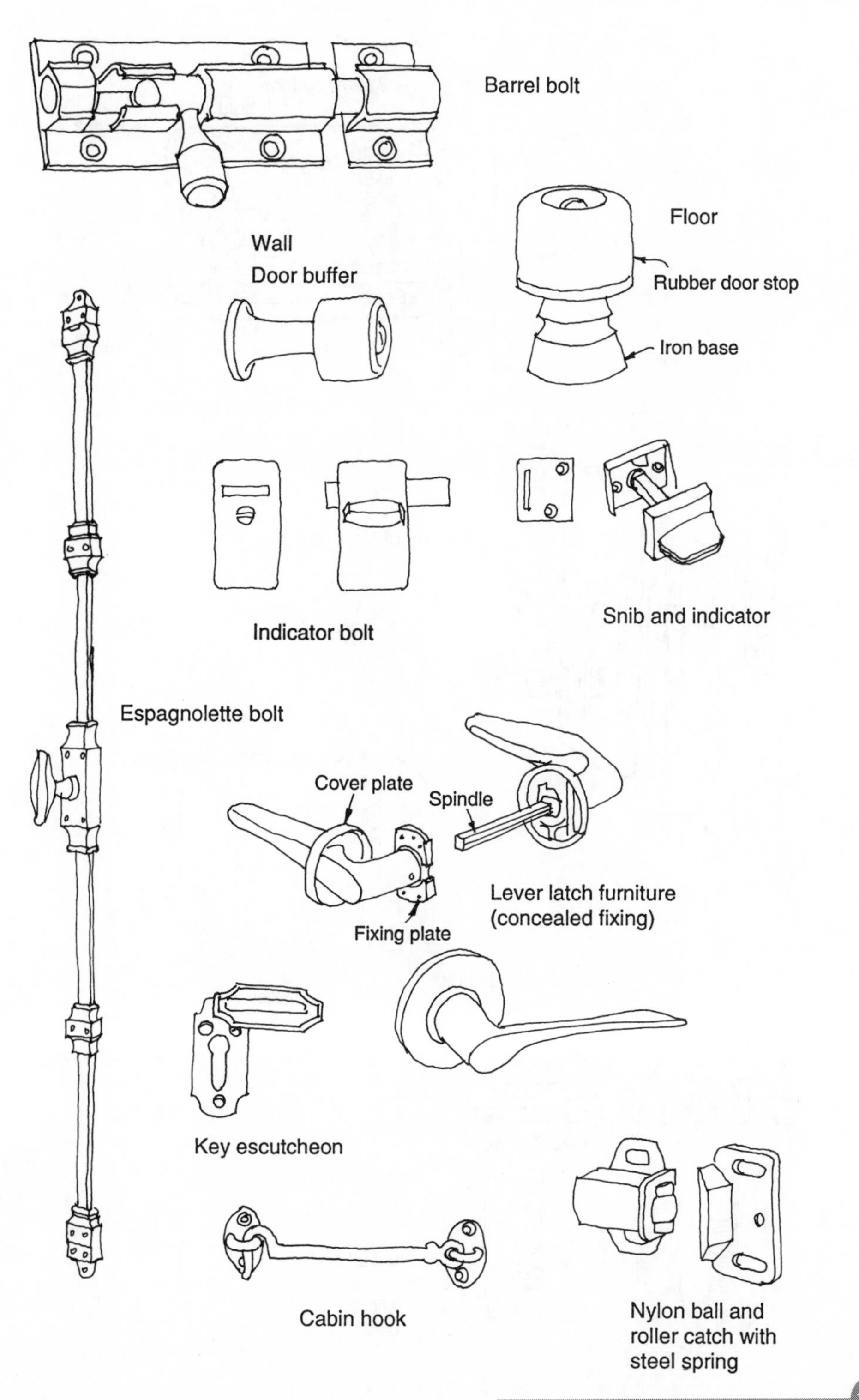
Barrel bolt
Floor
Wall
Door buffer
Rubber door stop
Iron base
Indicator bolt
Snib and indicator
Espagnolette bolt
Cover plate
Spindle
Lever latch furniture
(concealed fixing)
Fixing plate
Key escutcheon
Cabin hook
Nylon ball and
roller catch with
steel spring

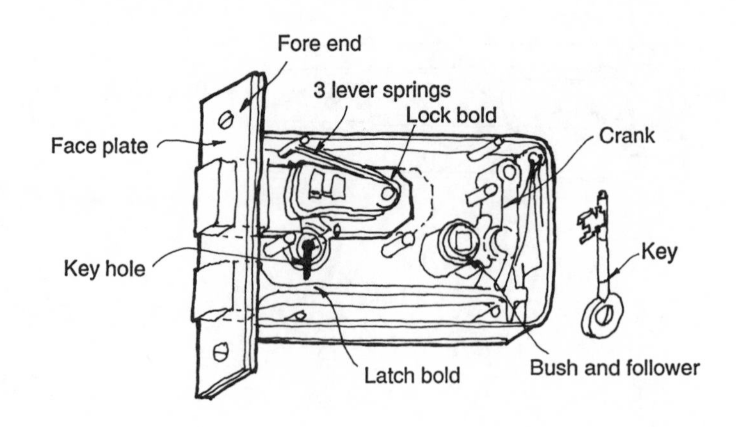
Fore end
3 lever springs
Lock bold
Face plate
Crank
Key hole
Key
Latch bold
Bush and follower

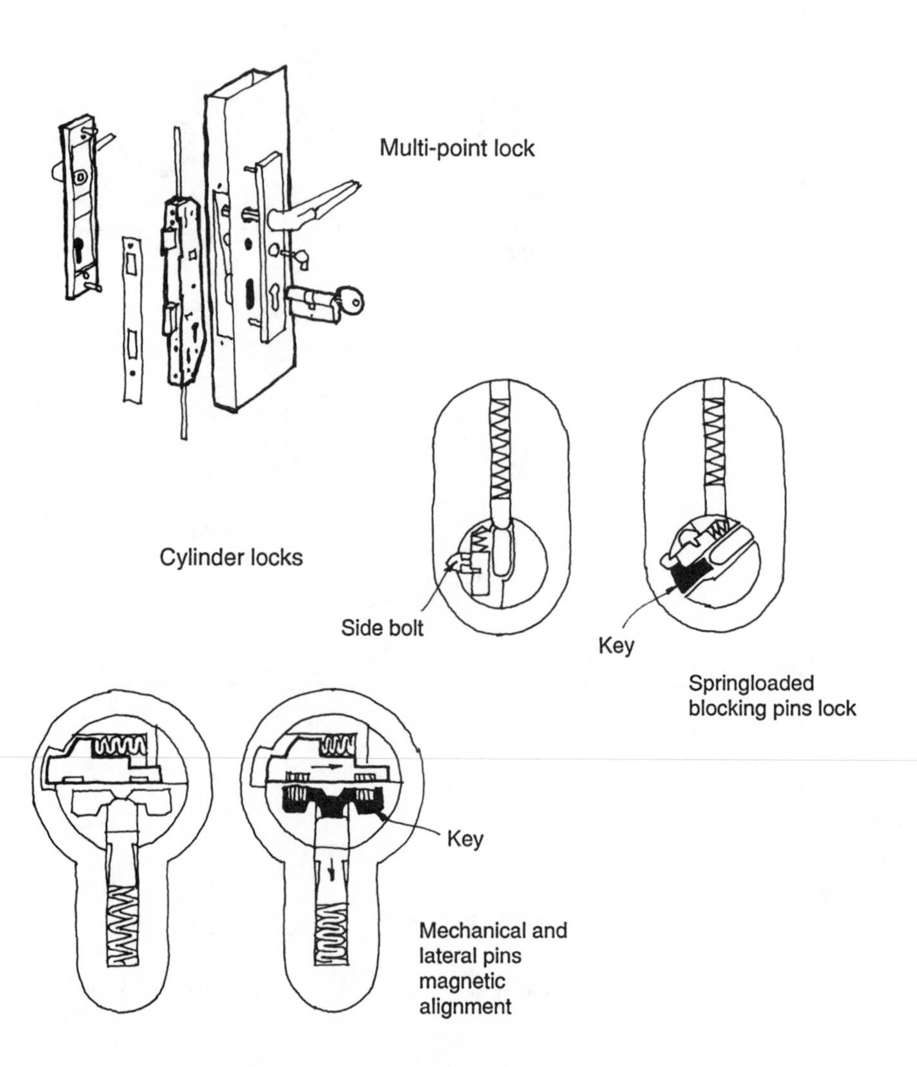
Multi-point lock
Cylinder locks
Side bolt
Key
Springloaded
blocking pins lock
Key
Mechanical and
lateral pins
magnetic
alignment

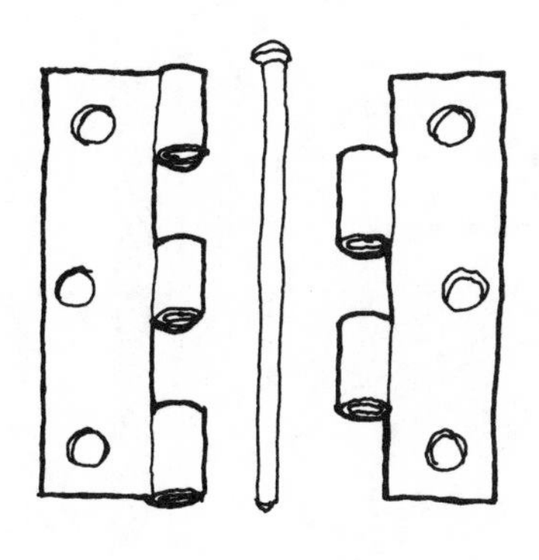

Loose pin

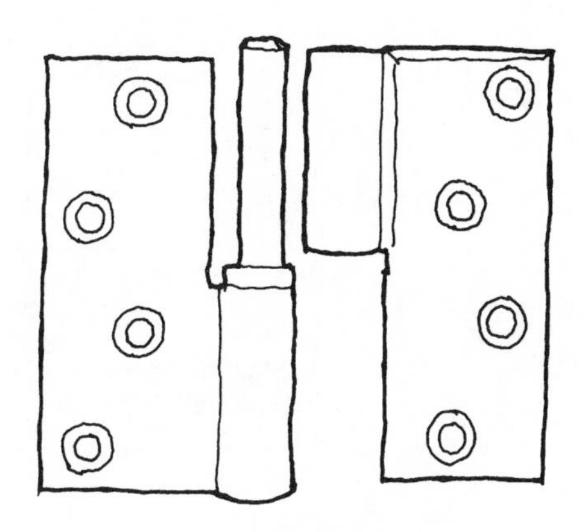

Lift-off butt

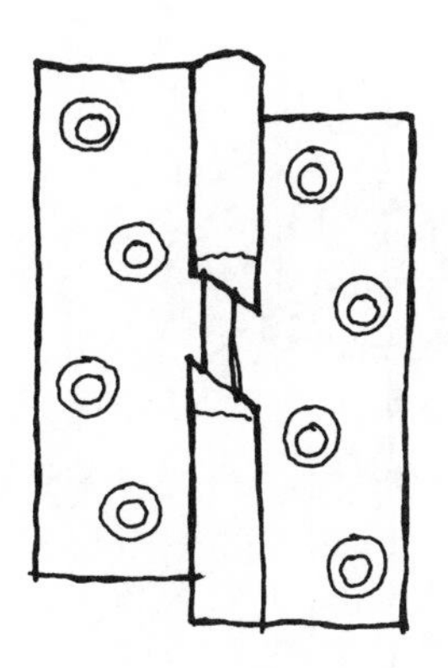

Rising butt

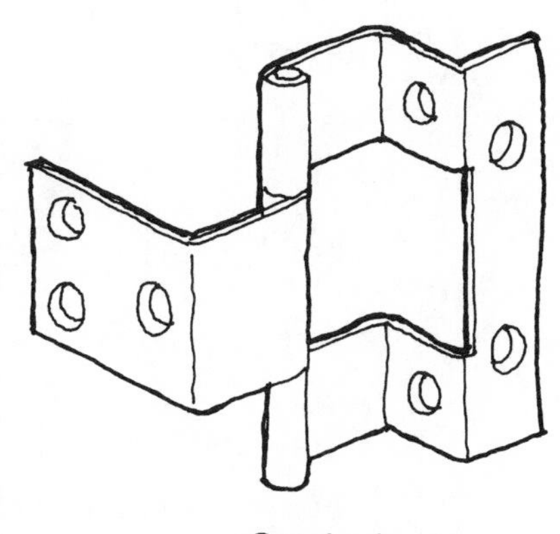

Cranked

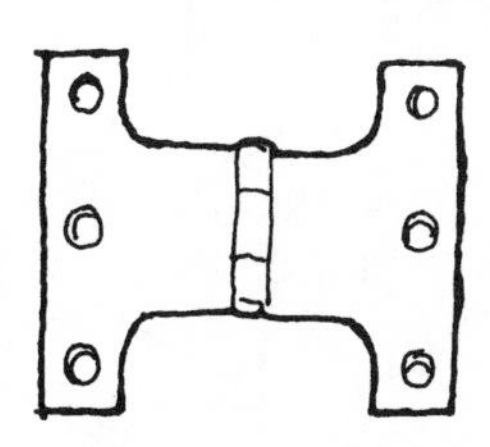

Parliament

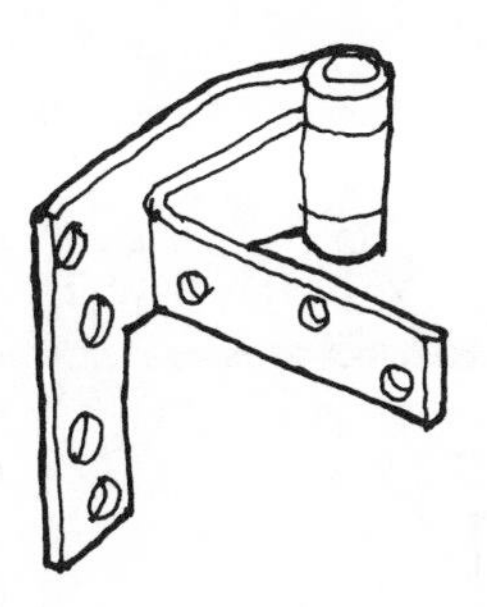

Offset-easyclean

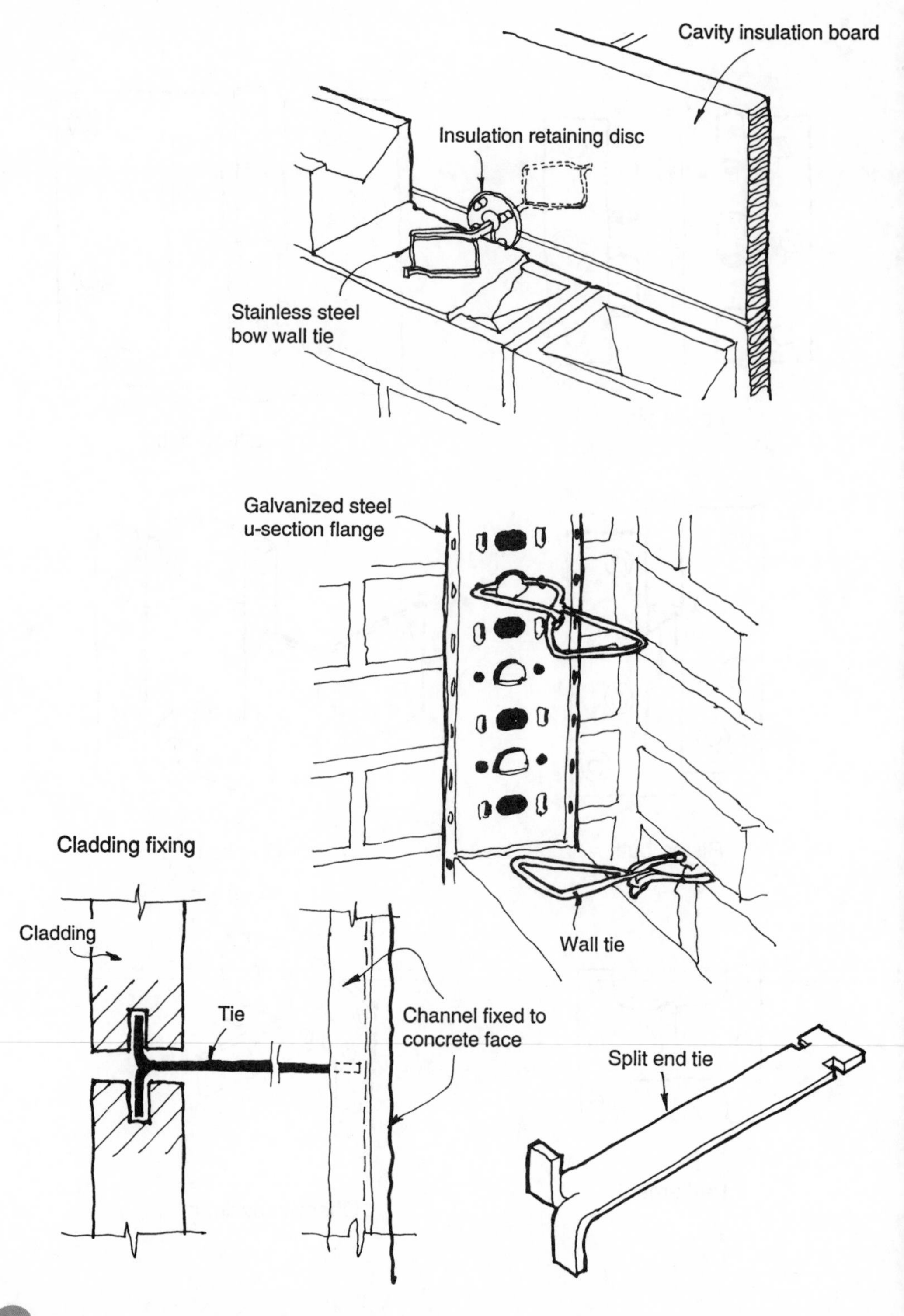
Cavity insulation board
Insulation retaining disc
Stainless steel
bow wall tie
Galvanized steel
u-section flange
Cladding fixing
Cladding
Tie
Channel fixed to
concrete face
Wall tie
Split end tie

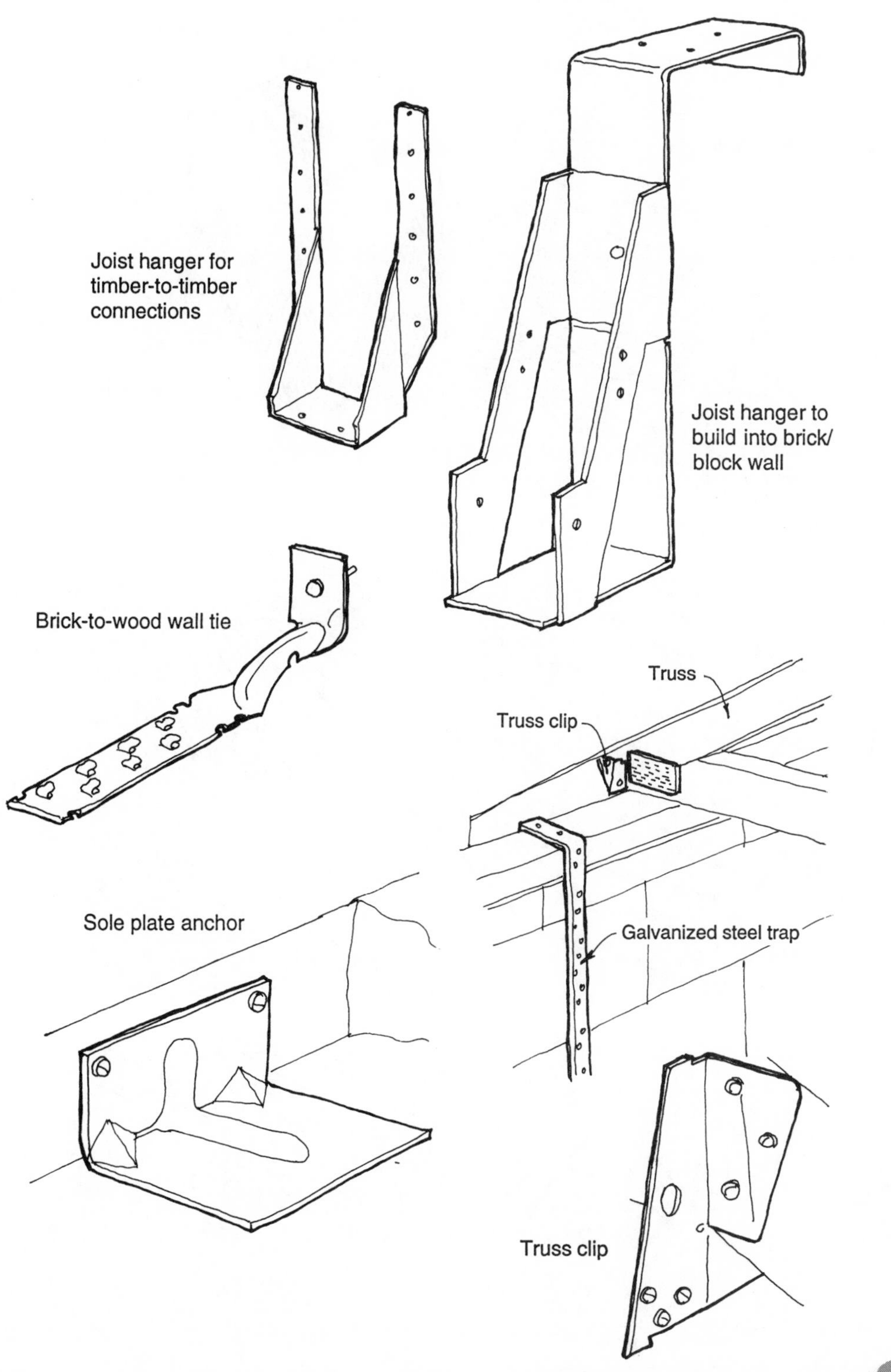
Joist hanger for timber-to-timber connections
Joist hanger to build into brick/ block wall
Brick-to-wood wall tie
Truss
Truss clip
Sole plate anchor
Galvanized steel trap
Truss clip

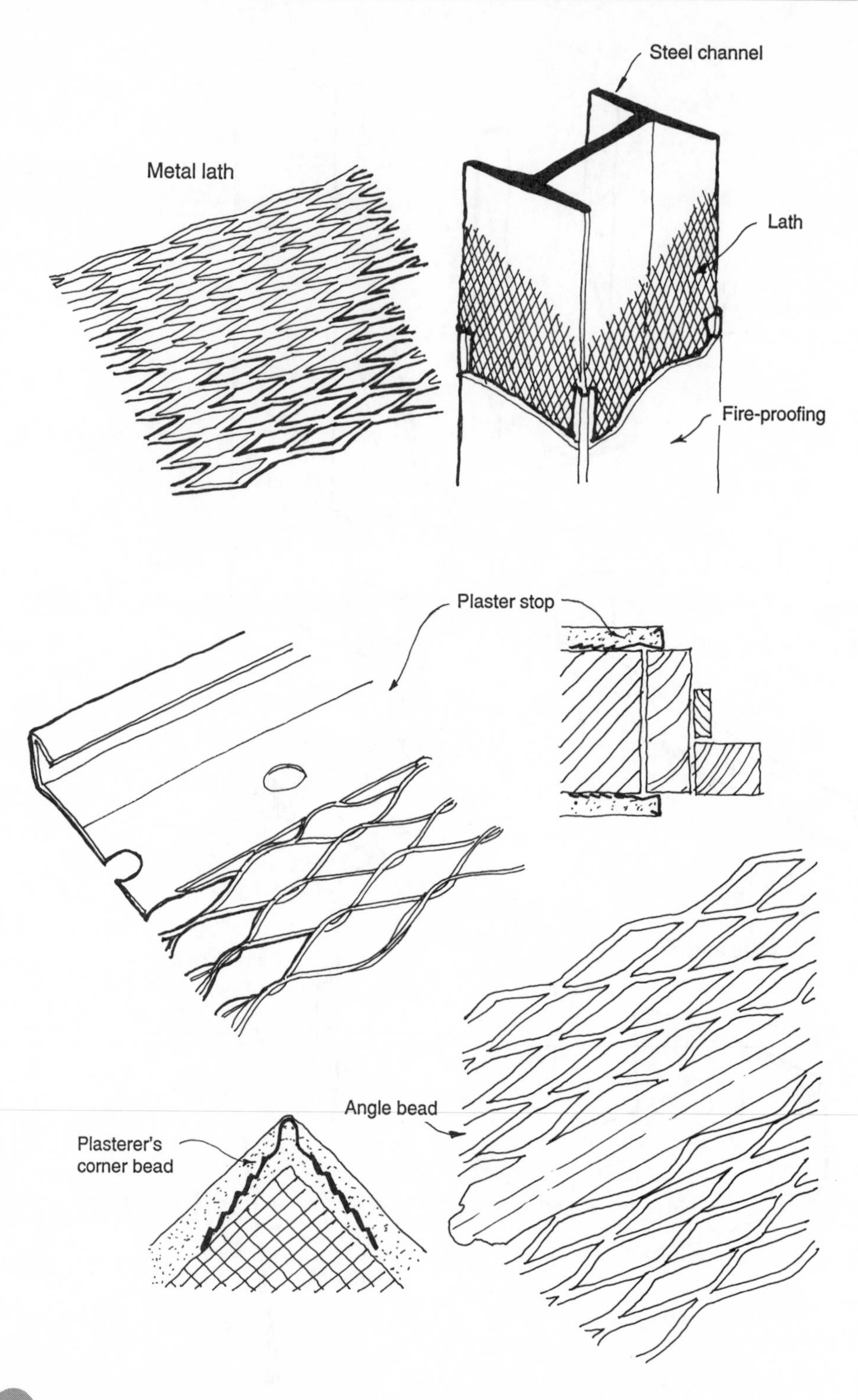
Steel channel
Metal lath
Lath
Fire-proofing
Plaster stop
Angle bead
Plasterer's
corner bead

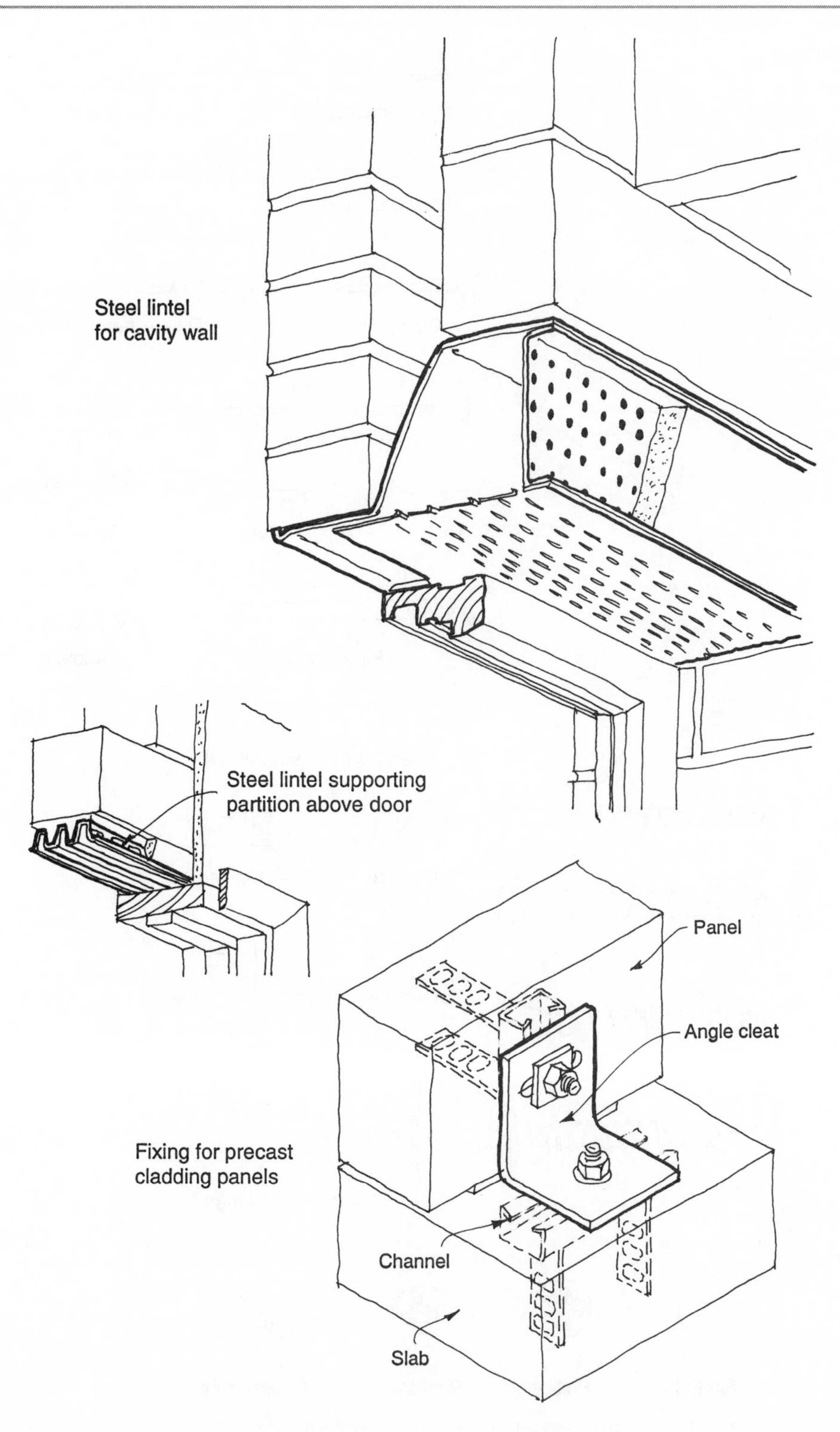
Steel lintel
for cavity wall
Steel lintel supporting
partition above door
Panel
Angle cleat
Fixing for precast
cladding panels
Channel
Slab

Nails and screws

Nails

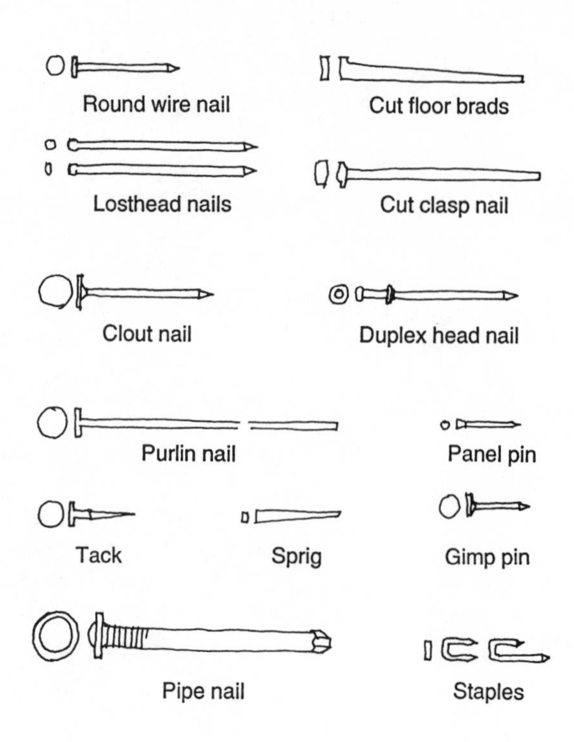

Screws

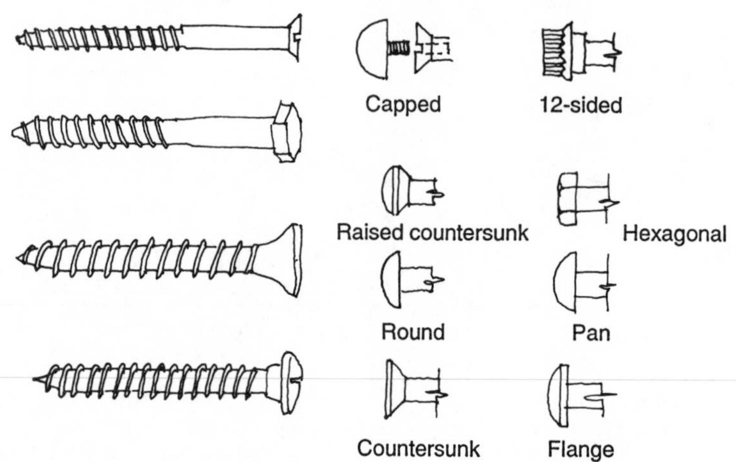

Plan

Slotted

Philips

Posidrive

Linread torx

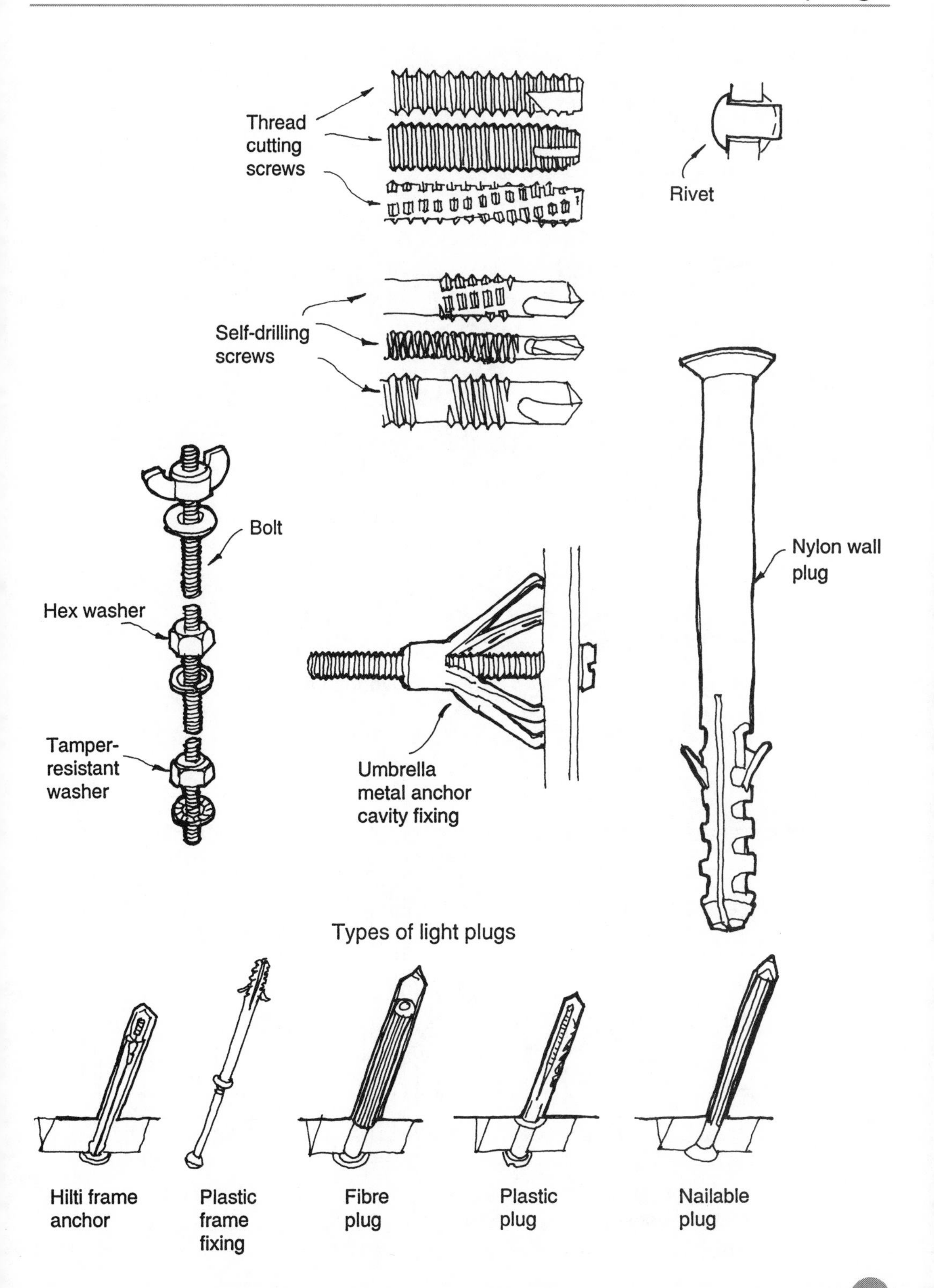
Thread
cutting
screws
Rivet
Self-drilling
screws
Bolt
Nylon wall
plug
Hex washer
Tamper-
resistant
washer
Umbrella
metal anchor
cavity fixing
Types of light plugs
Hilti frame
anchor
Plastic
frame
fixing
Fibre
plug
Plastic
plug
Nailable
plug

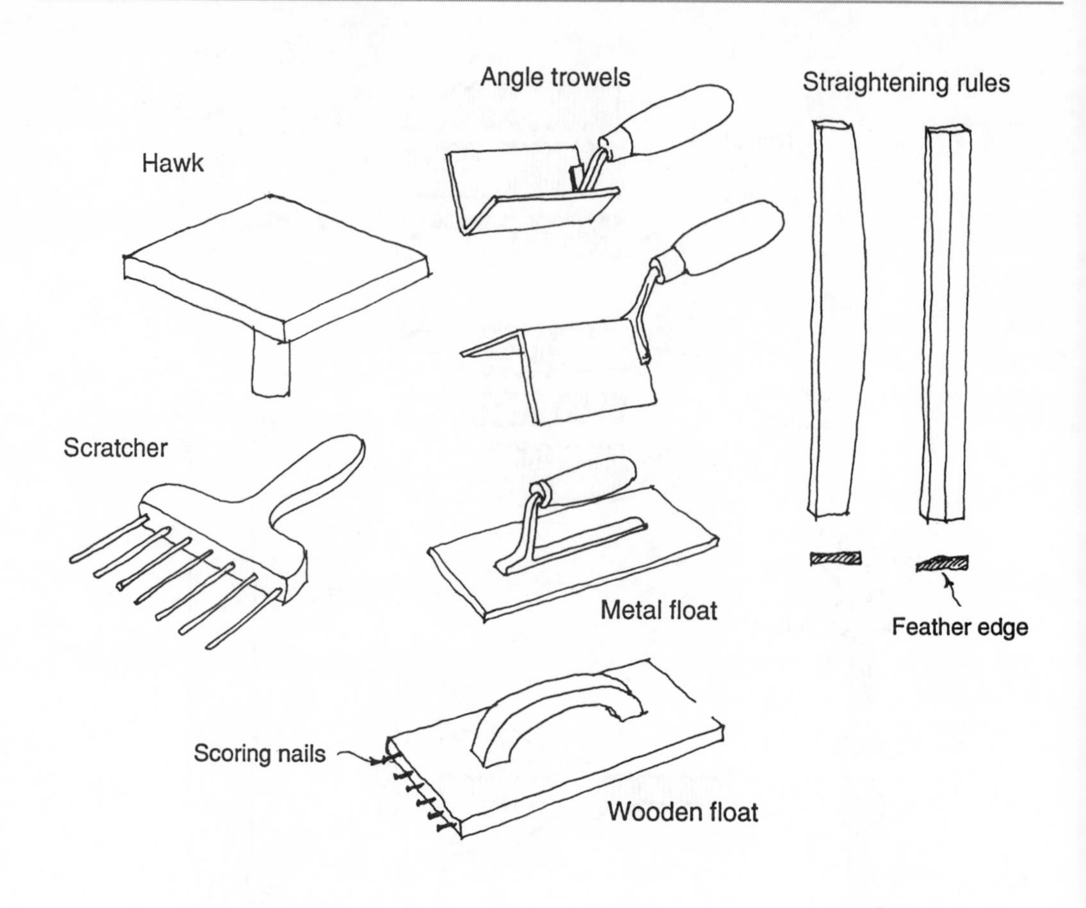
Angle trowels
Straightening rules
Hawk
Scratcher
Metal float
Feather edge
Scoring nails
Wooden float

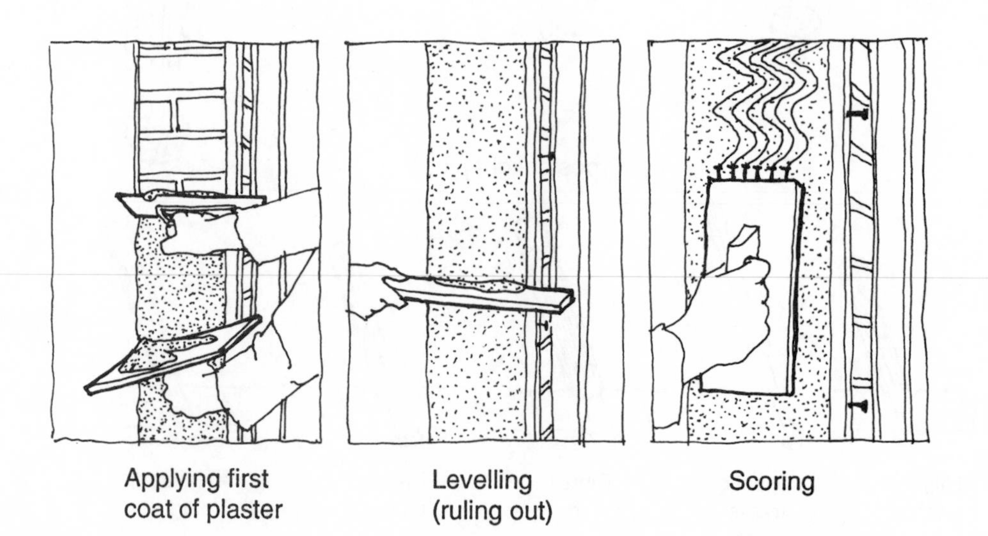
Applying first coat of plaster
Levelling (ruling out)
Scoring

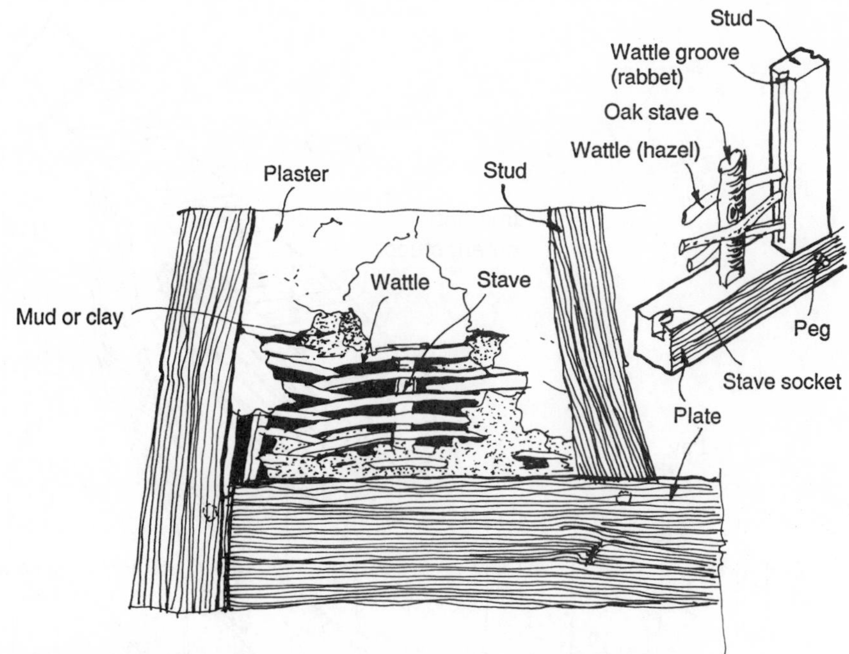

*Wattle is daubed both sides with a mixture of clay, dung and chopped straw

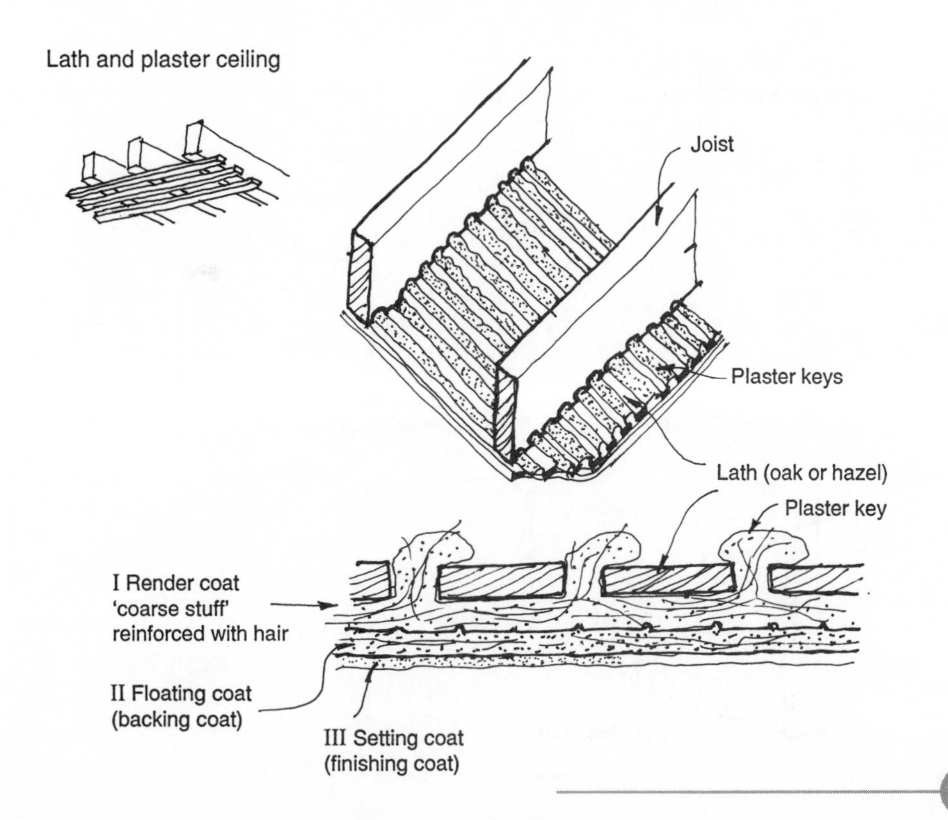

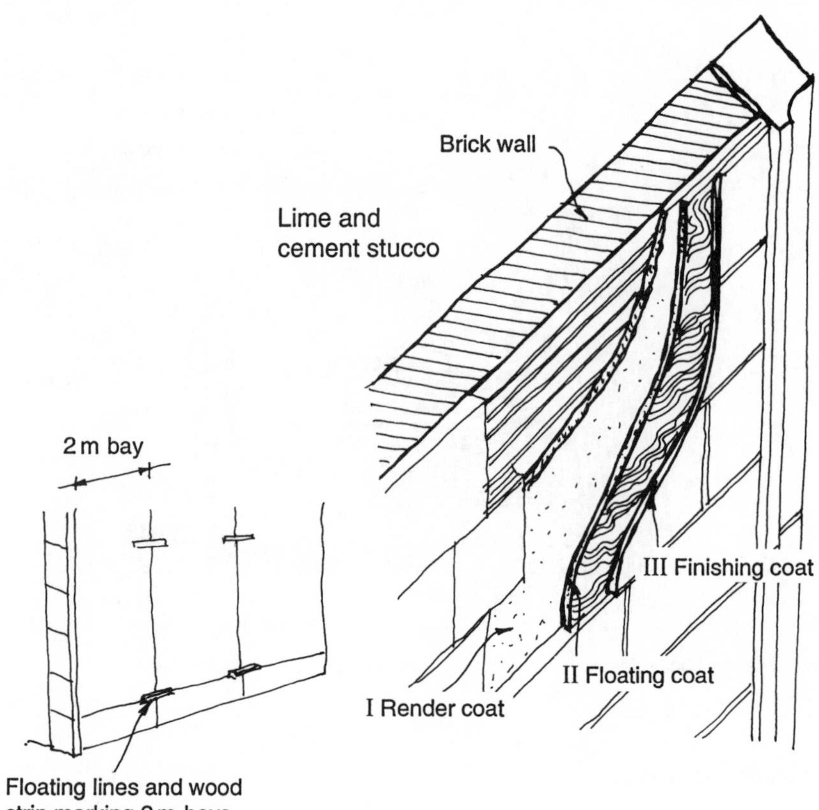
Brick wall
Lime and
cement stucco
2 m bay
III Finishing coat
II Floating coat
I Render coat

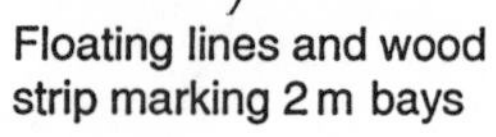
Floating lines and wood
strip marking 2 m bays

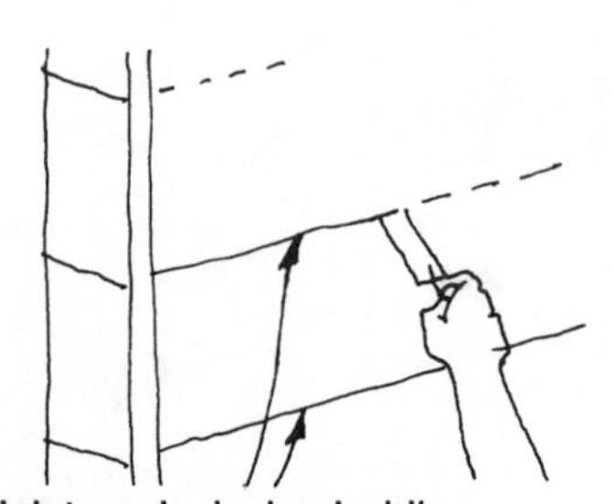
Joint marks incised while
finishing coat still soft

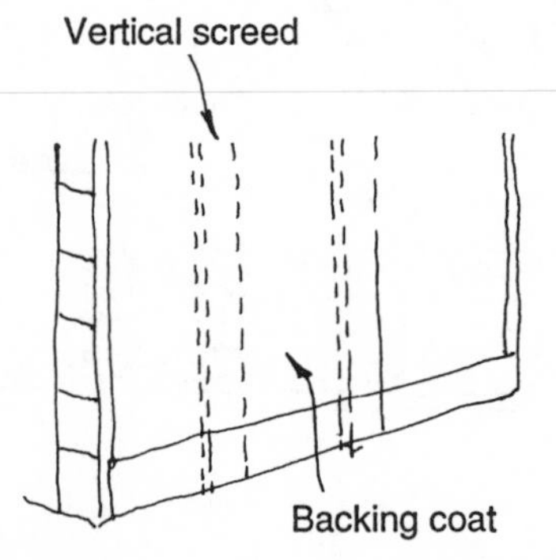
Vertical screed
Backing coat

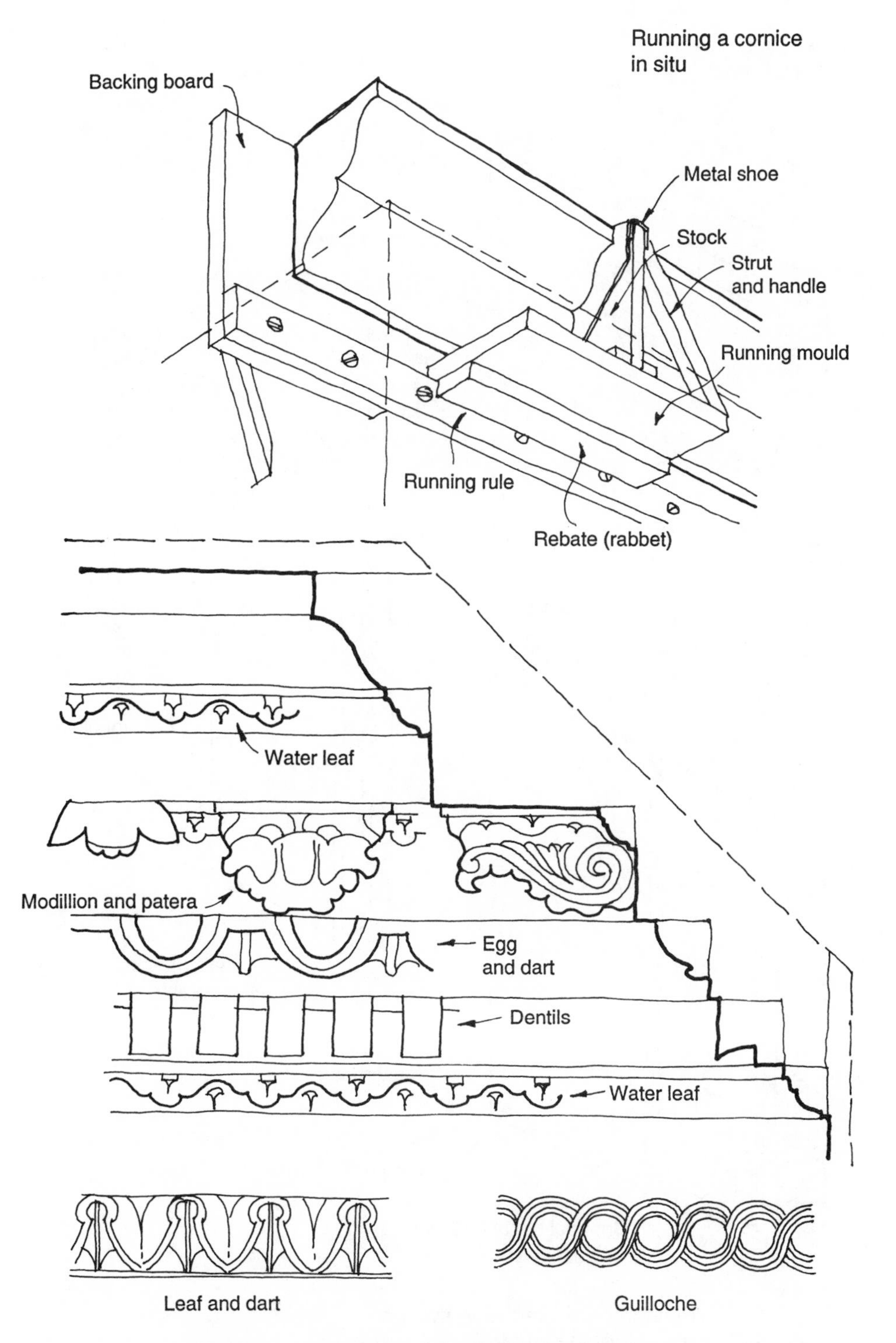

Cornice ornamentation (planted in wet stucco)

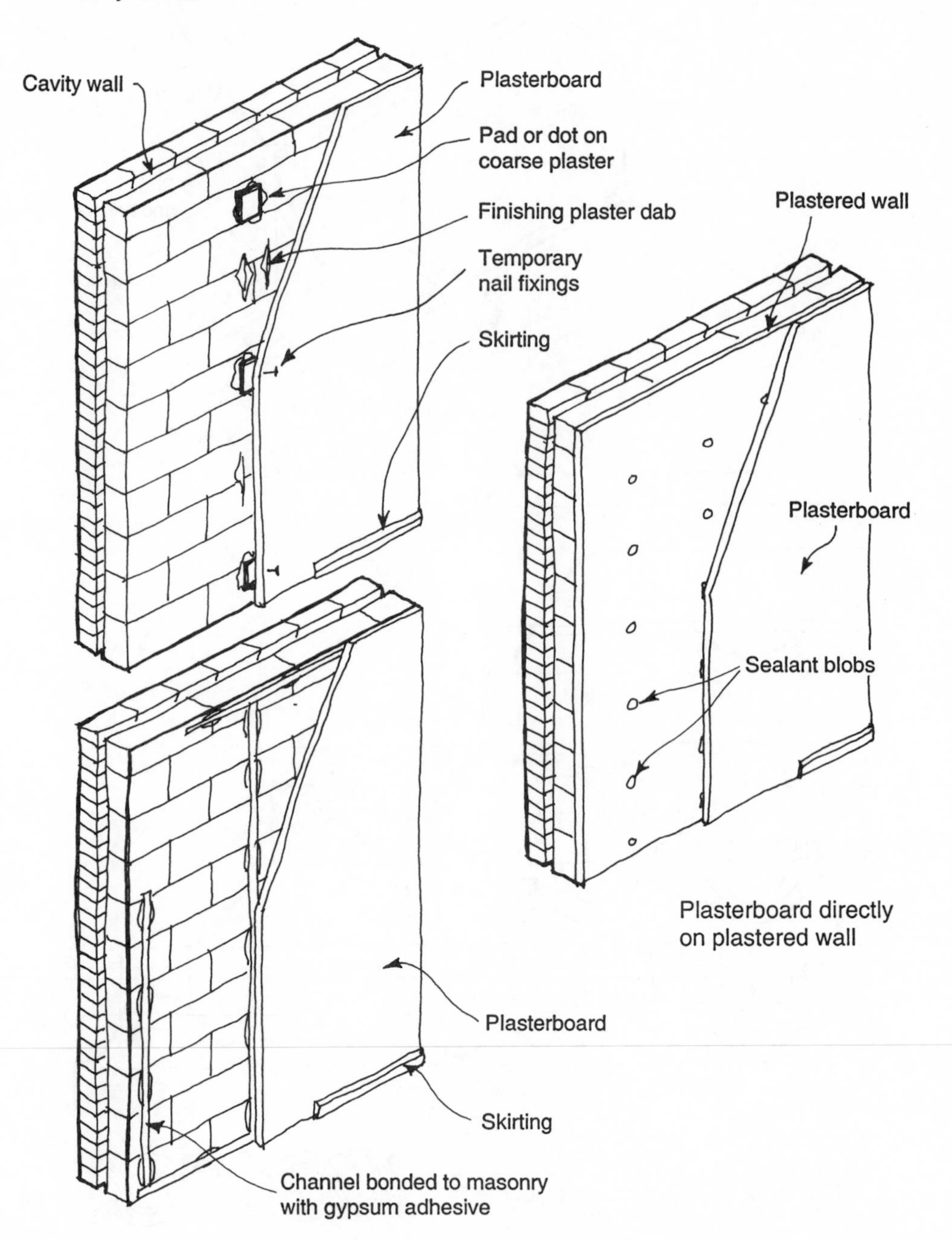
Plasterboard fixed
directly to wall
Cavity wall
Plasterboard
Pad or dot on
coarse plaster
Finishing plaster dab
Temporary
nail fixings
Skirting
Plastered wall
Plasterboard
Sealant blobs
Plasterboard directly
on plastered wall
Plasterboard
Skirting
Channel bonded to masonry
with gypsum adhesive
Plasterboard fixed to lightweight
metal furring channels

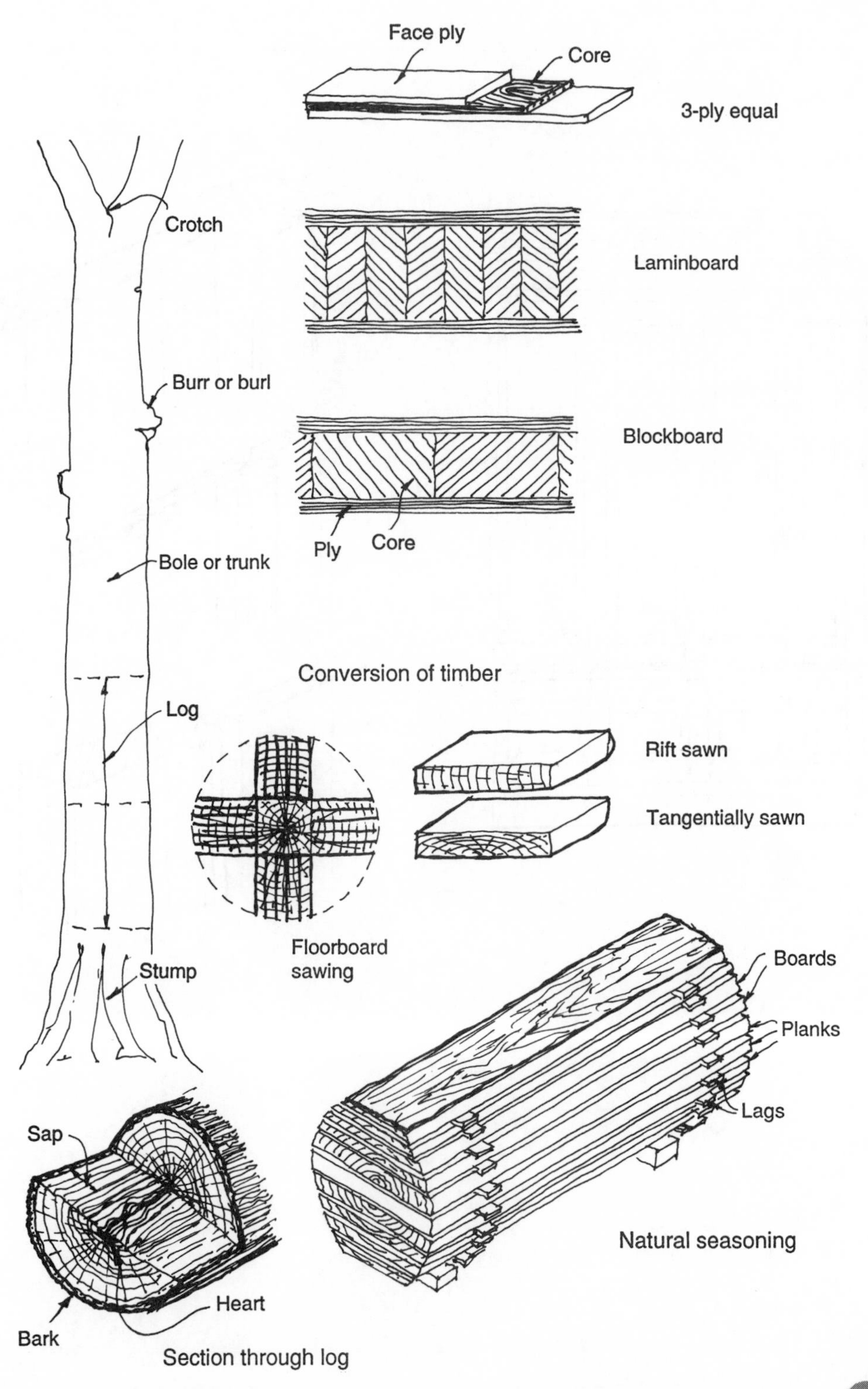
Face ply
Core
3-ply equal
Crotch
Laminboard
Burr or burl
Blockboard
Ply
Core
Bole or trunk
Conversion of timber
Log
Rift sawn
Tangentially sawn
Floorboard sawing
Stump
Boards
Planks
Lags
Sap
Natural seasoning
Heart
Bark
Section through log

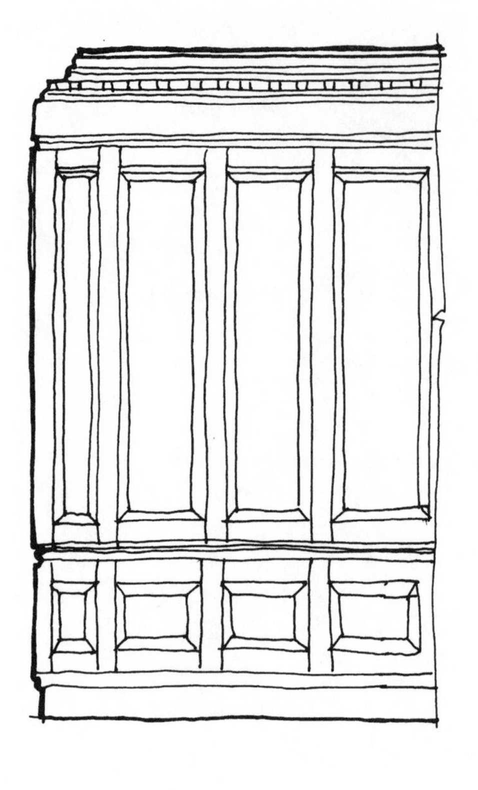

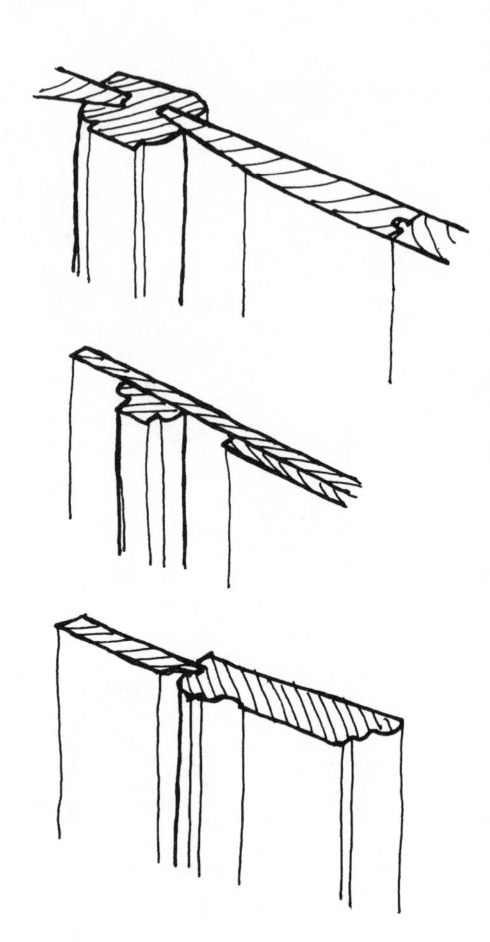

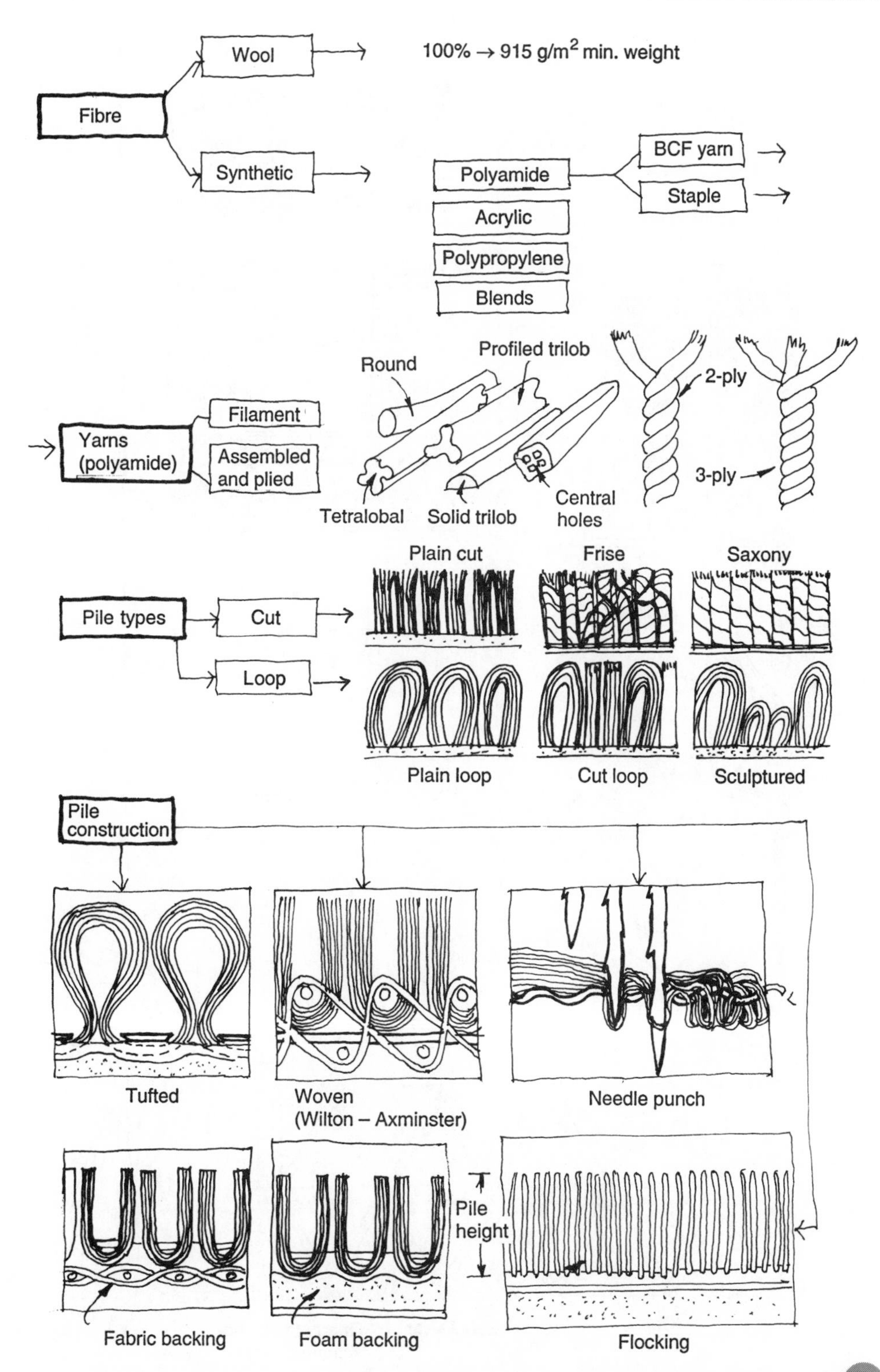
Wool
100% → 915 g/m² min. weight
Fibre
Synthetic
Polyamide
Acrylic
Polypropylene
Blends
BCF yarn
Staple
Yarns (polyamide)
Filament
Assembled and plied
Round
Profiled trilob
Tetralobal
Solid trilob
Central holes
2-ply
3-ply
Pile types
Cut
Loop
Plain cut
Frise
Saxony
Plain loop
Cut loop
Sculptured
Pile construction
Tufted
Woven (Wilton – Axminster)
Needle punch
Fabric backing
Foam backing
Pile height
Flocking

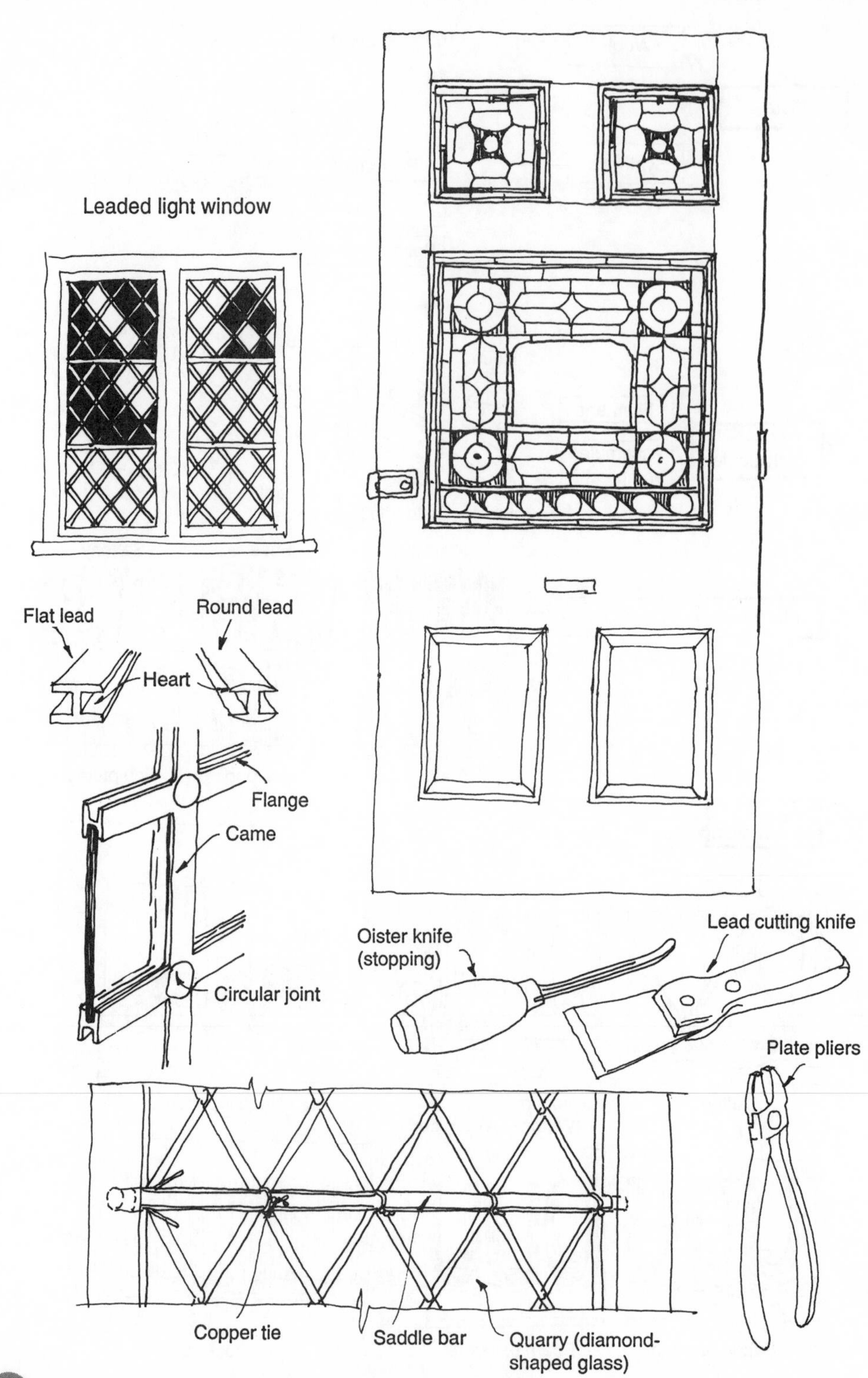
Leaded light window
Flat lead
Round lead
Heart
Flange
Came
Circular joint
Oister knife (stopping)
Lead cutting knife
Plate pliers
Copper tie
Saddle bar
Quarry (diamond-shaped glass)

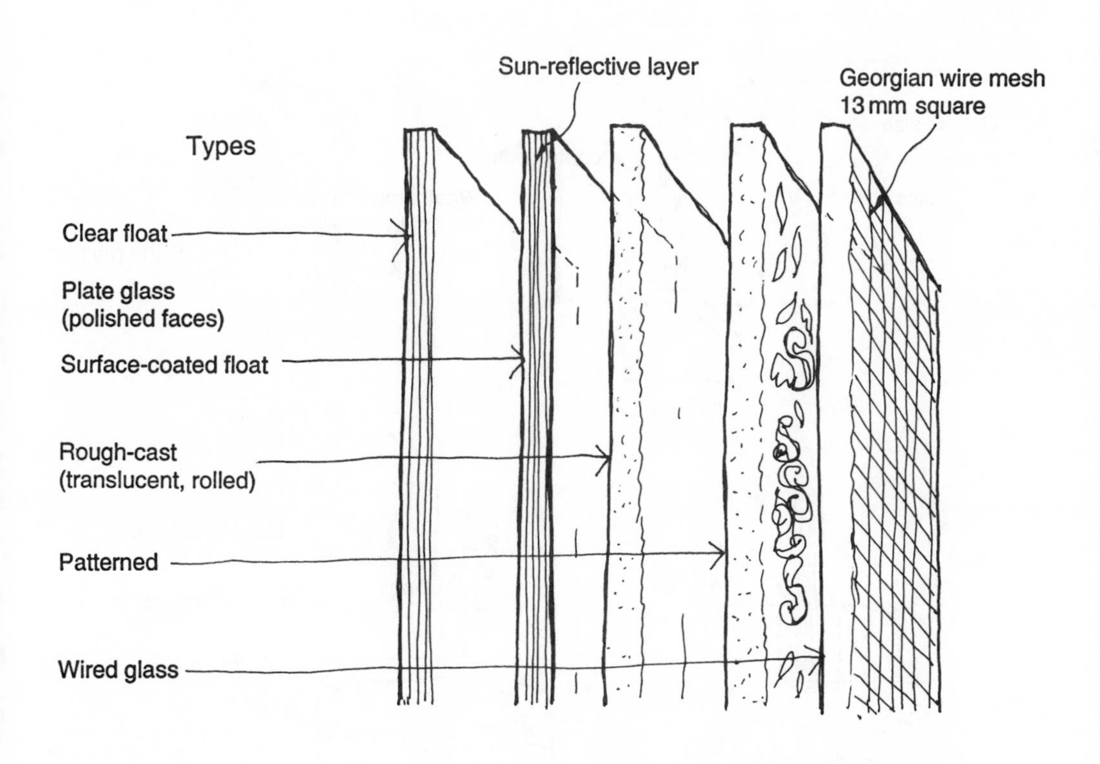
Sun-reflective layer
Georgian wire mesh
13 mm square
Types
Clear float
Plate glass
(polished faces)
Surface-coated float
Rough-cast
(translucent, rolled)
Patterned
Wired glass

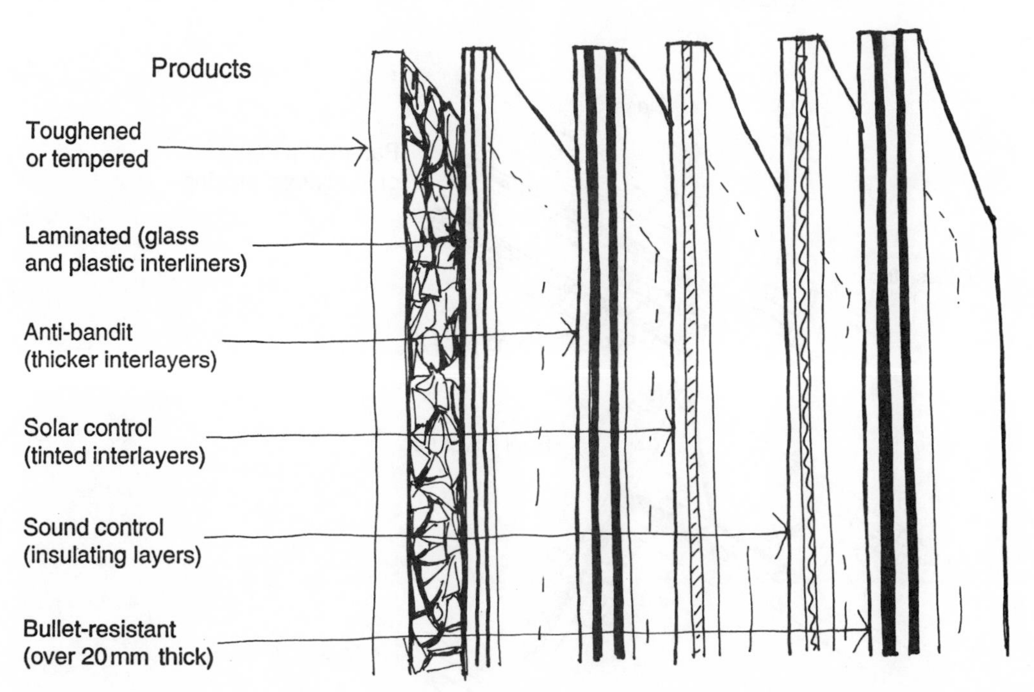
Products
Toughened
or tempered
Laminated (glass
and plastic interliners)
Anti-bandit
(thicker interlayers)
Solar control
(tinted interlayers)
Sound control
(insulating layers)
Bullet-resistant
(over 20 mm thick)

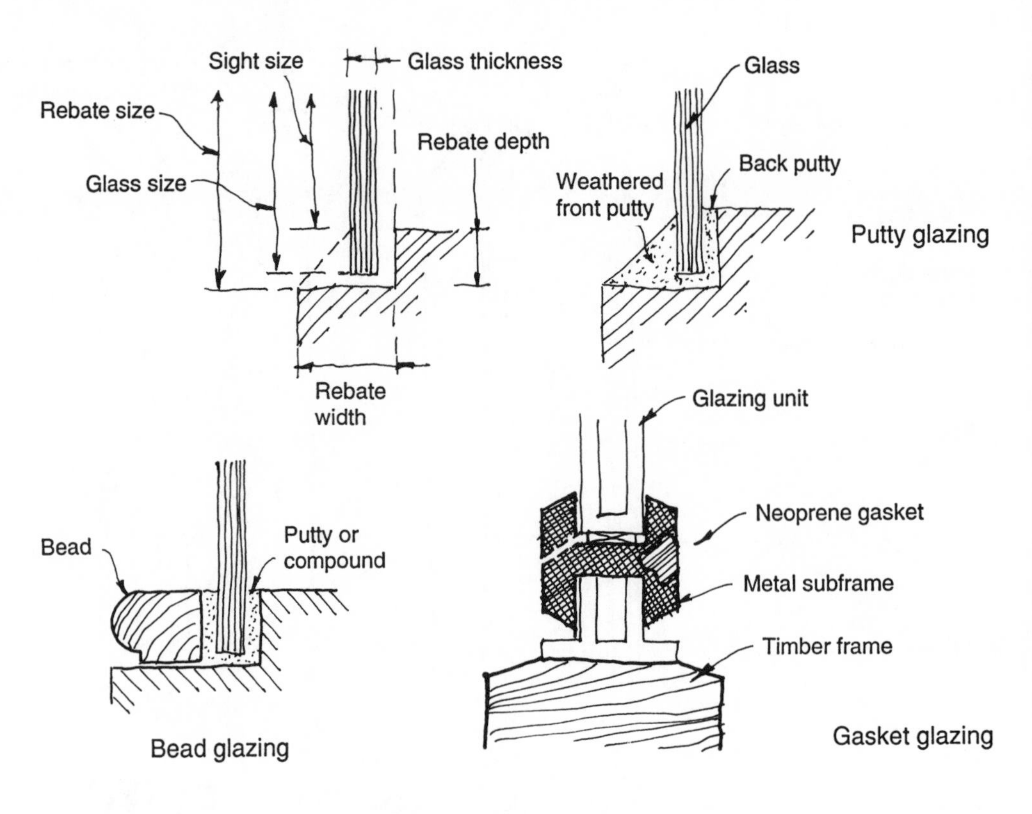
Sight size
Glass thickness
Rebate size
Rebate depth
Glass size
Rebate
width
Glass
Back putty
Weathered
front putty
Putty glazing
Glazing unit
Neoprene gasket
Bead
Putty or
compound
Metal subframe
Timber frame
Bead glazing
Gasket glazing

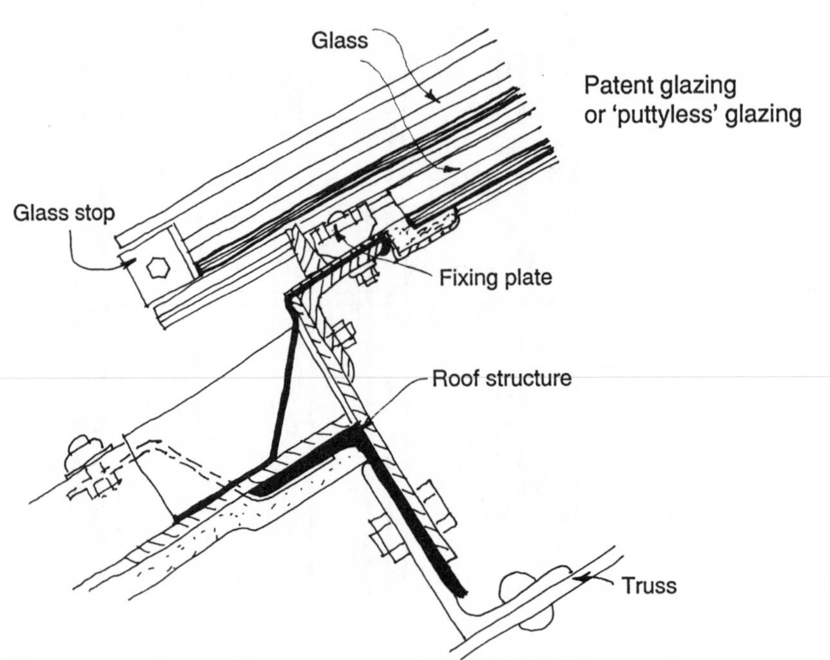
Glass
Patent glazing
or 'puttyless' glazing
Glass stop
Fixing plate
Roof structure
Truss

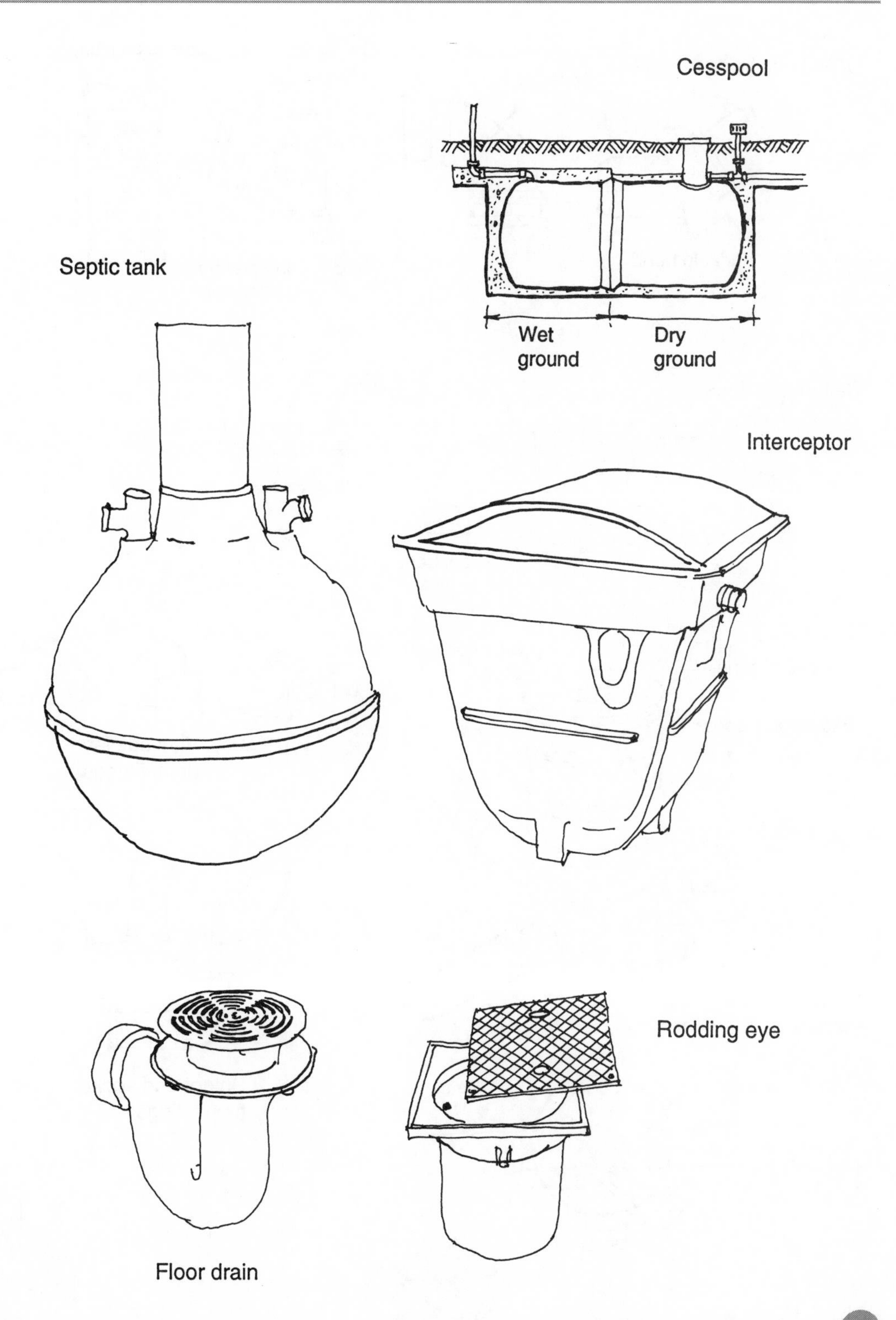
Cesspool
Wet ground
Dry ground
Septic tank
Interceptor
Rodding eye
Floor drain

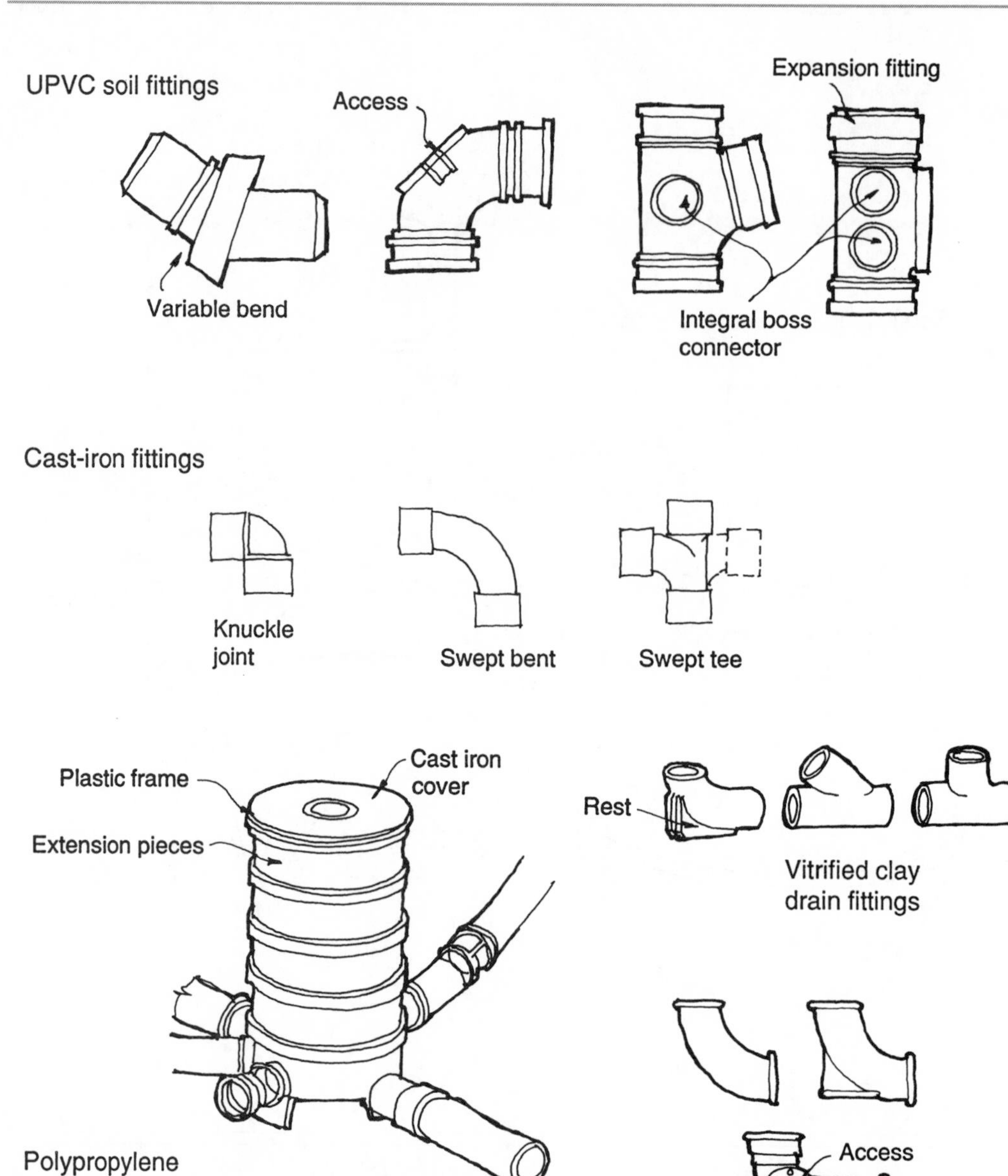
UPVC soil fittings
Access
Expansion fitting
Variable bend
Integral boss connector
Cast-iron fittings
Knuckle joint
Swept bent
Swept tee
Plastic frame
Cast iron cover
Extension pieces
Rest
Vitrified clay drain fittings
Polypropylene inspection chamber
Access
Spigot and socket type fittings
Plan

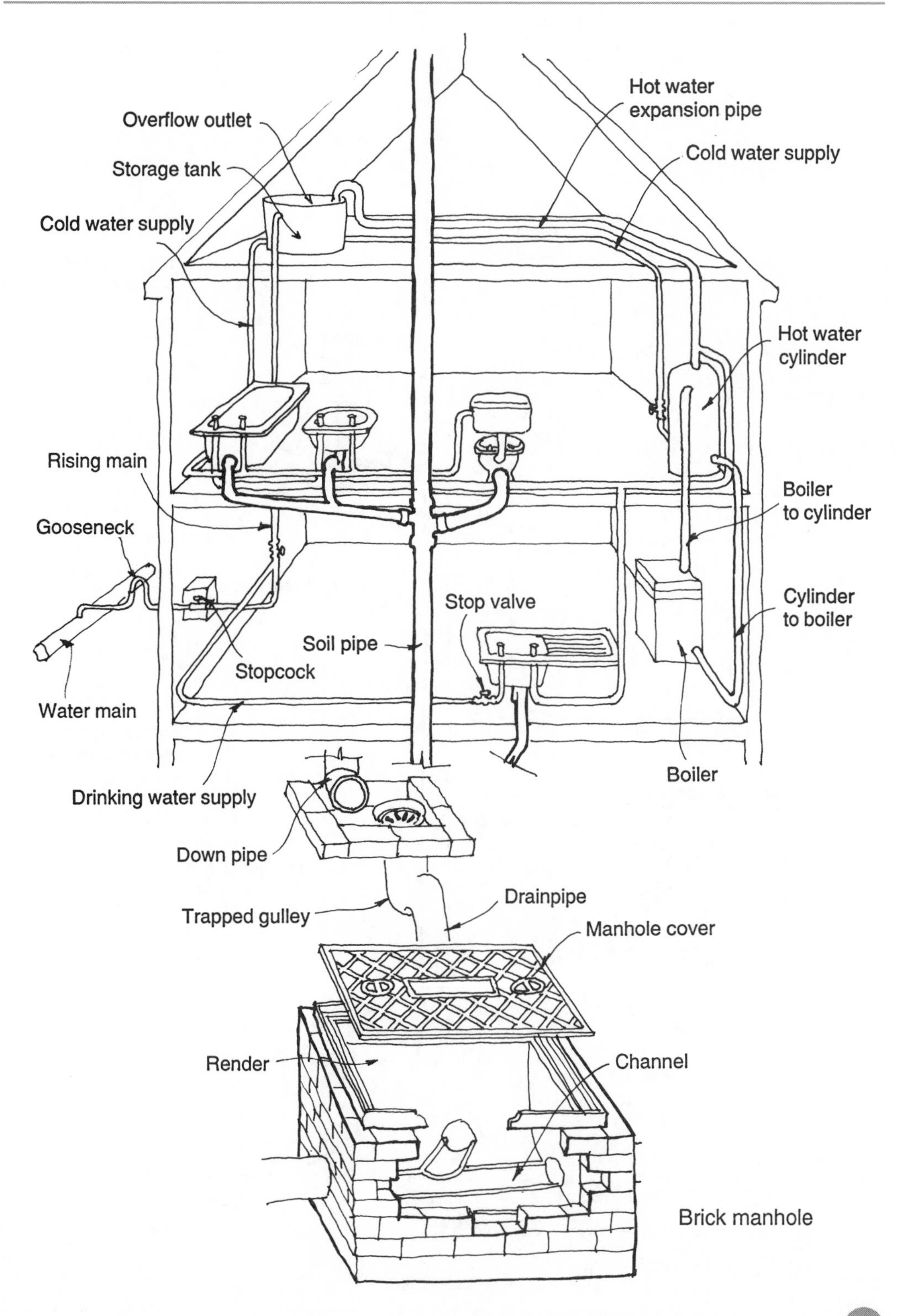
Overflow outlet
Storage tank
Cold water supply
Hot water expansion pipe
Cold water supply
Hot water cylinder
Rising main
Gooseneck
Boiler to cylinder
Cylinder to boiler
Stop valve
Soil pipe
Stopcock
Water main
Drinking water supply
Boiler
Down pipe
Trapped gulley
Drainpipe
Manhole cover
Render
Channel
Brick manhole

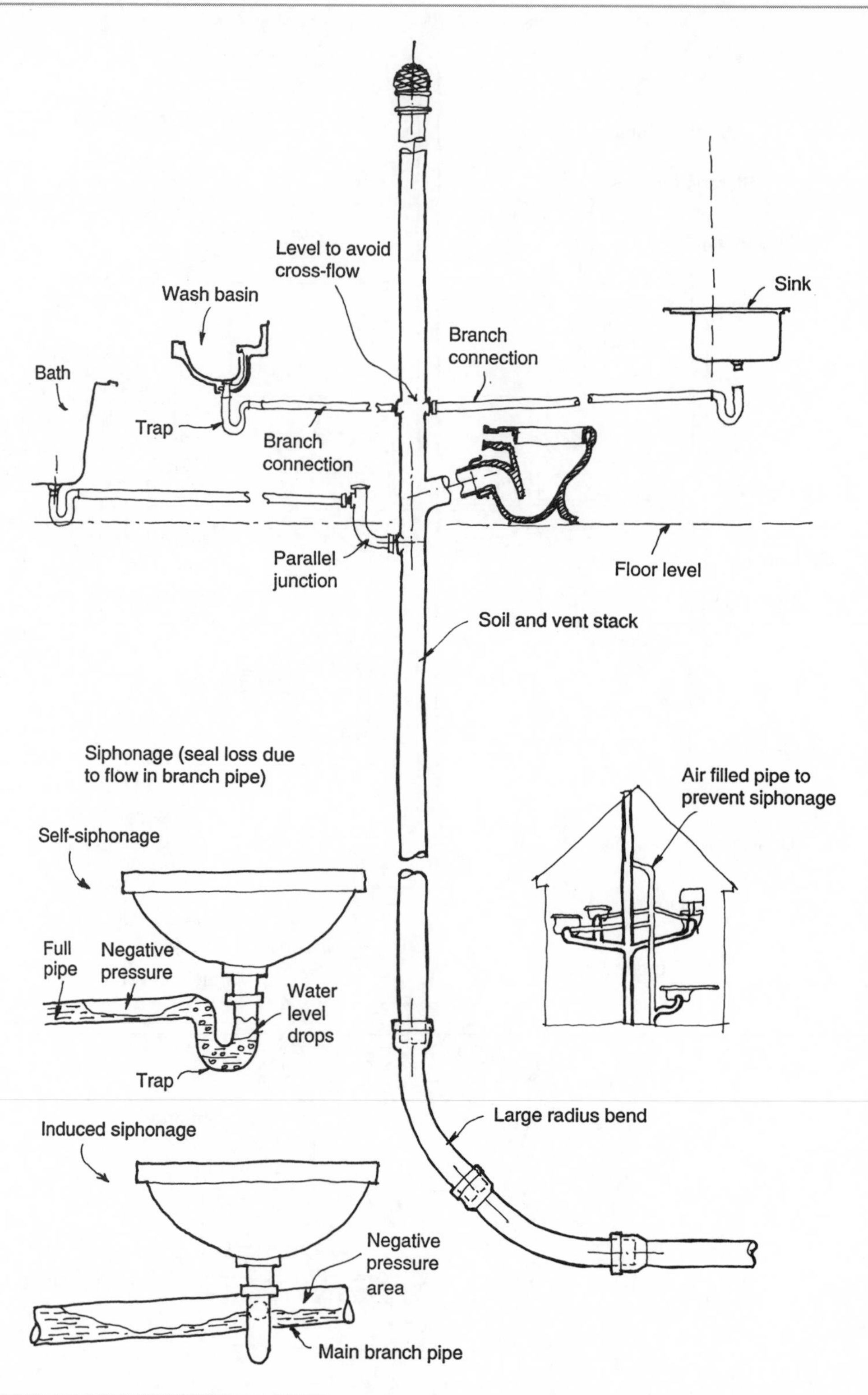
Level to avoid cross-flow
Wash basin
Sink
Branch connection
Bath
Trap
Branch connection
Parallel junction
Floor level
Soil and vent stack
Siphonage (seal loss due to flow in branch pipe)
Air filled pipe to prevent siphonage
Self-siphonage
Full pipe
Negative pressure
Water level drops
Trap
Large radius bend
Induced siphonage
Negative pressure area
Main branch pipe

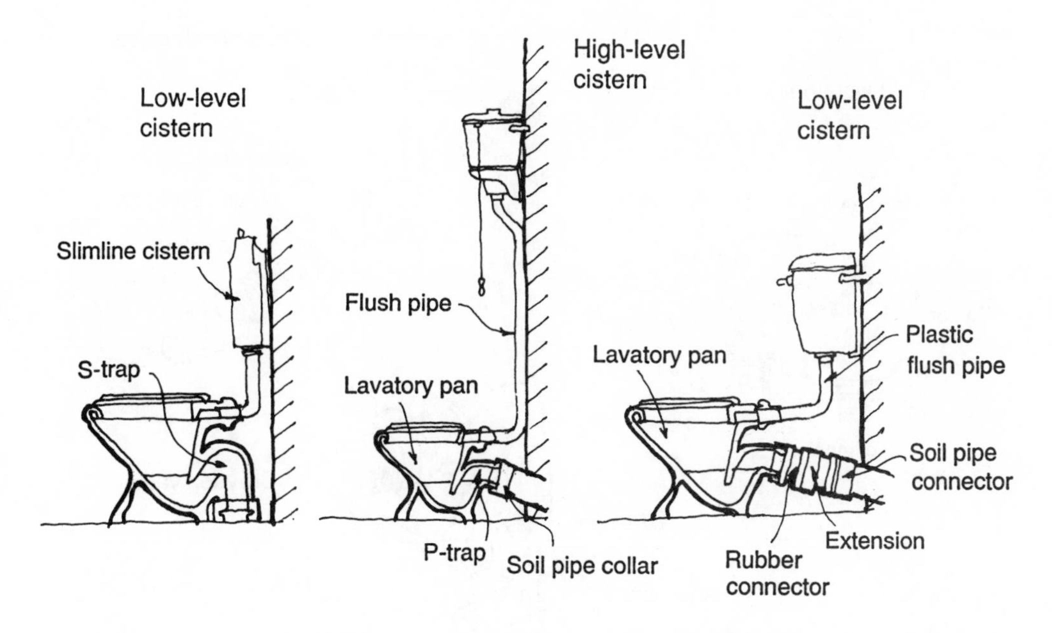
Low-level
cistern
High-level
cistern
Low-level
cistern
Slimline cistern
Flush pipe
S-trap
Lavatory pan
Lavatory pan
Plastic
flush pipe
Soil pipe
connector
P-trap
Soil pipe collar
Extension
Rubber
connector

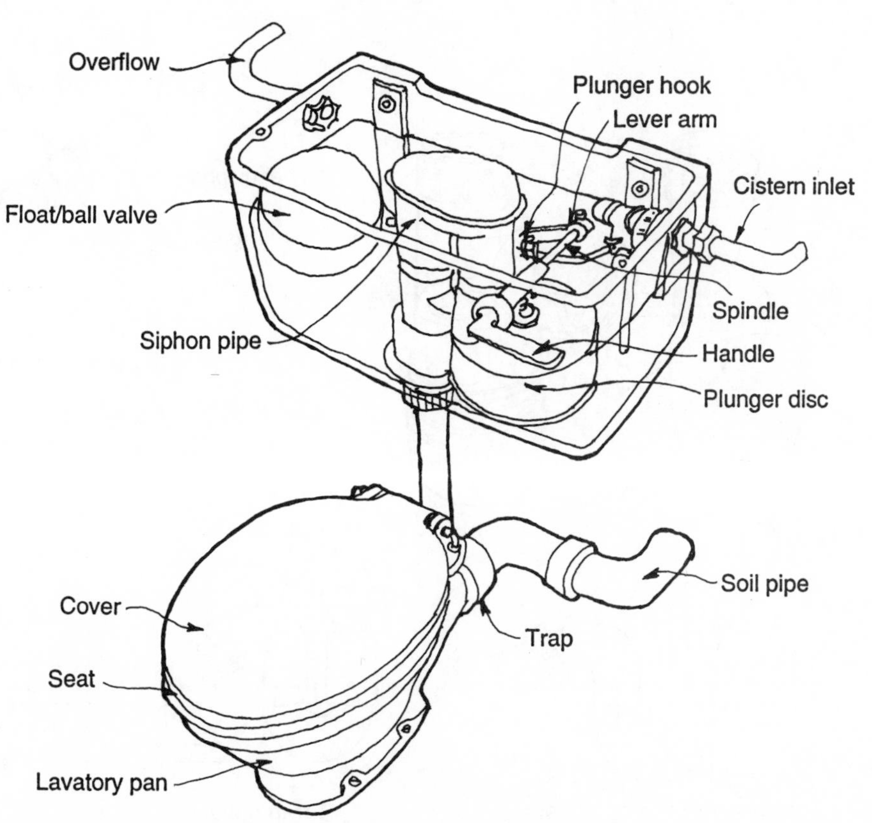
Overflow
Plunger hook
Lever arm
Cistern inlet
Float/ball valve
Spindle
Siphon pipe
Handle
Plunger disc
Soil pipe
Cover
Trap
Seat
Lavatory pan

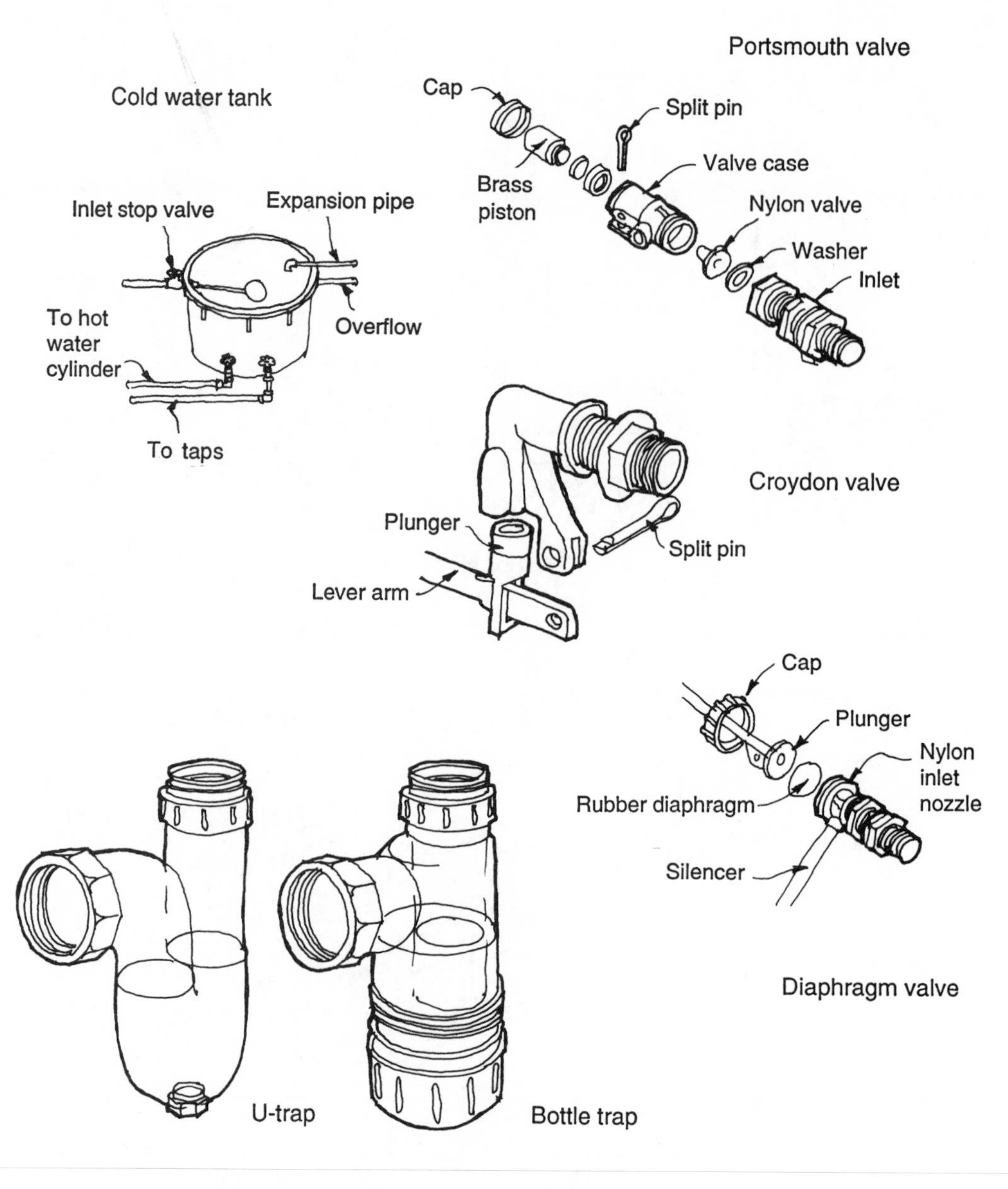

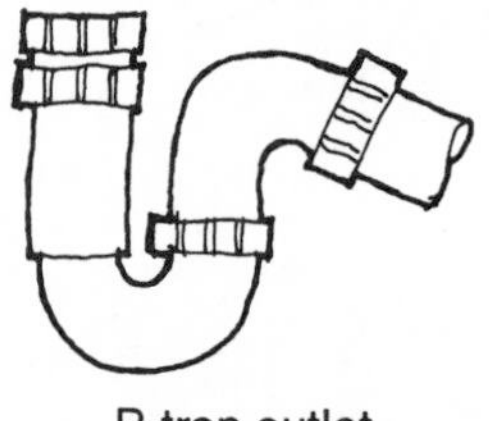

P-trap outlet

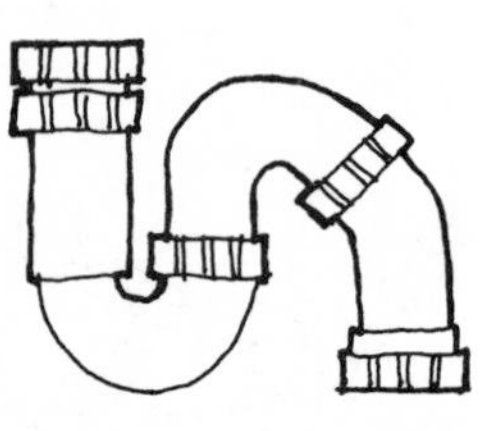

S-trap outlet

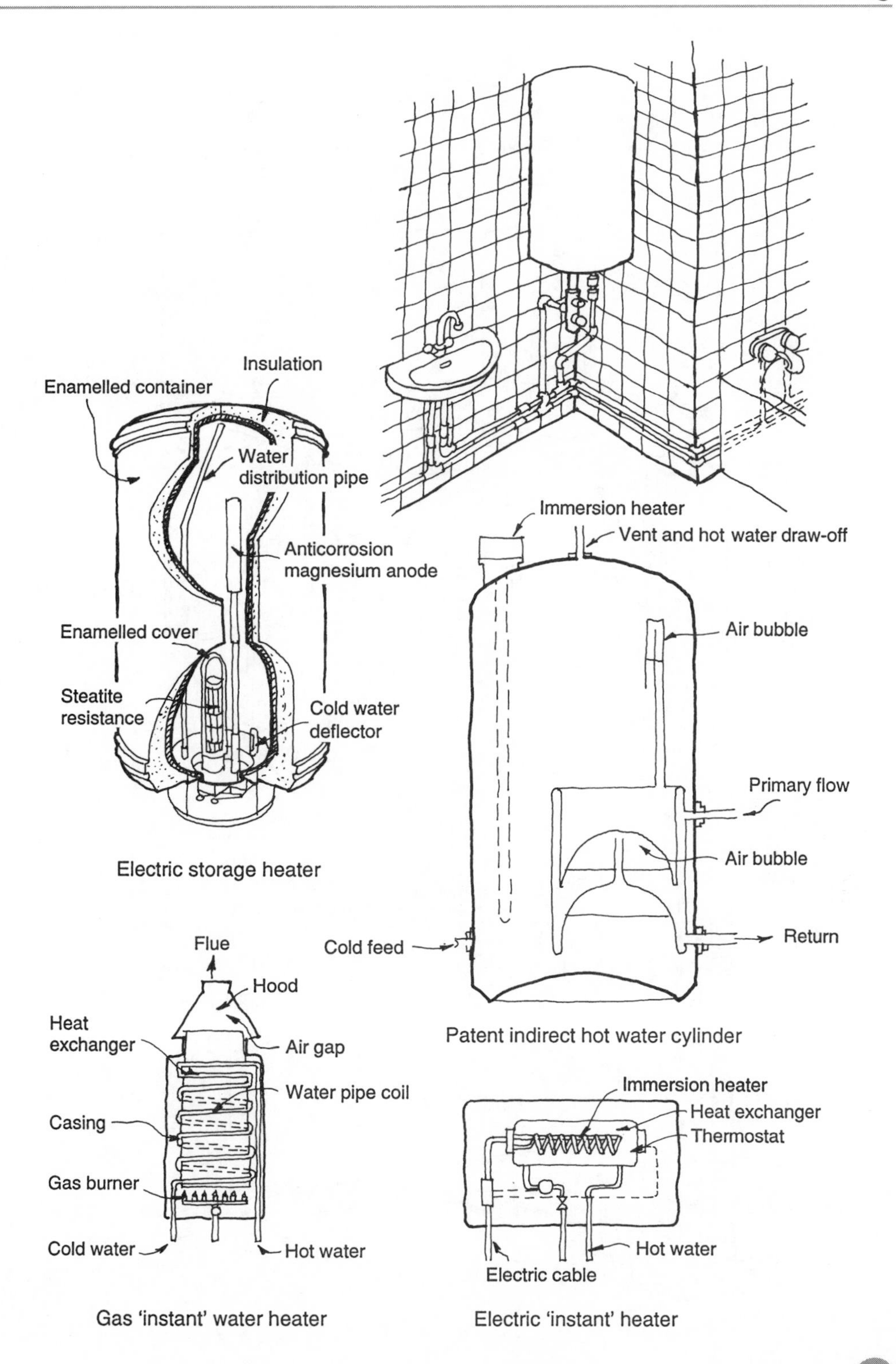

Electric storage heater

Patent indirect hot water cylinder

Gas 'instant' water heater

Electric 'instant' heater

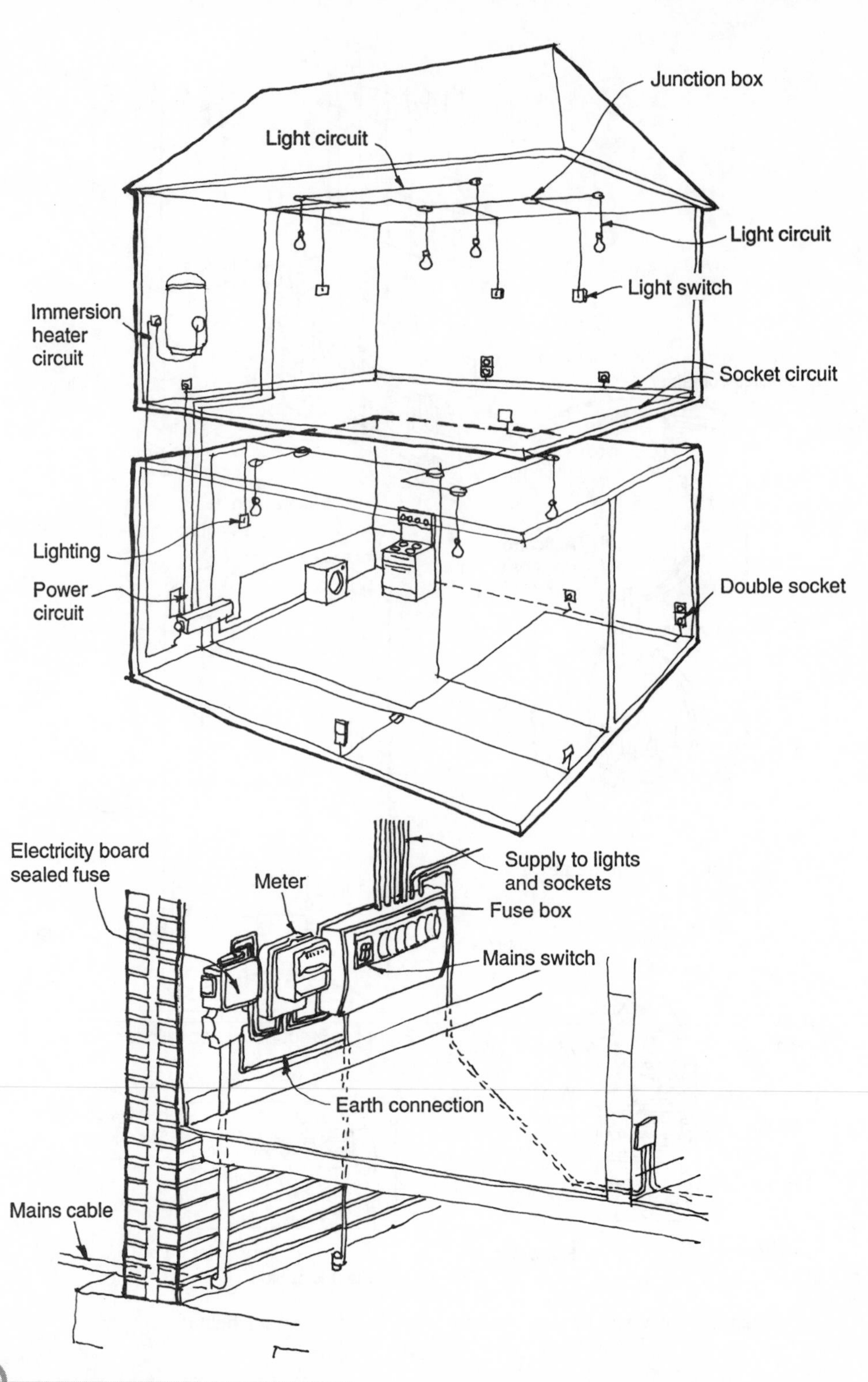
Junction box
Light circuit
Light circuit
Light switch
Immersion heater circuit
Socket circuit
Lighting
Power circuit
Double socket
Electricity board sealed fuse
Meter
Supply to lights and sockets
Fuse box
Mains switch
Earth connection
Mains cable

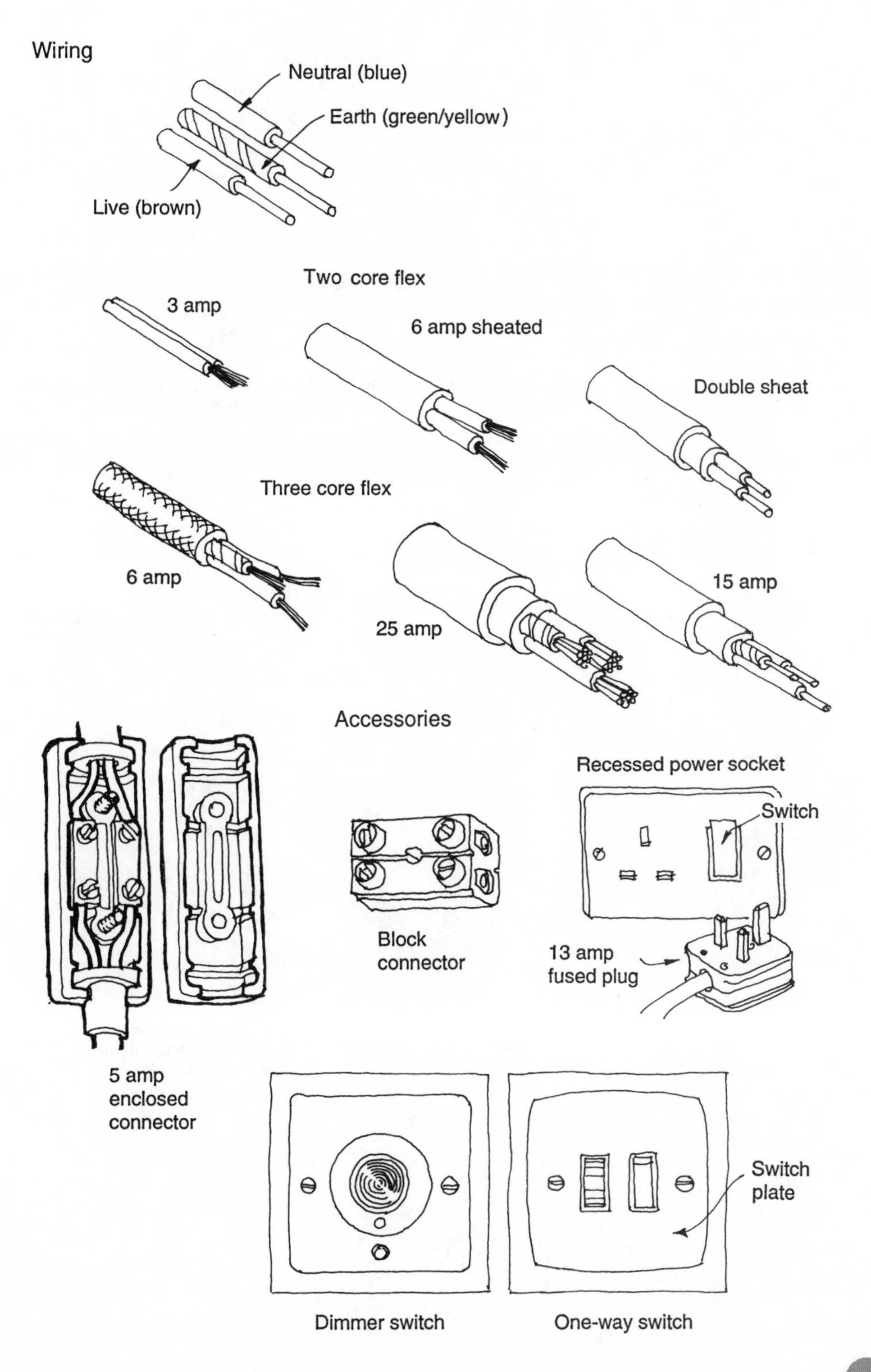
Wiring
Neutral (blue)
Earth (green/yellow)
Live (brown)
Two core flex
3 amp
6 amp sheated
Double sheat
Three core flex
6 amp
25 amp
15 amp
Accessories
Recessed power socket
Switch
Block connector
13 amp fused plug
5 amp enclosed connector
Switch plate
Dimmer switch
One-way switch

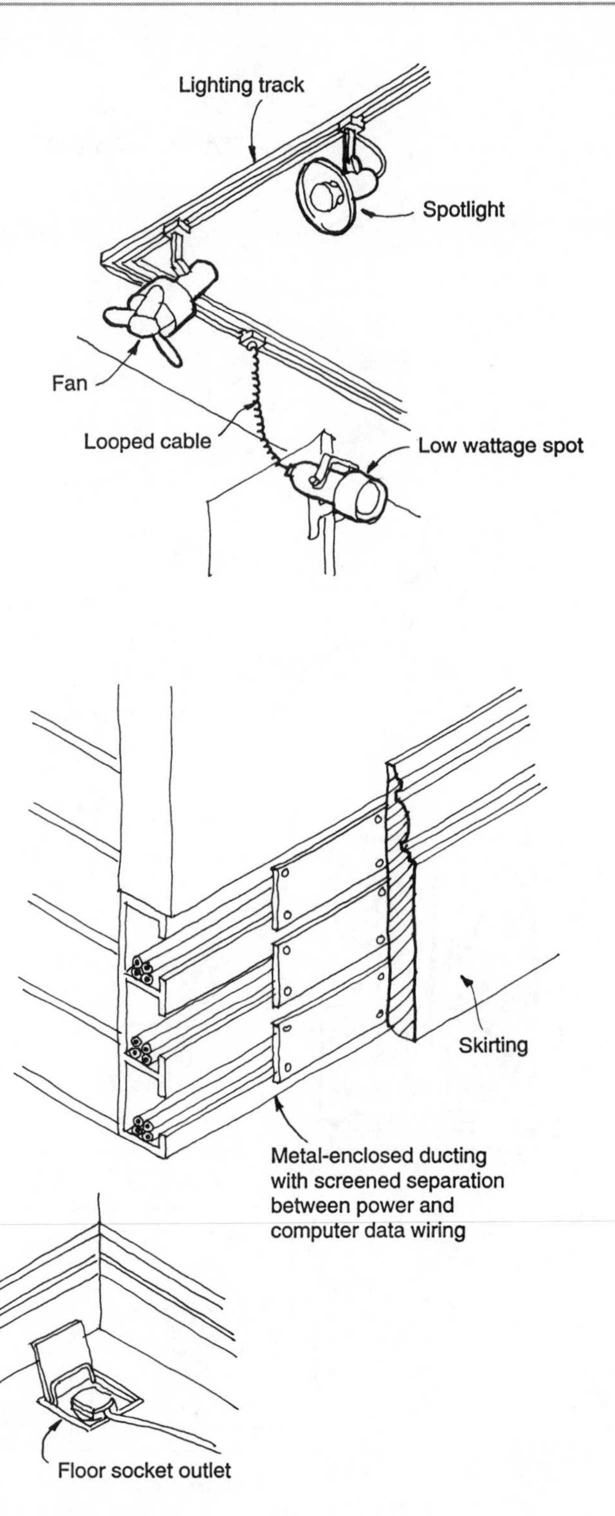
Lighting track
Spotlight
Fan
Looped cable
Low wattage spot
Skirting
Metal-enclosed ducting
with screened separation
between power and
computer data wiring
Power track
Safety plug
Floor socket outlet

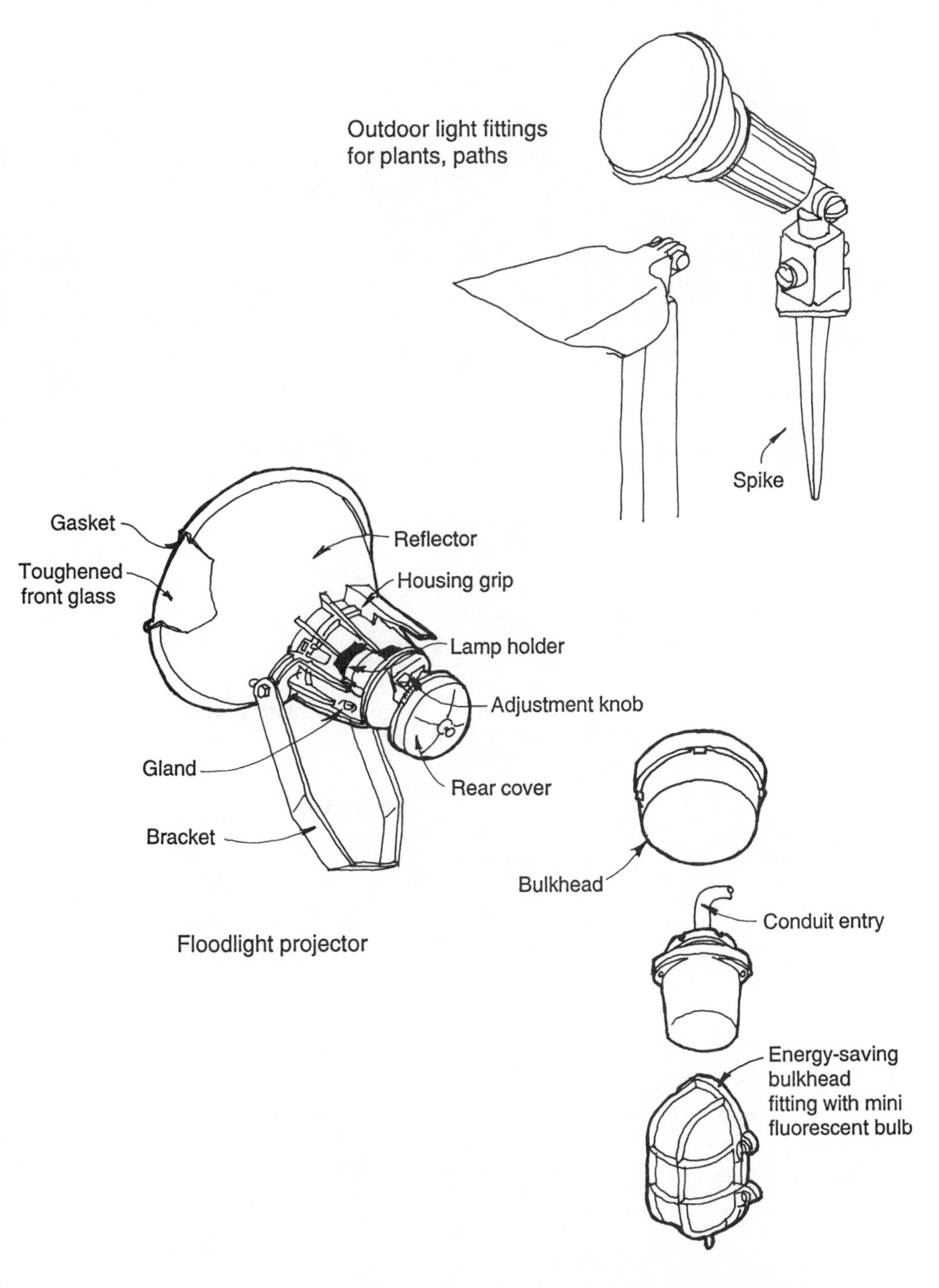

Floodlight projector

Outdoor lighting fittings

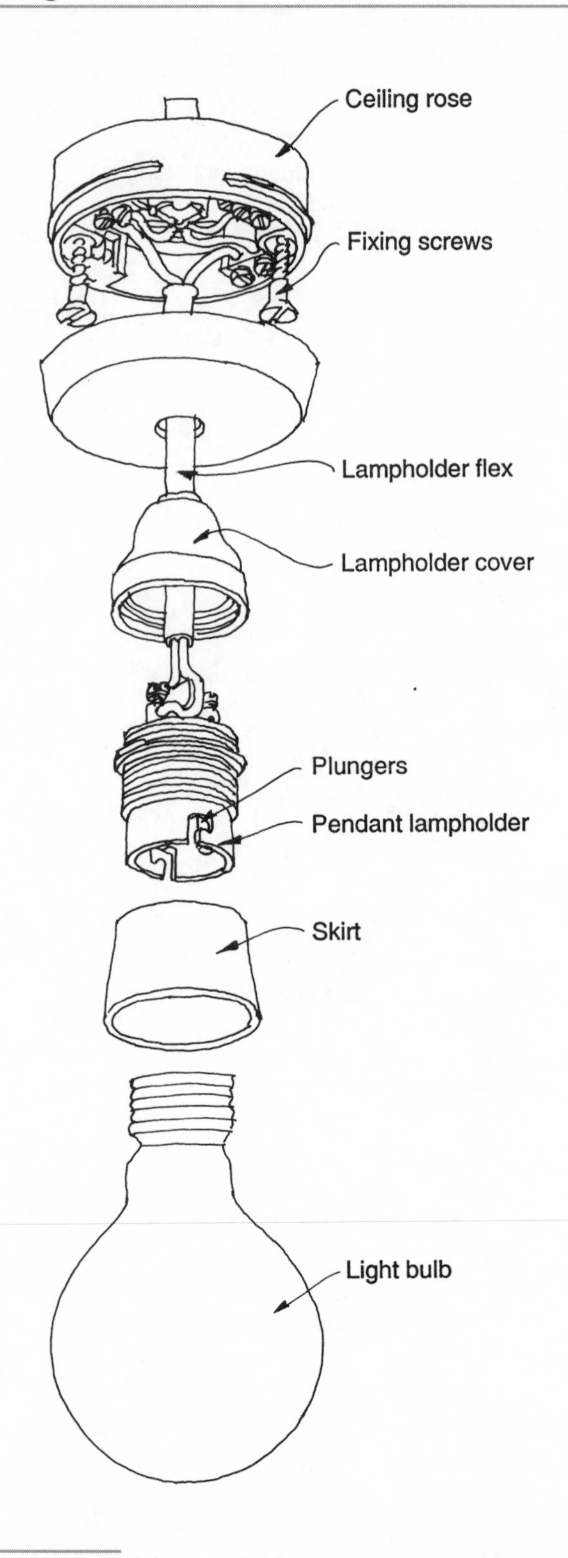
Ceiling rose
Fixing screws
Lampholder flex
Lampholder cover
Plungers
Pendant lampholder
Skirt
Light bulb

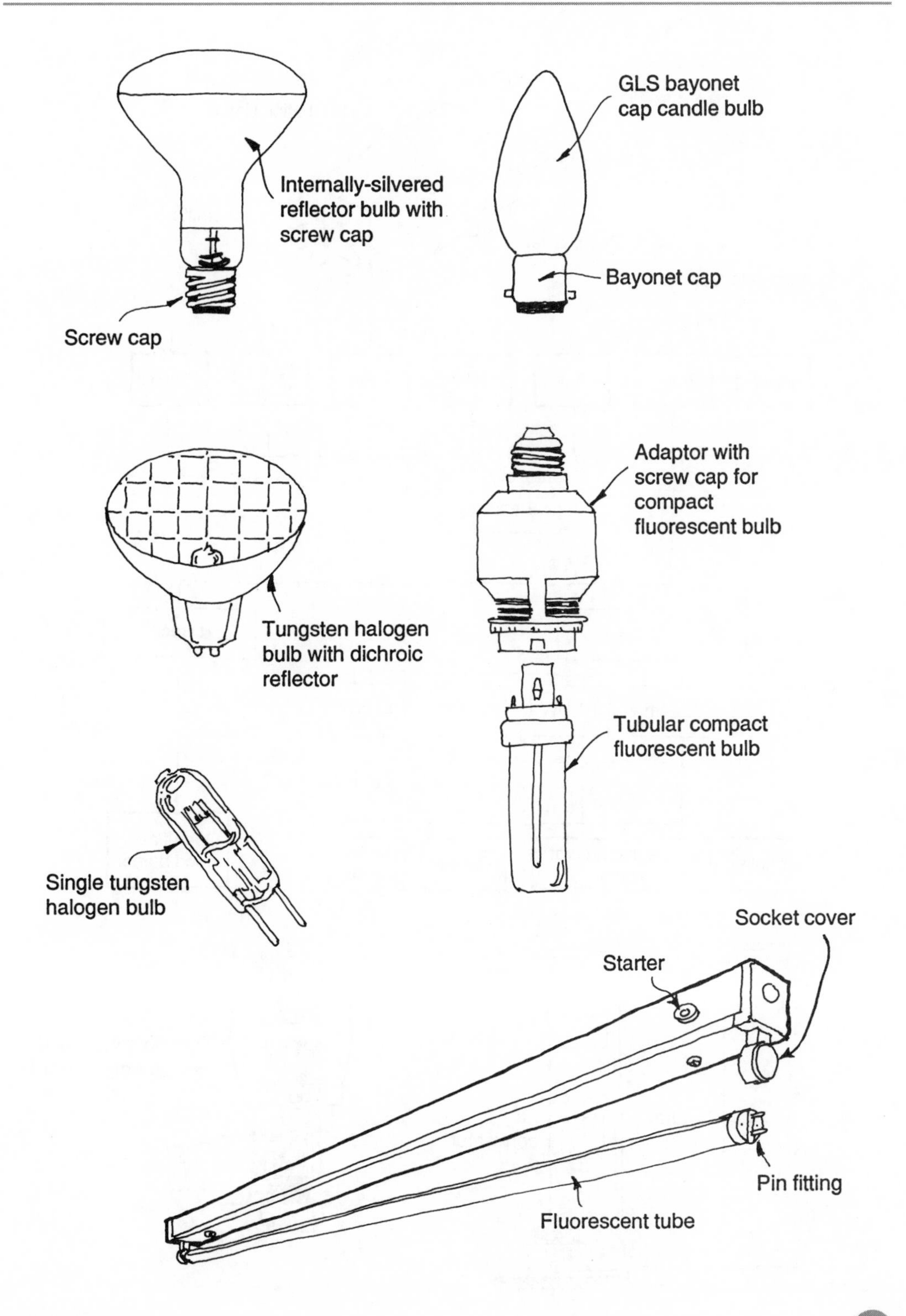
Internally-silvered reflector bulb with screw cap
Screw cap
GLS bayonet cap candle bulb
Bayonet cap
Adaptor with screw cap for compact fluorescent bulb
Tungsten halogen bulb with dichroic reflector
Tubular compact fluorescent bulb
Single tungsten halogen bulb
Socket cover
Starter
Pin fitting
Fluorescent tube

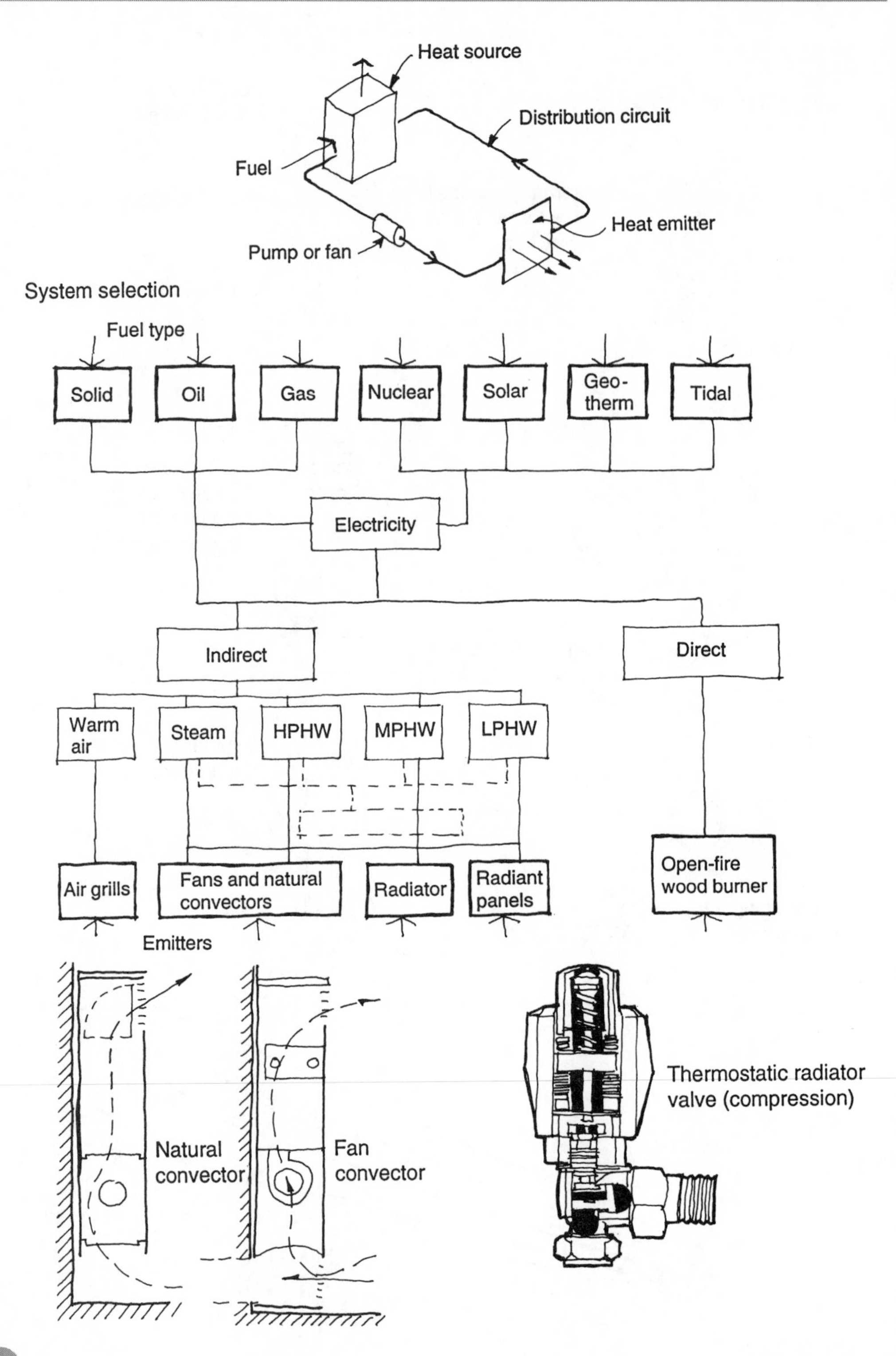
Heat source
Distribution circuit
Fuel
Heat emitter
Pump or fan
System selection
Fuel type
Solid
Oil
Gas
Nuclear
Solar
Geo-therm
Tidal
Electricity
Indirect
Direct
Warm air
Steam
HPHW
MPHW
LPHW
Air grills
Fans and natural convectors
Radiator
Radiant panels
Open-fire wood burner
Emitters
Natural convector
Fan convector
Thermostatic radiator valve (compression)

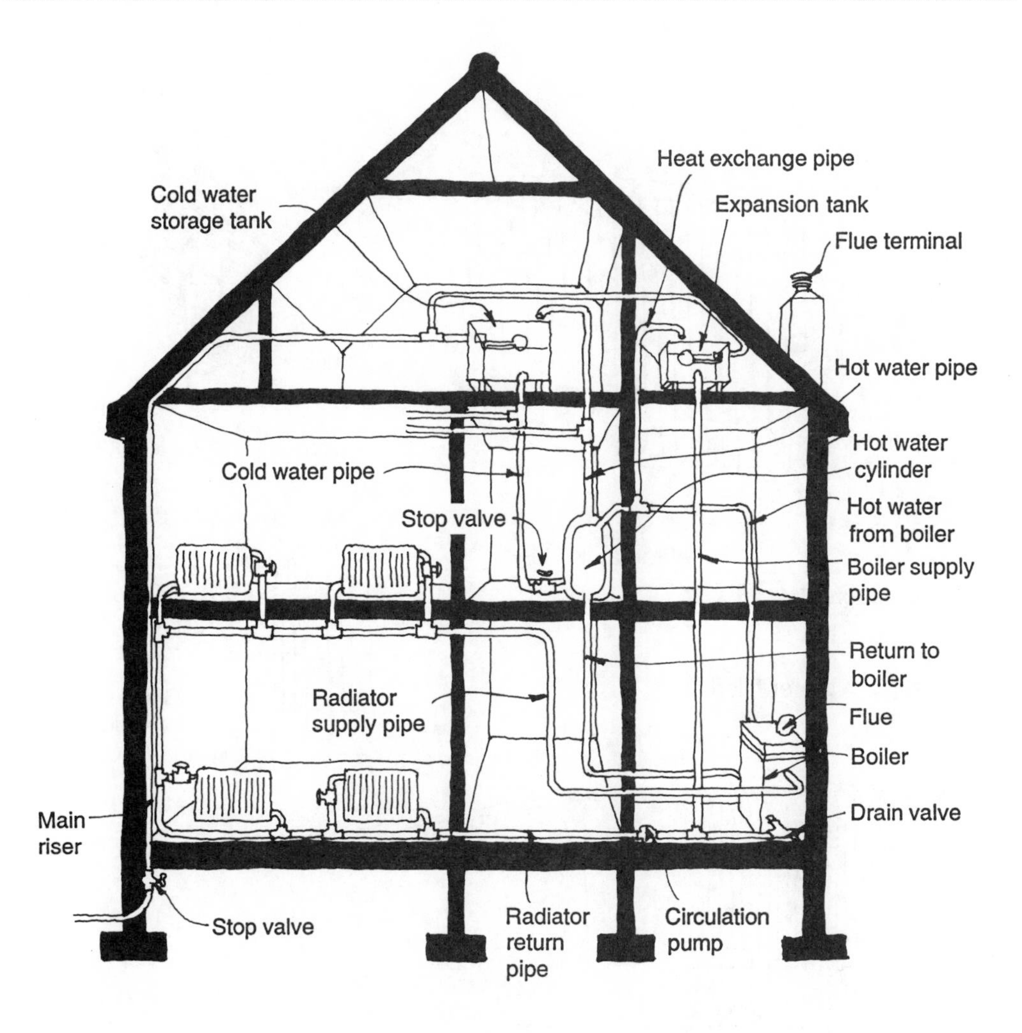

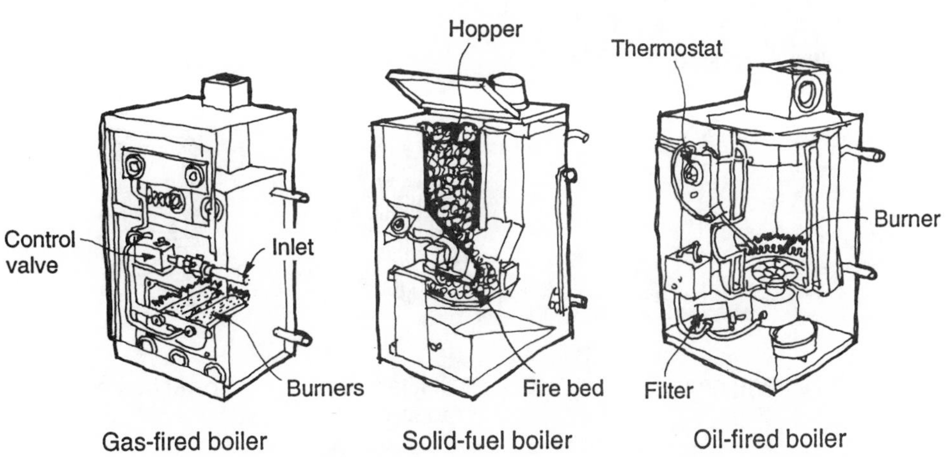

Gas-fired boiler Solid-fuel boiler Oil-fired boiler

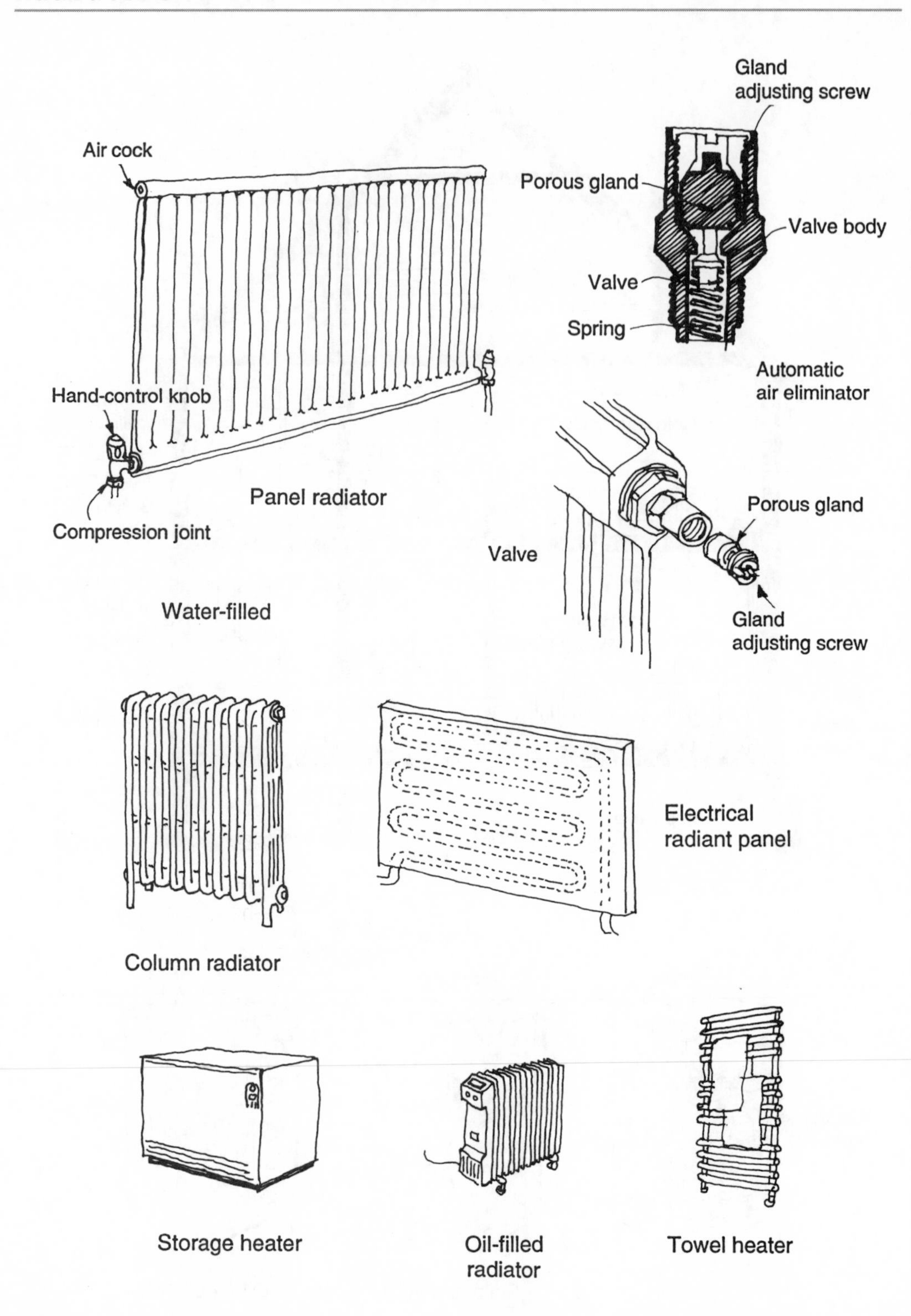
Gland
adjusting screw
Air cock
Porous gland
Valve body
Valve
Spring
Automatic
air eliminator
Hand-control knob
Panel radiator
Porous gland
Compression joint
Valve
Water-filled
Gland
adjusting screw
Electrical
radiant panel
Column radiator
Storage heater
Oil-filled
radiator
Towel heater

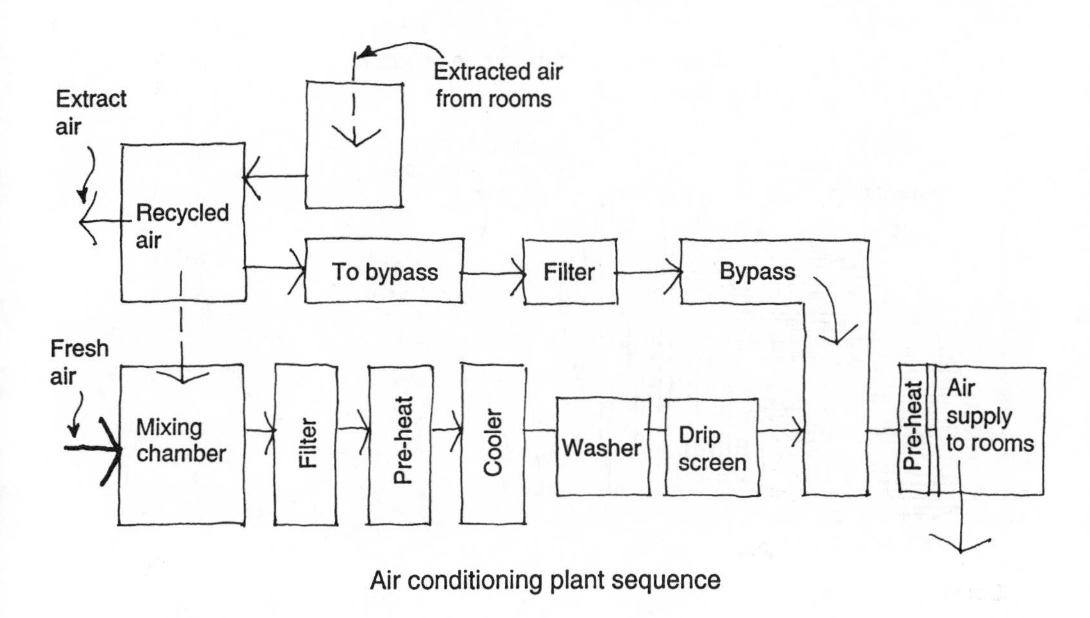

Air conditioning plant sequence

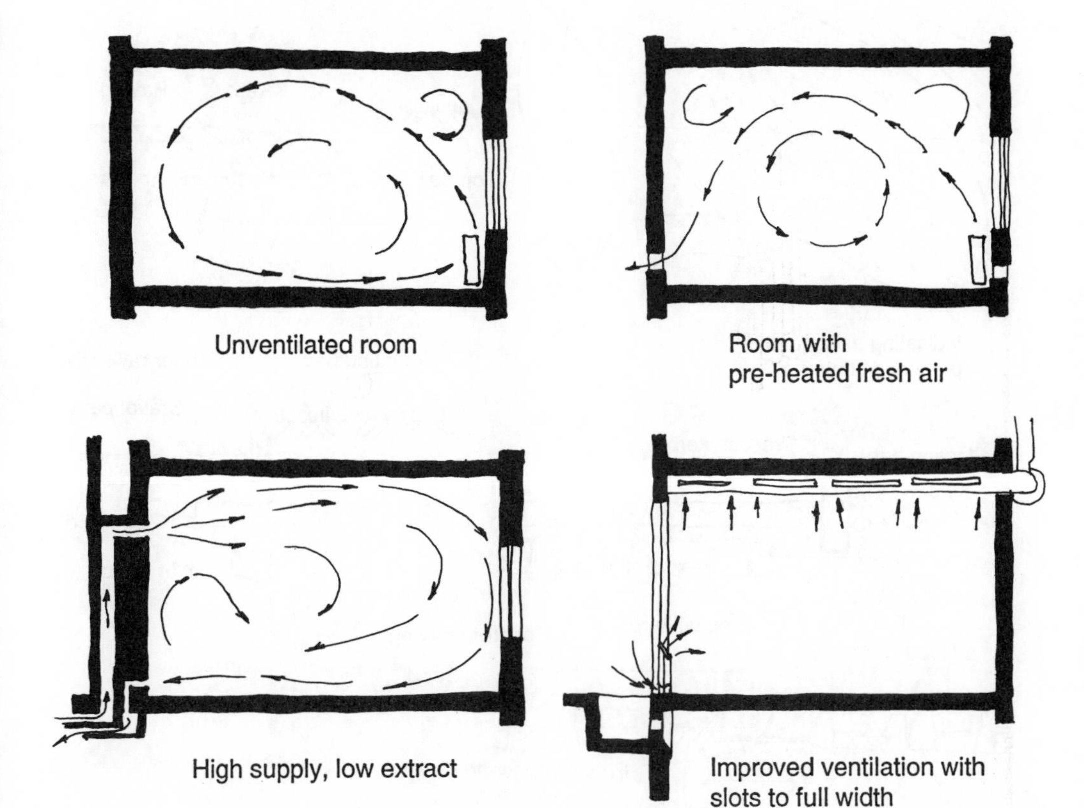

Unventilated room

Room with
pre-heated fresh air

High supply, low extract

Improved ventilation with
slots to full width

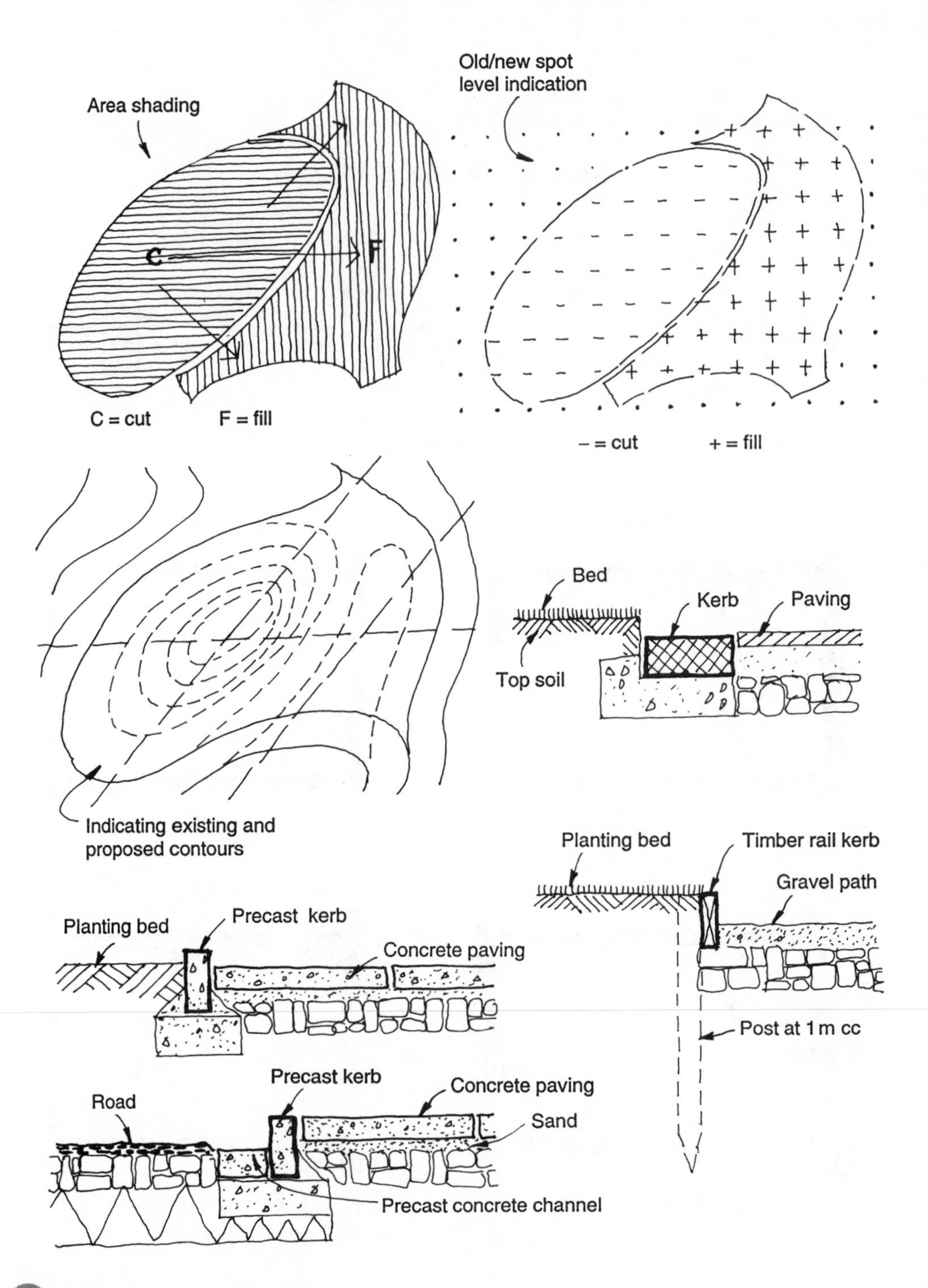
Graphic presentation of earthwork
Area shading
C
F
C = cut
F = fill
Old/new spot level indication
– = cut
+ = fill
Indicating existing and proposed contours
Bed
Kerb
Paving
Top soil
Planting bed
Timber rail kerb
Gravel path
Post at 1 m cc
Planting bed
Precast kerb
Concrete paving
Road
Precast kerb
Concrete paving
Sand
Precast concrete channel

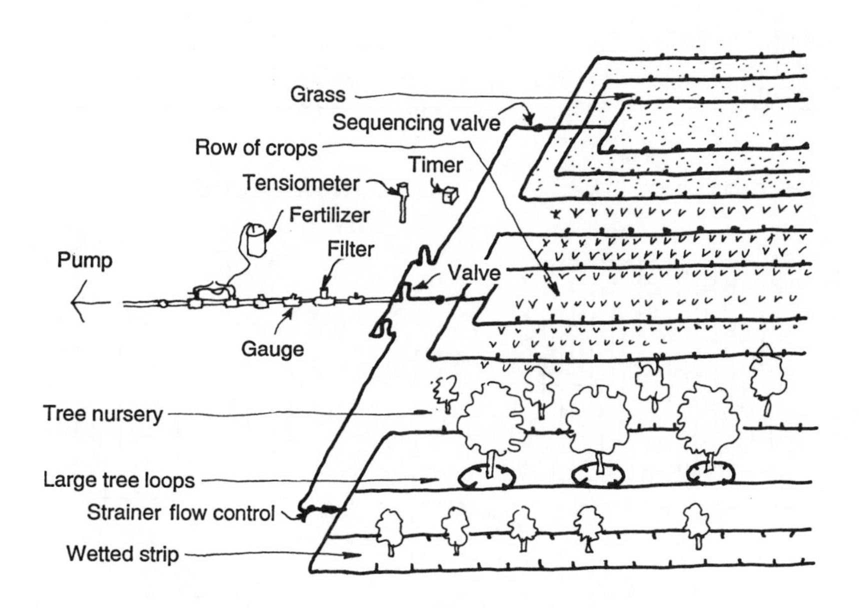
Irrigation system
Grass
Sequencing valve
Row of crops
Timer
Tensiometer
Fertilizer
Filter
Pump
Valve
Gauge
Tree nursery
Large tree loops
Strainer flow control
Wetted strip

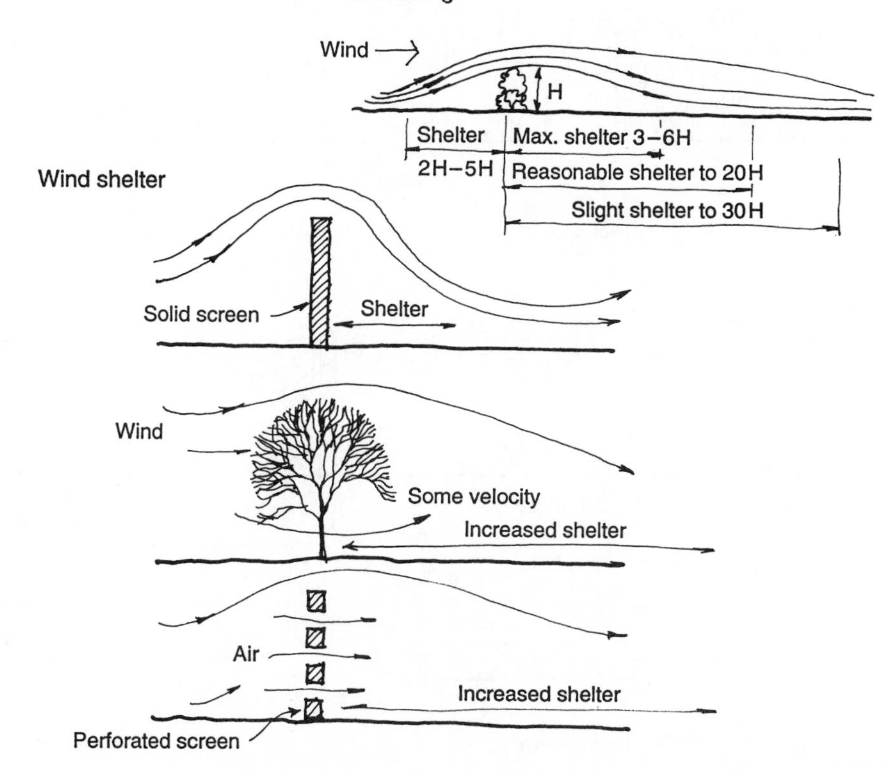
Screening
Wind
H
Shelter
2H–5H
Max. shelter 3–6H
Reasonable shelter to 20H
Slight shelter to 30H
Wind shelter
Solid screen
Shelter
Wind
Some velocity
Increased shelter
Air
Increased shelter
Perforated screen

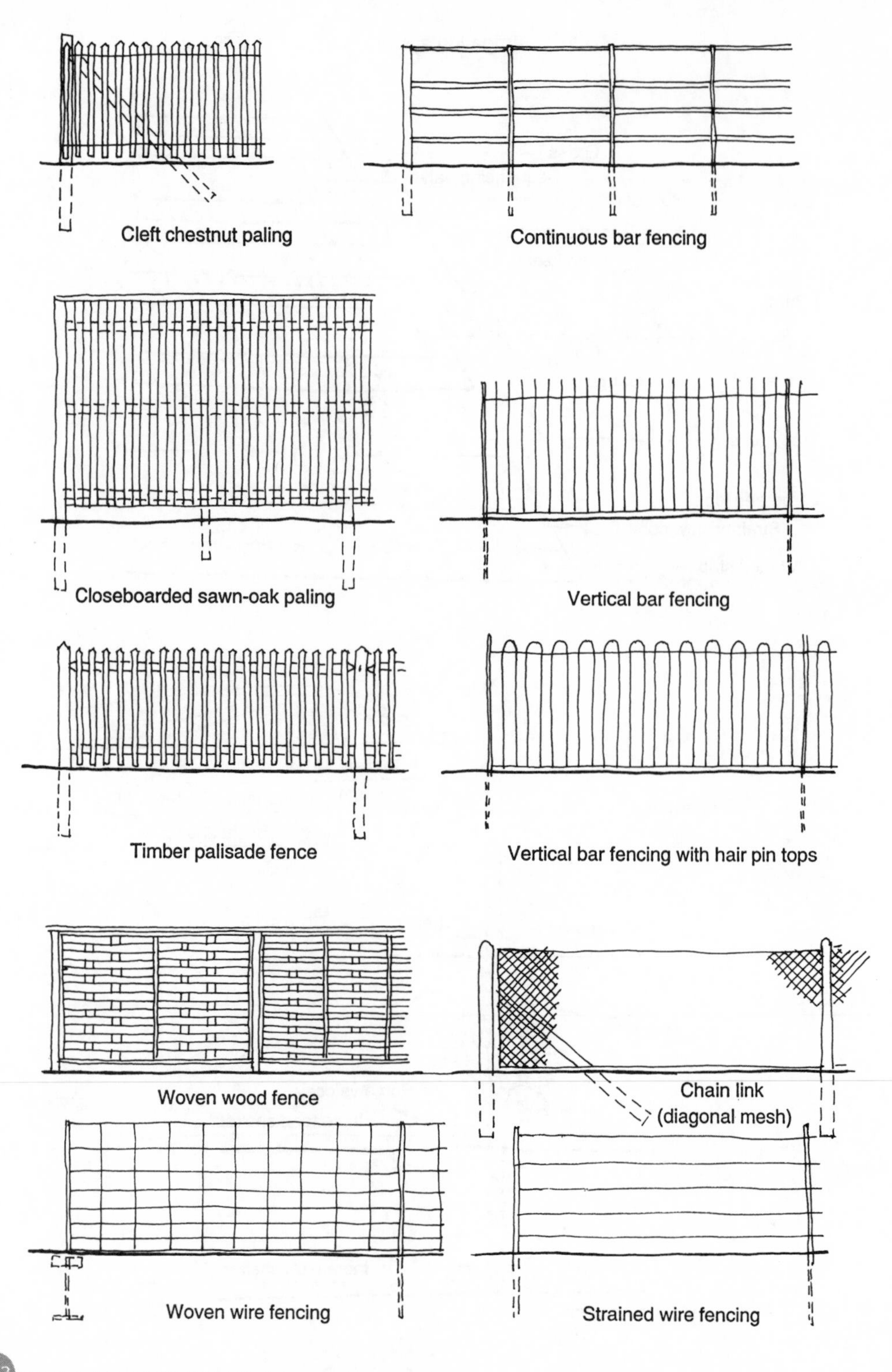
Cleft chestnut paling
Continuous bar fencing
Closeboarded sawn-oak paling
Vertical bar fencing
Timber palisade fence
Vertical bar fencing with hair pin tops
Woven wood fence
Chain link
(diagonal mesh)
Woven wire fencing
Strained wire fencing

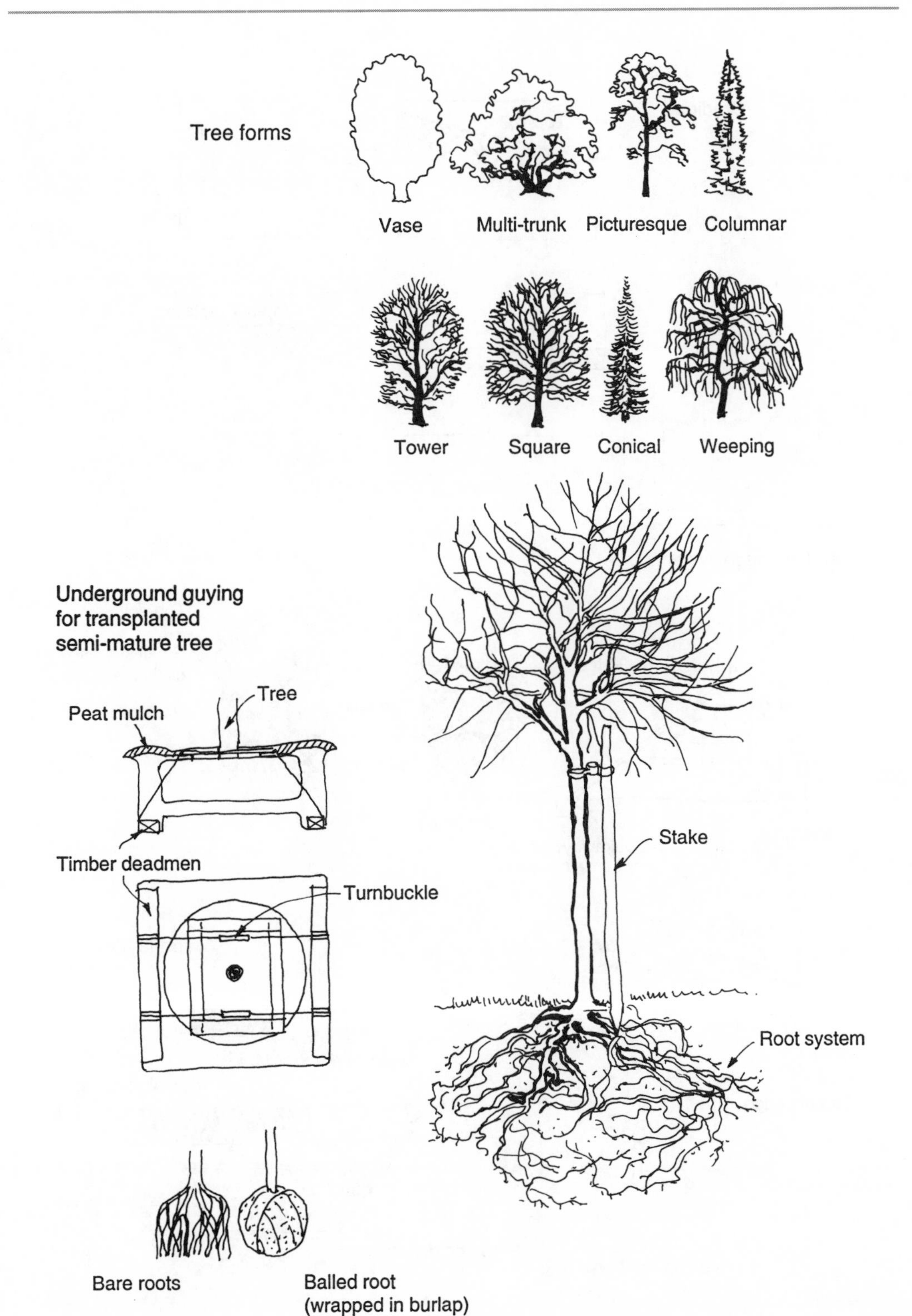
Tree forms
Vase
Multi-trunk
Picturesque
Columnar
Tower
Square
Conical
Weeping
Underground guying
for transplanted
semi-mature tree
Tree
Peat mulch
Timber deadmen
Turnbuckle
Stake
Root system
Bare roots
Balled root
(wrapped in burlap)

Grading, turf laying

Grading

Top-soil

Dig

Fil

Top-soil replaced

Land drainage systems

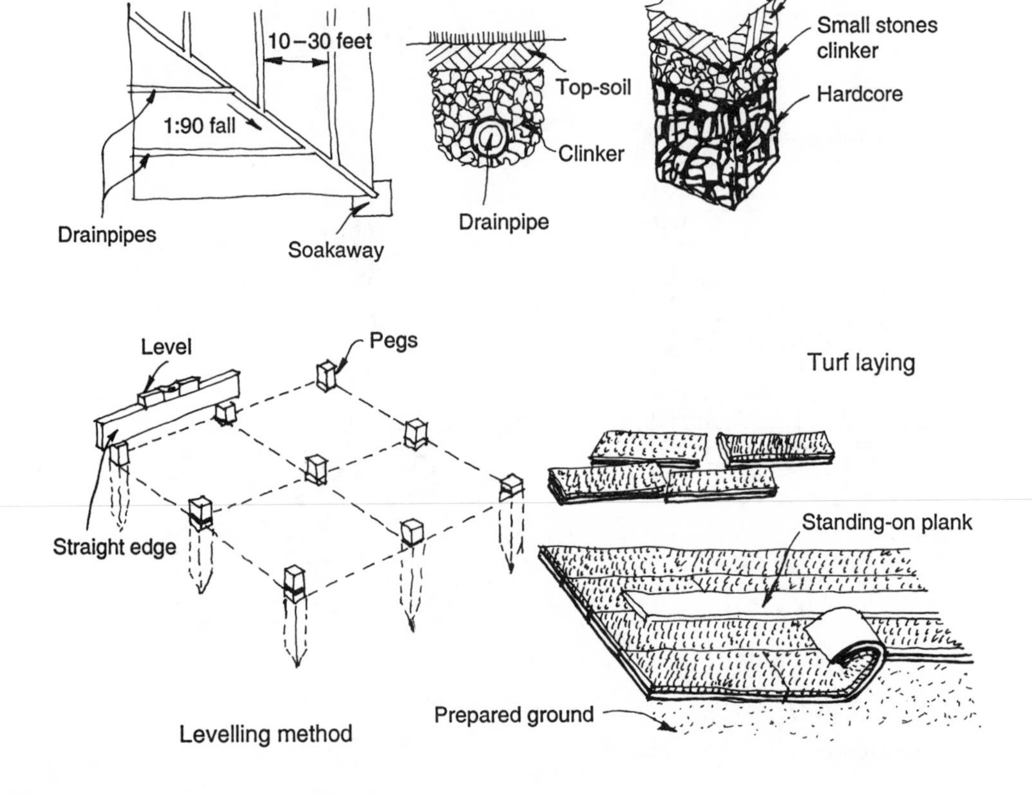

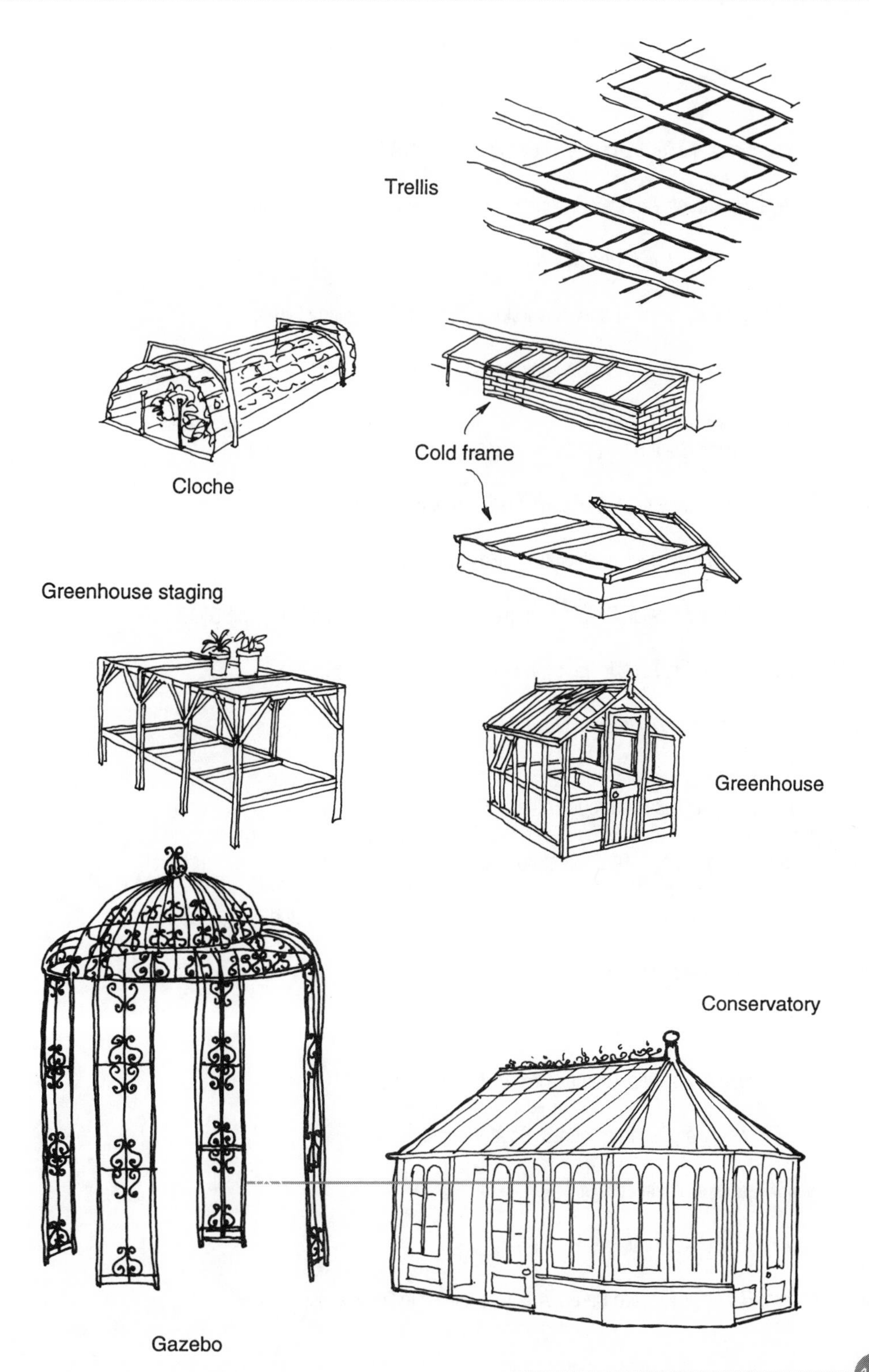
Trellis
Cloche
Cold frame
Greenhouse staging
Greenhouse
Conservatory
Gazebo

References

Architect's Data, Ernst Neufert, Crosby Lockwood Staples (1970).

Architect's Legal Handbook, Anthony Speaight and Gregory Stone, Architectural Press (1998).

Building Construction Vols I, II, III and IV, W.B. McKay, Longmans (1955).

The Building Design Easy Brief, Henry Haverstock, Morgan Grampian (1987).

Building Construction Handbook, R. Chudley, Laxton's (1988).

The Care and Conservation of Georgian Houses, Architectural Press with Edinburgh New Town Conservation Commitee, Paul Harris Publishing (1978). Architectural Press (1980).

Drawing Office Practice for British Standard 1192, Architects and Builders *(1953).*

English Historic Carpentry, Cecil A. Hewett, Phillimore (1980).

Farms in England, Peter Fowler, Royal Commission on Historic Monuments, HMSO (1983).

Handbook of Urban Landscape, Cliff Tandy, Architectural Press (1975).

History of the English House, Nathaniel Lloyd, Architectural Press (1975).

Mitchell's Building Series,
Structure and Fabric 1, Jack Stroud Foster (1973).
Structure and Fabric 2, Jack Stroud Foster and Raymond Harrington (1976).
Components, Harold King (1983) Batsford Academic and Education.

Modern Practical Masonry, E. G. Warland, Sir Isaac Pitman & Sons Ltd., 2nd edn (1953).

Modulor Le Cobusier, Faber & Faber (1951).

New Metric Handbook, Edited by P. Tutt and D. Adler, Architectural Press (1979).

The Parish Churches of England, Charles Cox, B. T. Batsford (1954).

The Penguin Dictionary of Building, John S. Scott, Penguin (1982).

Repair Manual Reader's Digest (1976).

Specification 1 – 6 Architectural Press (1987).

Traditional Farm Buildings Richard Harris, Arts Council Exhibition Catalogue (1982).

Index

Index

Index

Index

Index